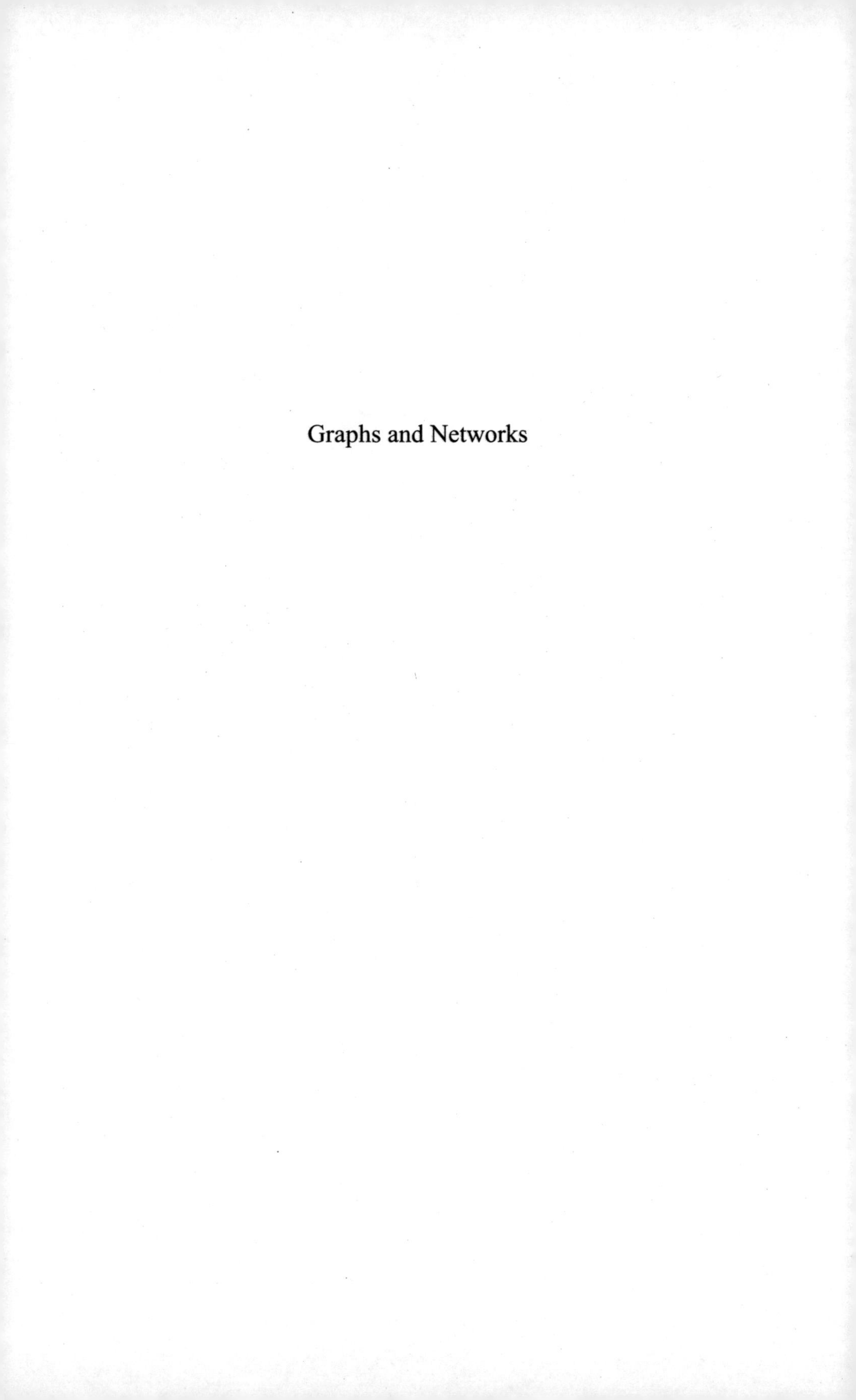

Graphs and Networks

Graphs and Networks

Edited by
Philippe Mathis

First published in France by Hermes Science.Lavoisier entitled "Graphes et réseaux : modélisation multiniveau"

First Published in Great Britain and the United States in 2007 by ISTE Ltd.

First South Asian Edition 2007

ISTE Ltd.
6 Fitzory Square
London WJT 5DX
UK

ISTE USA
4308 Patrice Road
Newport Beach, CA 92663
USA

www.iste.co.uk

ISBN 10: 1 905209 08 8
ISBN 13: 978 1 905209 08 8

Library of Congress Cataloging-in-Publication Data

Grahpes et réseaux. English.
Graphs and networks/edited by Philippe Mathis.
p.cm.
Includes index.
ISBN 10: 1 905209 08 8
ISBN 13: 978 1 905209 08 8
1. Cartography--Methodology. 2. Graph theory. 3. Transport theory. I. Mathis, Philippe. II. Title.
GA102.3.G6713 2007
388.01'1--dc22

2007006800

British Library Cataloguing-in-Publication Data
A CIP record of this book is available from the British Library
ISBN 978 1 905209 08 8

Printed and bound in India by Brijbasi Art Press Limited, I-72, Sector-9, Noida, U.P., India.

Table of Contents

Preface

This work is focused on the use of graphs for the simulation and representation of networks, mainly of transport networks.

The viewpoint is intentionally more operational than descriptive: the effects of transport characteristics on space are just as important to the planner as the transport itself.

The present work is based on the research conducted at Tours during the 1990s by various PhD students who have become researchers, teaching researchers or professionals.

The book is structured in three parts following an introductory chapter which contains a reminder of the necessary definitions from graph theory and of the representation problems.

The first part presents the traditional applications of graph theory in network modeling and the improvements required for their use as a planning tool.

The second part tackles the problem of the representation of graphs and exposes a certain number of innovations as well as deficiencies.

The third part considers the prior achievements and proposes to develop their theoretical justifications and fill in some gaps.

Philippe MATHIS

Introduction

Strengths and Deficiencies of Graphs for Network Description and Modeling

The focus of this book is on networks in spatial analysis and in urban development and planning, and their simulation using graph theory, which is a tool used specifically to represent them and to solve a certain number of traditional problems, such as the shortest path between one or more origins and destinations, network capacity, etc. However, although transportation systems in the physical sense of the term are the main concern and will therefore form the bulk of the examples cited, other applications, such as player, communication and other networks, will, nevertheless, be taken into account and the reader is welcome to transfer the presented results to others domains.

All of the examples presented below essentially correspond to a decade of research and some 10 PhDs of the Modeling group of the Graduate Urban Development Studies Center of Tours. Obviously, these will be supplemented by other contributions.

In network modeling, which is a field stemming from operational research, a certain form of empiricism tends to dominate, in particular in the intermediate disciplines between social sciences and hard sciences, such as urban development, which, essentially, borrow their tools. However, their specific needs are barely taken into account by fundamental disciplines, such as mathematics, or more applied ones, such as algorithmics, undoubtedly simply because the dynamics of research are very different. We will try to contribute to the mitigation of this difficulty.

Introduction written by Philippe MATHIS.

The modeling and description of networks using graphs: the paradox

The aim of this work is, among other things, to highlight a paradox and to try to rectify it. This paradox, once identified, is relatively simple. Since Euler's time [EUL 1736, EUL 1758] it has been known how to efficiently model a transport network by using graphs, as he demonstrated with the famous example of the Königsberg bridges and, following the rise of Operations Research in the 1950s and 1960s, a number of optimization problems have been successfully resolved with efficiency and elegance.

According to Beauquier, Berstel and Chrétienne: "graphs constitute the most widely used theoretical tool for the modeling and research of the properties of structured sets. They are employed each time we want to represent and study a set of connections (whether directed or not) between the elements of a finite set of objects" [BEA 92]. For Xuong [XUO 92]: "graphs constitute a remarkable modeling tool for concrete situations" and we could cite numerous further testimonies.

The power of the method increased considerably with the fulgurating development of computers and microcomputers[1]. However, although graphs are a powerful tool for the modeling and resolution of certain problems, they otherwise appear unable to represent and describe precisely and without implicit assumptions a network of paths on the basis of elements which are needed for the calculation such as, for example, minimal path or maximum flow, etc. Since on the basis of a matrix definition[2] of the graph, all the plots (i.e. representations) are equal and equivalent in graph theory.

We thus have a method that is simultaneously very simple and has great algorithmic efficiency, but is otherwise deficient, unless it were only to model a network represented on a roadmap, on which basis it delivers knowledgeable and powerful calculations. It does not satisfy the two essential criteria of all scientific work: reproducibility and comparability, particularly with respect to network modeling and the production of charts and/or synthesized images.

At first we propose to show the effectiveness of graph theory in the field of calculation, which we could quickly call of optimization. Then, we propose to demonstrate that the practice of modelers anticipated the theorization with pragmatism and efficiency, and, finally, to suggest some solutions and research paths to establish and generalize what has been conjectured by usage.

---

1 Has the generation of 50 year-olds not also been called the Hewlett-Packard generation? Its ranks remember calculations with a ruler, with logarithmic tables or with the electromechanical four operations machine, etc.

2 See below for the definition of the adjacency matrix, often referred to as *associated matrix* in the works from the 1960s, and of the incidence matrix.

Strength of graph theory

Simplicity of the graph

A graph can be defined as a finite set of points called vertices (i.e. nodes) and a set of relations between these points called edges (i.e. arcs).

Graph theory relates primarily to the existence of relationships between vertices or nodes and, in the figure that represents the graph, the localization of nodes is unimportant unless otherwise specified, and only the existence of a relationship between two nodes counts.

Formally, the graph $G = (V, E)$ is a pair consisting of:

– a set $V = \{1,2,\ldots, N\}$ of vertices;

– a set E of edges;

– a function f of E in $\{\{u, v\} \mid u, v \in V, u \neq v\}$.

An element (u, v) of VxV may appear several times: the arcs $e1$ and $e2$, if they exist, are called multiple arcs if $f(e1) = f(e2)$. The graph will then be a *multigraph* or *p-graph*, where the value of p is that of the greatest number of appearances of the same relation (u, v), i.e. the number of arcs between u and v.

If the arcs are directed, we will then talk of a *directed graph* or *digraph*. If the arcs are undirected, we are dealing with a *simple graph* that can be a multigraph[3].

The graph G is similarly characterized by the number of vertices, the cardinal of the set X, which is called *order* of the graph.

The total number of arcs between two nodes has a precise significance with regard to the definition of the graph only if: $p \neq 1$.

When $p > 1$, the number of relations between two nodes i and j may be between 0 and p. The graph is then called *p-graph* and *multigraph* when the arcs are undirected.

In order to know the number of pairs of connected nodes it is therefore necessary to have the precise definition of the relations, i.e. an integral description of E which is generally expressed in the shape of a file or a table[4].

---

3 See below the definition of the simple undirected graph and the multigraph.
4 See section 11.1.2.

If the graph admits loops, i.e. arcs, whose starting points and finishing points are at the same node, and it admits multiple arcs, we call it a *pseudo-graph*, which is the most general case.

Graph theory only takes into account the number of nodes and the relationships between them but does not deal with the vertices themselves. The only exception to this rule is the characteristic of source and/or wells which is recognized at nodes in certain cases, such as during the calculation of the maximum flow for Ford-Fulkerson [FOR 68], etc.

However, merely taking into account the existence of nodes, their number and the relationships between them in graph theory is insufficient for network modeling. A better individual description of network vertices is an important problem that graph theory must also tackle to enable certain microsimulations, such as the study of flows and their directions within the network crossroads, or the capacity of the said crossroads, etc.

Thus, graph theory only deals with relationships between explicitly defined elements which are limited in number. Indeed, in order to determine certain traditional properties of graphs, such as the shortest paths, the Hamiltonian cycle, etc., the number of nodes must necessarily be finite.

The graphic representation of G is extremely simple: "it is only necessary to know how the nodes are connected" [BER 70]. The localization of the nodes in the figure, i.e. implicitly on the plane, the representation or drawing of the graph do not count, nor does the fact that the latter has two, three or *n* dimensions.

This offers great freedom in representing a graph. On the other hand, for the reproduction of a transport network, for example, and if we wish the result to resemble the observation, in short, if we want to approximate a map, this representation will have to be specified. This is done by associating to it the necessary properties or additional constraints, so that the development process of the representation can be repetitive and the result reproducible (for example, definition of the coordinate type attributes for the nodes), which is what Waldo Tobler requires for maps.

Simplicity of the methods of definition and representation of graphs

Let us consider the *associated matrix* or *adjacency matrix* A of graph G. It is the Boolean matrix $n \times n$ with 1 as the (i, j)-ith element when u and v are adjacent, i.e. joined together by a edge or a directed or undirected arc and 0 when they are not [COR 94].

Other authors [ROS 98] generalize this notation by accepting the loop (by noting it 1 at the (i, i)-ith position) and multiple arcs, thus considering that the adjacency matrix is then not a zero-one or boolean matrix because the (j, i)-ith element of this matrix is equal to the number of arcs associated to $\{ui, vi\}$. In this case, all the undirected graphs, including multigraphs and pseudo-graphs, have symmetric adjacency matrices.

The problem of the latter notation is that it can be difficult to distinguish, unless we define beforehand a valuated adjacency matrix when the valuations are expressed as integers and small numbers.

The list of adjacency

The use of the adjacency matrix is very simple. However, it may be cumbersome, in particular in the case of a large graph whose nodes are only connected by several arcs which is, for example, the case of a road network or a lattice on a plane. In this case, the matrix proves very hollow and the majority of the boxes are filled with zeros. To optimize the calculation procedures we then use methods which make it possible to remove these zero values and to only retain the existing arcs.

One of the simplest ways of describing a graph, in particular by using a machine, is to enumerate all its arcs when there are no multiple ones or to enumerate them by identifying [MIN 86] those whose origin and destination are identical when we are dealing multigraphs or directed p-graphs, which constitutes an arcs file[5]. The writing can be simplified by using an *adjacency list.*

This adjacency list specifies the nodes which are adjacent to each node of the graph G. We can even consider for a Boolean adjacency list of a p-graph or of a multigraph that the number of times where the final node is repeated indicates the number of arcs resulting from the origin node and leading to the destination node, *half a bipolar degree.* If the description of the graph is not only Boolean, it might then be necessary to identify each arc between the same two nodes, in particular, by their possible valuation, weighting or another characteristic, such as a simple number.

The incidence matrix

For a graph without loop, the values of the incidence matrix "vertices-edges" $\Delta(G)$ are defined [BEA 92] by:

– 1 if x is the origin of the arc;

5 See in Chapter 12 an example of time-lag graphs.

– -1 if x is the end of the arc, 0 otherwise.

In order to avoid confusion let us recall that it is completely different from the "node-node" adjacency matrix whose valuation is equal to 1 when the two nodes considered are connected by an arc. It is this latter matrix, which in certain works is referred to as the associated matrix.

The algorithmic ease has already been underlined and the methods of description of graphs listed above, which are naturally usable by a machine, do nothing but amplify it.

The adjacency matrix enables a simple usage of numerous algorithms, as well as numerous indices, as we will be able to see. It also makes it possible to use sub-tables, etc. However, the description by using an adjacency list enables a greater processing speed due to the absence of zero values tests and the possibility of using pointers[6].

Hereafter we will establish that with some supplements this description of graphs enables us to describe representations and reproducible plots, and that it is sufficiently flexible to extend the formalism of graphs to other fields.

Glossary of graph theory for the description of networks

The definitions of graph theory are commonly allowed and scarcely leave ground for ambiguity. However, certain terms have evolved through time, just as it happens in any active field. We propose to develop the representations of graphs by considering them as strictly belonging to the theory and to express other representations in the form of graphs. Therefore, we must now specify the definitions of the most used terms.

Indeed, since the fundamental work of Berge [BER 70] was published in France 30 years ago a certain number of definitions have evolved through use (see below).

---

6 It can be defined as the address of an element.

Directed graph

A directed graph (*V,E*) consists of a set of vertices *V* and a set of edges *E*, which are pairs of the elements of *V* [ROS 98].

"Pseudographs form the most general type of undirected graphs, since they can contain multiple loops and arcs. Multigraphs are undirected graphs that may contain multiple arcs but not loops. Finally, simple graphs are undirected graphs with neither multiple arcs, nor loops" [ROS 98].

Arc and edge

An arc is a directed relation between two nodes (*U, v*) of the set of nodes of G. An edge is always an undirected arc between two nodes (*U, v*) of G.

Adjacency

Adjacency defines the contiguity of two elements. Two arcs are known as adjacent if they have at least one common end. Two nodes are adjacent if they are joined together by an arc of which they are the ends. The nodes *u* and *v* are the final points of the arc $\{u, v\}$.

Incidence

Incidence defines the number of arcs, whose considered node is the origin (incidence towards the exterior: out-degree) or the destination (incidence towards the interior: in-degree). Since the degree of a node is equal to the number of arcs of which it is the origin and/or destination, each loop is counted twice.

Regular graph

When all the nodes have the same degree, the graph is known as *regular.*

Degree of a node

The degree of a node in an undirected graph is the number of arcs incidental to this node, except for a loop that contributes twice to the degree of this node. The degree of this node is noted by deg(*v*).

Symmetric graph

A graph is known as symmetric, if each node is the origin and destination of the same number of arcs.

The adjacency matrix of a symmetric graph is symmetric.

Complete graph

A complete graph is a graph where each node is connected to all the other nodes by exactly one arc. A complete graph with n nodes is noted by Kn. A complete directed graph is a digraph where each node is connected to all the others by two arcs of opposite directions.

Subgraph

A subgraph is defined by a subset $A / A \subset V$ of nodes of G and by the set of arcs with ends in $A / UA \subset U$, $GA = (A, UA)$. For example, the graph of the Central region is a subgraph of France. It is fully defined by an adjacency submatrix.

Partial graph

A partial graph is defined by a subset of arcs $H \subset E / GS = (V,E)$. A partial graph may be a monomodal graph of a multimodal graph as well as a graph of trunk roads within the graph of all the roads in France. The adjacency matrix of a partial graph has the same size as the adjacency matrix of the complete graph.

For example, if the partial graph is a modal graph (i.e. defined by a specific means of transport), the adjacency matrix of the complete graph (i.e. of the transportation system) is the sum of all the adjacency matrices of the partial graphs (various means of transport).

Partial subgraph

A partial subgraph combines the two characteristics mentioned above: it is formed by a subset of nodes and a subset of arcs GSA = (A,V) such as, for example, the partial subgraph of TGV cities.

Chain

A chain is a sequence of arcs, such that each arc has a common end with the preceding arc and the other end is in common with the following one. The cardinal of the considered set of arcs defines the length of the chain. In a transport network where the arcs are, by definition, directed, the chain only makes sense only if the arcs are symmetric, i.e. directed both ways.

Path

A path is a chain where all the arcs are directed in the same way, i.e. the end of an arc coincides with the origin of the following one.

Circuit

A circuit is a path whose origin coincides with the terminal end.

Cycle

A chain is called a cycle if it starts and finishes with the same node.

Eulerian cycle

A Eulerian cycle in a graph G is a simple cycle containing all the arcs of G. A Eulerian chain in a graph G is a simple chain that contains all the arcs of G.

A chain is known as Hamiltonian if it contains each node of the graph only once.

Connected graph

An undirected graph is connected if there is a chain between any pair of nodes.

A directed graph is known as connected if there is there a path between any pair of nodes.

Strongly connected graph

A graph is described as strongly connected if, for any pair of nodes, there exists a path from the origin node to the destination node.

In other words, in a strongly connected graph it is possible to go from any point to any other point and to return from it, which is one of the essential properties of a transport network.

Quasi-strongly connected graph

A graph is known as quasi-strongly connected if for any pair of nodes u, v, there is a node t, from which a path going to u and a path going to v start simultaneously.

A strongly connected graph is thus quasi-strongly connected.

Bi-partite graph

A graph G is bi-partite if the set V of its nodes can be partitioned into two non-empty and disjoined sets V1 and V2 in such a manner that each arc of the graph connects a node of V1 to a node of V2 (so that there is no arc of G connecting either two nodes of V1 or two nodes of V2)[7].

A joint or pivot

A node is a joint if upon its suppression the resulting subgraphs are not connected.

Isthmus

An isthmus is an edge or an arc whose suppression renders the resulting partial subgraphs unconnected.

Articulation set

By extension, a set $UA \subset U$ is an articulation set if its withdrawal involves the loss of the connectivity of the resulting subgraphs G.

List 1. *Essential definitions*

7 See below the K3,3 graph.

Description, representation and drawing of graphs

For the majority of authors the term representation indicates the description of the graph by the adjacency matrix and the adjacency list or the incidence matrix and the incidence list, as well as that the graphic representation of the considered graph in the form of a diagram, whose absence of rules we have seen[8].

For representations in the form of a list or a matrix table we will use the term *description*, possibly by specifying computational description and by mentioning the possible attributes of the nodes, such as localization, form, modal nature[9], valuations[10] of the arcs, etc.

For graphic, diagrammatic representation we will use the term (graphic) representation or drawing of the graph.

This notation appears more coherent to us since, in the first case, we describe the graph by listing all of the nodes and arcs, possibly with the attributes of the nodes and the characteristics of the arcs: modal nature, valuation, capacity, etc., which are necessary for computational calculation. For the computer the representation of arcs has neither sense nor utility.

On the other hand, in the second case, we carry out an anthropic representation of the graph, possibly among a large number of available representations according to constraints that we set ourselves, such as planarity, special frame of reference, isomorphism with a particular graph, or geometrical properties that we impose on a particular plot, such as linearity of arc, etc.

Isomorphic graphs

The simple graphs G1 = (V1,E1) and G2 = (V2,E2) are isomorphic if there is a bijective function f of U1 in U2 with the following properties: u and v are adjacent in G1 if and only if $f(u)$ and $f(v)$ are adjacent in G2 for all the values of u and v in E1. Such a function f is an isomorphism.

---

8 See section 1.1.1.1.
9 Here the term indicates the means of transport which is possibly assigned to the arc: terrestrial, such as a car, a truck, a train, or maritime, by river or air.
10 Indicates a qualifying value allotted to an arc or an edge, such as distance, duration, cost, possibly modal capacity, etc. Two arcs stemming from the same node and having the same node as destination can have different valuations, for example, distance by road and rail between two cities.

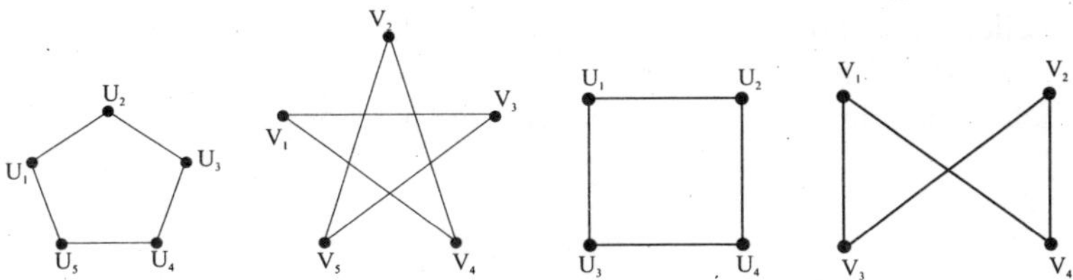

Figure 1. *Example of isomorphic graphs*

Plane graph

A plane graph is a graph whose nodes and arcs belong to a plane, i.e. whose plot is plane. By extension, we may also speak of a plot on a sphere, or even on a torus.

Two topological graphs that can be led to coincide by elastic strain of the plane are not considered distinct.

All the graph drawing are not necessarily plane; they can be three-dimensional like the solids of Plato, or like a four-dimensional hypercube traced in a three-dimensional space and projected onto a plane as the famous representation of *Christ on the Cross* of Salvador Dali.

Planar graph

It is said that a graph *G* is planar if it is possible to represent it on a plane, so that the nodes are distinct points, the arcs are simple curves and two arcs only cross at their ends, i.e. at a node of the graph.

The planar representation of *G* on a plane is called a topological planar graph and it is also indicated by *G*.

Any planar graph can be represented by a plane graph, but the reciprocal is not necessarily true.

Saturated planar graph

A planar graph is described as saturated when no arc can be added without it losing its planarity. In a saturated planar graph the areas delimited by arcs are triangular.

Christaller's transport network (Figure 3) [CHR 33] is a plane graph based on triangular grids; it is neither planar nor saturated because some arcs do not only cut across each other at the nodes and some areas are quadrangular.

Figure 2. *European*[11] *quadrimodal graph*

11 Graph plotted by CESA *Geographical position working group 1.1 Study Program on European Spatial Planning*, December 1999. An extended version integrates the *ferry boat* into this graph which represents four modes of transport.

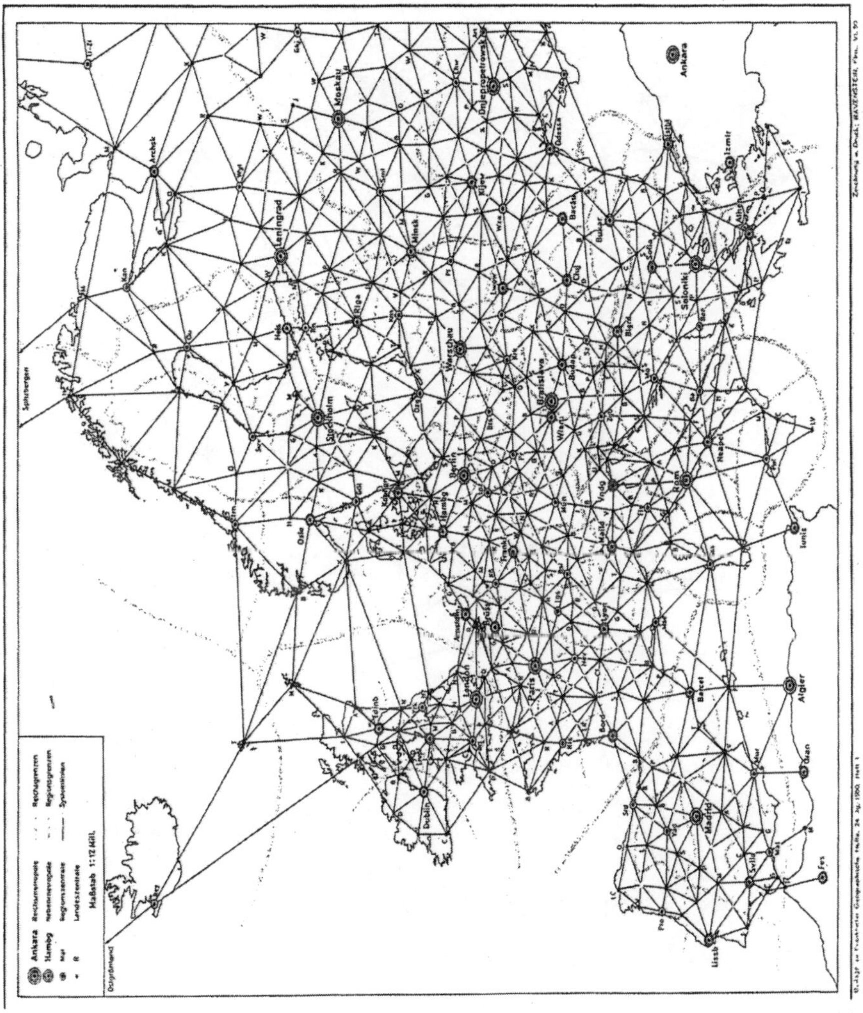

Figure 3. *Walter Christaller's network*

Figure 4. *Multimodal graph of France*[12]

---

12 Planar multimodal saturated graph of France. Plotted by CESA, this graph is truly multimodal because, due to the zoom method of Laurent Chapelon (see below), in its extended version it can integrate regional, departmental, agglomeration and even intra-urban graphs with their specific modes (1999).

The search for the planarity of graphs led to famous publications and numerous algorithms. Among the best known results let us quote two traditional properties:

– Euler's formula:

- let G be a connected planar simple graph with e edges and v vertices,

- let r be the number of regions (or areas) in a planar representation of G.

Then $r = e - v + 2$.

– Kuratowski's theorem: a graph is not planar if and only if it contains a homeomorphic subgraph with K3,3 or K5 (see Figure 5).

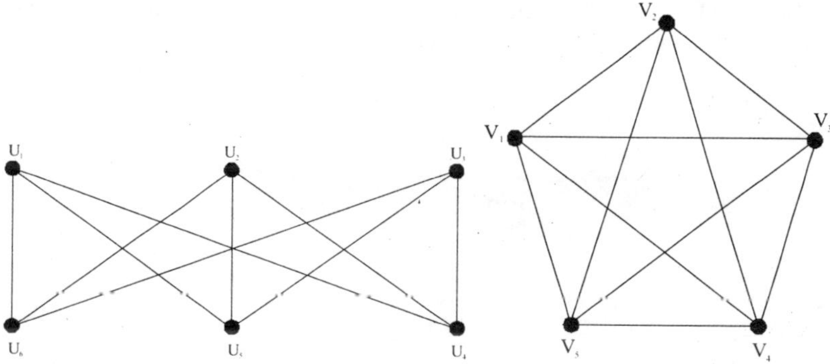

Figure 5. *K3,3 graph and K5 graph [ROS 98, p. 479 and 419]*

We add two definitions to these traditional ones in order to specify graph plottings with more than two dimensions.

Graph with geographical reference

A graph with geographical reference (GGR) is either a plane graph or a graph plotted on a simple surface, for which it is possible to perform a bijection between the nodes of the plotting and those of its projection onto a plane. In other words, no node or face of a GGR must have a double point as a projection.

The typical example will be a digital terrain model like that in Figure 6 [BRU 87]: this graph, whose areas are quadrangular and thus not necessarily planar, belongs to a surface defined in a three-dimensional space which is qualified by the use of the 2.5D in landscape analysis[13].

Graph with spatial reference

We will also call a graph with space reference, or GSR, a graph plotted on a convex surface or in a three-dimensional space or more, such as, for example, the solids of Plato but also the GSR like those used by Kamal Serrhini to define co-visibilities [SER 00][14]. In this case, when the graph is projected onto a plane there does not have to be a bijection between the drawing of the graph in the space and its plane projection because certain points, nodes or areas can be doubled[15]. Let us note that the use sometimes qualifies the graphs drew on a "planar" sphere by extension, since in this case the plane is considered as a sphere with infinite radius.

Figure 6. *Digital terrain model of the Mont Blanc [BRU 87]*

13 This qualification, i.e. dimension 2.5, is usual and established by the use in co-visibility. It should be specified that it is used to differentiate this type of GGR grid from the GSR defined hereafter, which, in turn, is not limited by the constraint of bijection. This qualification of dimension 2.5 has nothing to do with a non-integer dimension of the fractal type.

14 See below Part 2, Chapter 10.

15 For example, several points of a façade or a work are projected orthogonally onto the same point in the ground.

Dual graph

There exist many definitions of duality[16]. We will retain the one used by the majority of authors. The duality of a graph consists of associating each area of a graph called primal to a node of the dual graph. Berge provides the following definition for it [BER 87]: "let us consider a planar graph G, which is connected and without isolated nodes. We make it correspond to a planar graph G* in the following manner: inside every face s of G we place a node x * of G *; we make every edge e of G we correspond to an edge e* of G*, which connects the z nodes x* and y* corresponding to the faces s and t that are on both sides of edge e. The graph G* thus defined is planar connected and does not have an isolated node: it is called the dual graph of G" (see Figure 7).

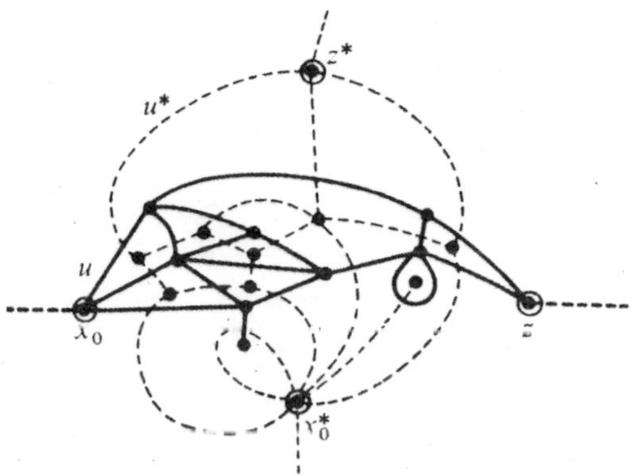

Figure 7. *Primal graph and its dual [BER 70]*

This definition is insufficient and, for example, Aldous and Wilson specify [ALD 00]: "let G be a connected graph. Then a dual is constructed from a *plane drawing* (italics added) of G, as follows…". The definition is more exact, as they demonstrate, by defining the dual of a convex three-dimensional polyhedron, as we examine in Part 3.

Similarly, in certain, even very simple, cases the dual is not of the same type as the primal in the sense that a primal 1-graph can have a p-graph as a dual, as we will see. This may present a problem in terms of graph description, since the adjacency matrix of the dual then has to be an extremely hollow *p*-dimensional matrix…

16 For example, the definition of the dual graph of a grid (adjacency matrix) provided by Pumain will not be retained here: [PUM 97], page 31.

The concept of dual graph is very rich, but insufficiently used. Indeed, it makes it possible to make areas correspond to networks and vice versa, which is one of the fundamental problems of synthesized images and of specialized network representations among others. Thanks to the duality and to what stems from it, it is possible to bypass some of the limitations facing modeling.

A particular type of graphs: the tree and tree structure

A tree is a connected undirected graph without a simple circuit. An undirected graph is a tree, if and only if each pair of nodes is connected by a simple and single circuit.

Tree and tree structure: we can call a particular node of the tree the root and then attribute a direction to each node in such a way that there is a only path from the root to each node. We then obtain a directed graph which is referred to as a tree structure (see Figure 8).

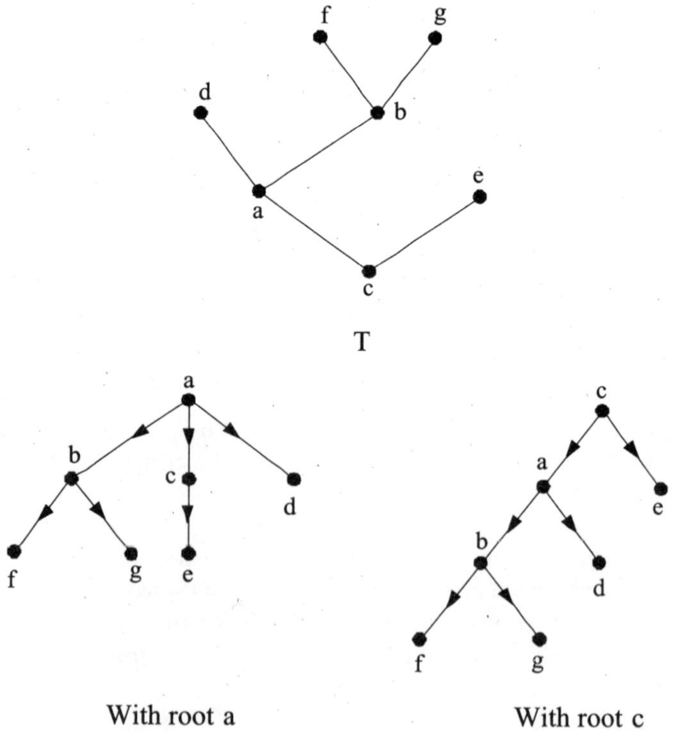

Figure 8. *Examples: a tree and two tree structures [ROS 98, p. 506]*

If v is a node other that the root, the father of v is the single node u, such that there is a directed arc from u to v. If u is the father of v, v is called the child of u. Nodes with the same father are called siblings. The ancestors of a node other than the root are nodes of the path leading from the root to this node. The descendants of a node v are the nodes that have v as ancestor. The node of a tree that does not have offspring is called a leaf. Nodes that have offspring are called internal nodes. A subtree is the subgraph comprised of a node of the tree, its descendants and all the arcs leading to its descendants.

A tree structure is described as *m-ary* if each internal node does not have more than m offspring. The tree is called a *complete m-ary* tree if each internal node has exactly m descendants. An m-ary tree with $m = 2$ is called a binary tree.

A tree structure will be quasi-strongly connected if there is a node u, from which we can reach all the others via a path. The node u is known as the root of the tree: it is the common ancestor of all the nodes.

An example of a tree structure in networks is a monopolar access map, such as those plotted by Laurent Chapelon[17], or Legrand's star map presented below (see Figure 9).

Somc additional thcorcms deserve being mentioned:

– a tree with n nodes contains $n-1$ arcs;

– a complete *m-ary* tree with i internal nodes contains $n = mi + 1$ nodes;

– a complete m-ary tree with:

- n nodes contains $i = (n-1)/m$ internal nodes and $l = [(m-1)\ n + 1]/m$ leaves,

- i internal nodes contains $l = (m-1)\ i + 1$ leaves,

- l leaves contains $n = (ml-1)/(m-1)$ nodes and $i = (l-1)/(m-1)$ internal nodes.

17 Chapelon Laurent; see Chapter 1.

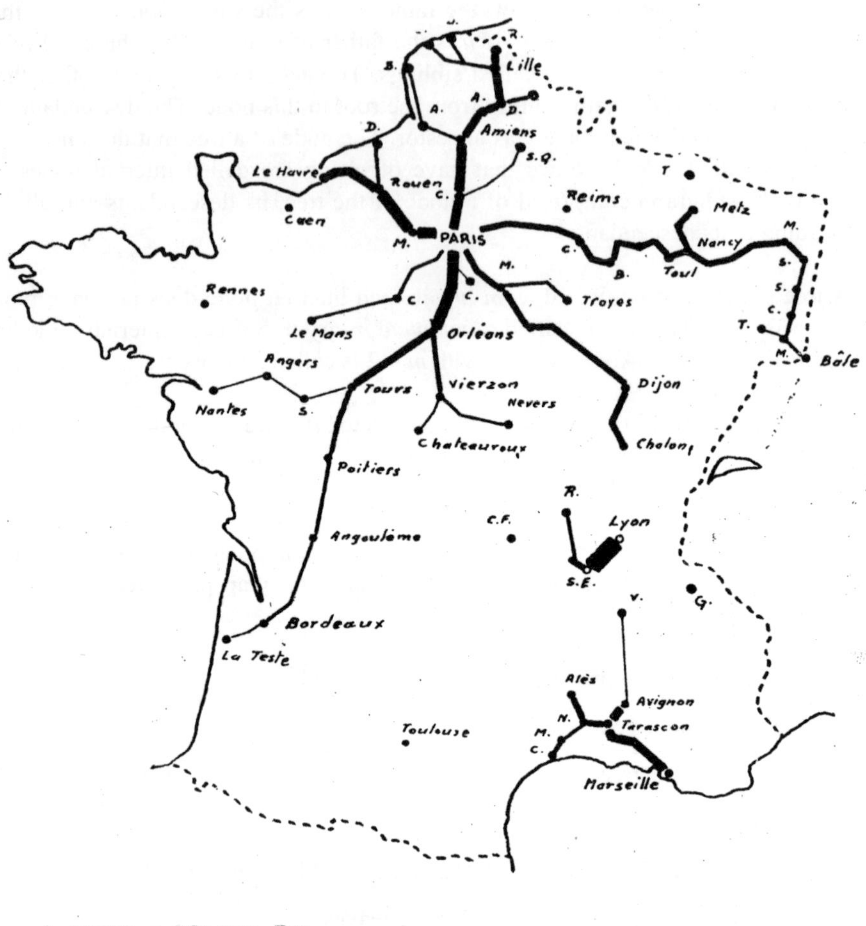

Figure 9. *Merchandise traffic by rail in France. Annual throughput by line section (in effective thousands of tons) in 1854. Extract from Renouard, Les transports des merchandises depuis 1850, Armand Colin*

The height or depth of a tree structure is the maximum level of the nodes or the length of the longest path between the root and any node. An m-ary tree structure of height h is known as balanced if all the leaves are at the levels *h* or *h-1*. There are at most m^h leaves in an m-ary tree of height *h*.

Spanning tree

Let G be a simple graph. A covering tree of G is a subgraph of G, which is a tree containing each node of G. Finally, a simple graph is connected if and only if it has a spanning tree.

The great algorithmic simplicity of graph theory is the heritage of operational research. Numerous problems that arise in the network simulation have been solved by operational research [FIN 01, KAU 64], which developed many algorithms first for a manual and then for a computerized resolution.

Among many publications that have made it possible to obtain effective algorithms it is necessary to cite Ford and Fulkerson for the flow algorithm in graphs, Dantzig and then Ford for the first shortest paths algorithms, etc. These works paved the way for many later achievements that have been encouraged and reinforced by developments in data processing, which provided the necessary means of computation.

Graph description methods for computer by list or by adjacency or incidence matrix, and all the methods derived from those easily lend themselves to the computerized processing and thus to the resolution of problems involving large graphs.

The use of these methods of computerized graph description has been made even simpler since the 1990s in the domain of transport networks by the development of data capture methods and the development of adjacency list-type files through the digitalization of maps or direct on-screen capture and modification (Chapelon, *Les logiciels MAP et NOD* [*MAP and NOD Software*]; see Chapelon, L'Hostis). The rapid development and generalization of GIS (geographical information systems) or of SRIS (spatial reference information systems) reinforce and accelerate this evolution.

Once an algorithm has been developed, the size of the graph or manageable network depends only on the data, the characteristics of the machine and the available computing time. However, it has to be said that the computing time of certain algorithms grows very quickly with the size of the graph. This is a problem of algorithm efficiency, of calculation complexity: "let us consider a given class of problems, whose size is given by the integer n. We say that a solution algorithm A has a complexity of the order $f(n)$ – noted by $Of(n)$ – if the asymptotic growth of the computing time $TA(n)$ according to n, which is the size of the problem, is of the order $f(n)$ at the most (where f is an increasing positive function of n, which is generally a polynomial function). An algorithm A is known as polynomial and of complexity $O(n^k)$ if, in the worst case scenario, the resolution time grows as the kth power of n, which is the size of the problem, when $n \to \infty$..." [MIN 86]. For example, the time needed for the research of the minimal paths between any pair of nodes in a graph by Floyd's algorithm [COR 94, p. 550] grows according to the cube of the number of nodes. In this algorithm that has an exceptionally beautiful symmetry of expression, there are three overlapping loops and when the number of nodes doubles, the computing time increases eightfold. It is an algorithm of complexity $O(n^3)$, that is, a polynomial complexity [ROS 98, p. 97 *et seq.*].

On the other hand, although computers have progressed considerably in their capacities to rapidly produce high definition images, this problem has practically not been tackled by graph theory whose development logic remains largely marked by its history and dependant on the research logic of mathematicians and operational researchers who are practically unaware of this aspect.

However, the fastest method to apprehend networks and results expressed as networks is the synthesized image, i.e. an effective and precise graphic representation, which is reproducible and verifiable to the same extent as the optimization calculations of operational and therefore algorithmic research. In spatial analysis and urban development planning the synthesized images must become a measurement and diagnostic instrument, as in many other disciplines: medicine, physics, biology, etc.

The representation of networks by graphs

Network modeling by graphs constitutes a relatively recent application for an old requirement. Although the network in its current meaning is a relatively modern concept, the need to represent a set of roads and terrestrial or maritime routes, on the other hand, is very old and has become accentuated with time. Indeed, this need develops along with the progressive interpenetration of societies witnessing the development of their exchanges. Little by little, tradesmen, sailors, soldiers will feel a greater need for not being dependant on local guides, for no longer being limited by local knowledge and for being able to consider all of the characteristics of various displacements: difficulty, length, duration, risk, locally available resources, etc. That explains the development of cartographic representation methods rather than networks and spaces.

From functional representation to resemblance

One of the most ancient functional representations of a road network is the "Peutinger Table" (Figure 10 in the color plates section) from the 2nd or 4th century (which reached us thanks to a medieval copyist) which represents a map of ancient routes from city tc city with the indication of distances.

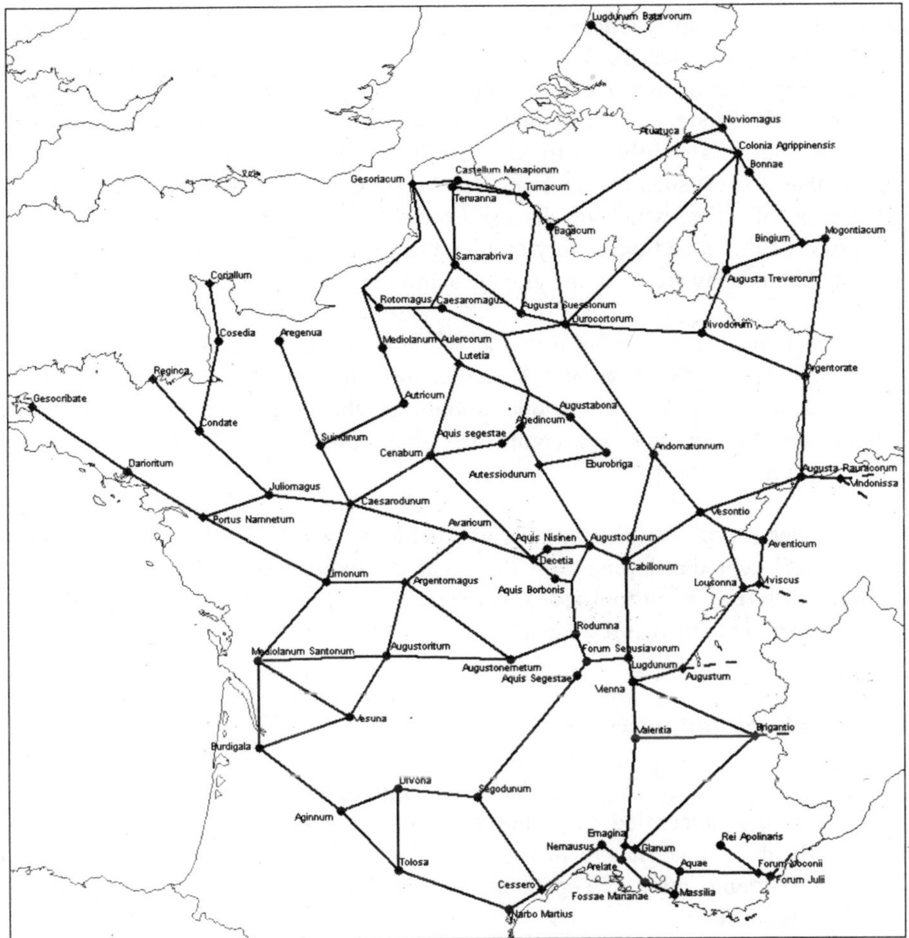

Figure 11. *Layout of the network of the Peutinger Table against the background of a modern map*

This Table is very interesting because it does not constitute a map but a functional representation corresponding to the drawing of a graph. It makes it possible to identify stages (poles or nodes of the graph), to calculate distances (valuation of the arcs) and, thus, the duration of the journey.

Indeed, we note that it strictly corresponds to the definition of a graph plotting provided by Berge: "it is only necessary to know how the nodes are connected". The Peutinger Table only slightly resembles a map and its representation against the background of a map[18], whose form corresponds to that of the territory as above, makes it possible to see the differences (see Figure 11). These two representations are isomorphic but neither one nor the other has the properties of reproducibility and comparability. We could think that the second is very close to having these properties, but it is simply the force of habit that makes us forget all the implicit notions that it uses, such as, for example, the localization of nodes identical to the localization of cities, which results from these "stages mentioned in the table...". The proof of this assertion is very simple and to observe it would suffice to ask a student, who is unaware of all the graphic semiology, to draw this map.

A millennium later, starting from the 13th and 14th centuries, the first portolans[19] established by the Genoese and Venetian and then by Arabic and Portuguese navigators, illustrated the famous pilot's logbooks thus showing that the description of the possible route must be accompanied by a representation, a graphic draw and that the text is not enough[20].

This need for cartographic representations supplementing the "nautical instructions" kept developing with the great discoveries and would constitute an extraordinarily important tool at sea and then on solid ground. Was it not Napoleon who considered the map as a weapon?

Increasingly scientific mapping

A considerable progress in precision, in semiological and mathematical rigor in the increasingly objectified maps has been carried out, among others, under the influence of the geometricians, of whom the Cassinis[21] were undoubtedly the most famous. The triangulation of a space thus corresponds to defining a saturated planar graph in this space.

---

18 This map has been plotted in 2001 by Olivier Marlet, a student of "Urban Sciences" DEA archeology option at Tours.

19 From the Italian *portulano*: pilot.

20 Indeed, if machine descriptions of graphs as adjacency matrix, for example, enable many calculations including those of certain "morphological" indicators, they do not make it possible for the human mind to visualize the specific network, which is described without additional data.

21 From 1696, which is the date of the first planisphere plotted by Gian Domenico Cassini, to the large map known as Cassini's with a scale of 1/86,400, which was introduced in 1789 to the French National Assembly by his great-grandson Jean Dominique Cassini. This is one of the first topographic maps based simultaneously on a complete triangulation of France and geodetic readings taken on the ground.

However, between the sea or terrestrial maps and the Peutinger Table there is a fundamental difference: maps have the aim of representing the territory (i.e., amongst other things, a surface) in the most precise possible way: the relief of the depths or altitudes, its form (plateaus, mountains, peaks, reefs, cliffs, crossing points, major and minor rivers, currents, etc.), whereas the Peutinger Table is only a functional representation and does not resemble road networks.

Even if the map resulting from the Cassinis features the road communication networks, its aim is different: one of its constraints is that of "resemblance" to the territory because it must describe it and not be limited to networks, to "paths and circuits" in the sense of graphs, but also make it possible for someone to orientate himself in this terrestrial or maritime space. It is a tool to define the position in space in the most precise possible way.

This constraint of resemblance between the terrestrial surface and a plane representation was the object of many mathematical works and the development of the most traditional projection methods as those of Mercator[22] and Lambert.

This objective of resemblance that the maps are given with the aim of representing the space or the territory implies the existence of homeomorphism between the space represented and the representation of space[23]. This property of homeomorphism doubles on maps due to a resemblance resulting from graphic semiology codes: the forests are in green, the rivers, the lakes and the seas in blue, etc. Even on a small scale Michelin map, the winding roads are represented with many turns indicating their sinuosity, etc. We are relatively far from Peutinger's functional representation, even though it features some zigzags, stylized mountains and urban monuments.

Already overloaded with stylized decorations, fantastic characters and animals to fill in the blanks and to compensate for the unknowns, the map is overloaded with descriptive details of the territory by means of graphic semiology: the map has to account for the territory and it must make it possible to apprehend it as a whole, its evolution and its significant details. We refer to this as the constraints of resemblance.

---

22 Mercator Gerhard 1521-1594 *Tabulae geographicae ad mentem Cl. Ptolemaei* (1578) – Mercator readopts the projection invented in the 1st century by Martin de Tyr. This projection on a cylinder tangent to the equator is improved by Lambert who uses a cone tangent to a central meridian line.

23 This increasingly strong homeomorphism risks creating confusion between the map and the territory, which Korzyski denounces in: *Science and Sanity, an Introduction to Non-Aristotelician Systems and General Semantics* (1933): "the map is not the territory".

Networks, roadmaps and graphs: the constraints of resemblance

The maps representing networks must give the same impression of immediate resemblance, thus enabling an instantaneous comprehension. Undoubtedly, it is for this reason that the layout of a road on the Michelin map evokes the real layout of the road: straight where it is actually straight and curving when the layout is winding. Having said that, there is no strict relation between the two layouts because the width of the route is not connected with that of the representation but only bears a similarity: bolder lines for highways, thinner for minor roads, etc. Also, the smaller the scale of the map, the larger the disproportion becomes, which necessarily requires a simplified layout (Figure 12).

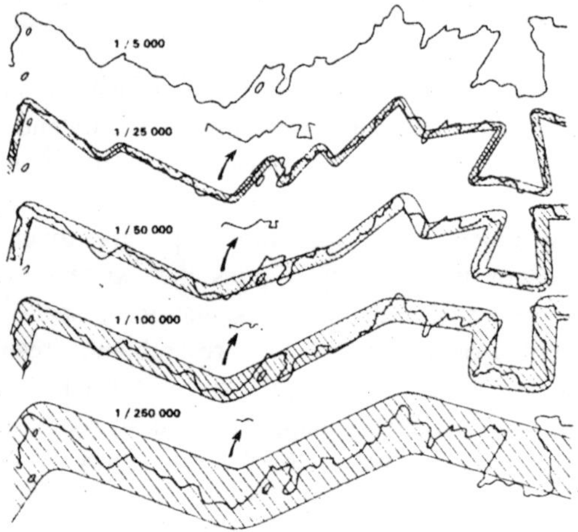

Figure 12. *Evolution of a coastal layout through the reduction of the scale of representation*

Networks, tree structures, flow charts: the constraints of hierarchy

The described networks can be of particular types, such as flow charts retracing a hierarchical system in a structure. The usage dictated by habit encourages a representation of this type of dissymmetric network in a pyramidal form, with the most important element in the hierarchy sitting atop and the subordinate levels following each other in order of decreasing importance.

These tree structures use a technique which is found in scientific applications, in physics and data processing: diagrams blocks ([CHO 64], p. 91 *et seq.*), which will hereafter be referred to as data flow charts [FAU 75], which are graphic translations of a program or of a part of a program.

Many domains are still insufficiently explored

Graph theory primarily developed in two research directions: mathematical research and operational research, according to the specific dynamics of these two fields. The concerns and the questions posed by mathematicians are very specific, but often have very little to do with those of the developers of transport networks, space analysts and urban planners.

The concerns of the operational researchers are much closer to them. Some are obviously common, but urban planners, as those for whom their work is meant, i.e. contractors and the public, have a visualization requirement that cannot be circumvented and a need for illustration that, except in rare cases, is not present for operational researchers.

Urban planners need representations that are repetitive and verifiable, as well as comprehensible for all the public, in particular, within the framework of public interest investigations. Moreover, we have seen that graph theory is absolutely not preoccupied with the representation in the sense of graphic plotting of a graph. In a certain manner Berge eliminates the problem by writing: "it is only necessary to know how the nodes are connected. The localization of the nodes in the figure, the representation or plotting of the graph do not count" [BER 70, BER 87].

It is neither considered nor even mentioned that the representation can be "plotted" on a plane or in a three-dimensional space or be the projection of one onto the other [CAU 76, CAU 86], except in the case of a planar graph and that of the construction of a dual graph, which can be plotted on a plane, a sphere or a torus [HEA 76, HEA 86, LHU 76, LHU 86, POI 76, POI 86]. The concerns of the founding fathers of graph theory have been simultaneously developed and reduced.

The drawing of graphs and their constraints, network image or images

This absence of rules of graph drawing is a considerable difficulty for the representation of modeled networks, whereas for network cartography there is a graphic semiology.

We pose as an axiom that the rules of the effective, material layout and of a representation form an integral part of graph theory and that the necessary and sufficient conditions for this realization to be reproducible and verifiable must be established.

We propose as a working hypothesis that this definition will enable us to integrate various geometrical results into graph theory, in particular, certain basic elements of fractals that will make it possible to provide the rules for the extension or reduction of graphs, to ground in theory the already operational concept of zoom by using the property of *self-similarity* or strict or probabilistic internal homothety and to provide global morphological indicators of networks, such as the fractal dimension [GEN 00].

The problem of graph extension and reduction and the scales of representation under constraints are very important problems in network modeling. We have mentioned the complexity of the algorithms for the resolution of traditional problems, such as that of the search for the shortest paths in a graph, and also mentioned that it was of the $O(n^3)$ level. It is then easily conceived that two problems present themselves acutely: that of the reduction of large graphs and that of the rule to be used for the process not to be left completely to the free interpretation of the modeler, but to be, as we have already stated, reproducible and verifiable; basically, to be more objective, more scientific.

This problem of reduction of graphs and of the size of graphs to be processed, even though it evolves rapidly with the performance of machines, *ipso facto* poses the problem of the levels of calculation or the scales in spatial analysis. The problem of the existence of various network apprehension and modeling scales immediately leads to the question of the existence of internal homothety between the various levels considered. Can they be treated in the same manner and up to which point, which limit?

We pose the working hypothesis that at a certain level the logic and the hypotheses change. For example, the reflection on the saturation of a transport system can be solved by calculating the capacity of one or several cross-sections of the network by using the Ford-Fulkerson theorem. However, this is done at the cost of very strong assumptions and, in particular, those according to which the node is strictly neutral, which is the general assumption in graph theory that only deals with the relationships between nodes and not with them as such. Consequently, the saturation of an arc is related to its capacity. This is obviously at the very least an abusive assumption and completely opposed to the reality of the flux of flows. Any car driver knows from experience that the saturation of an axis almost always results from the saturation of a crossroads, i.e. of a node, except for cases that constitute an accident, like the reduction of the number of lanes, and even in this last case, normally, the capacity of an arc should equal that of its narrowest portion.

Similarly, the total disinterest of the theory in the vertex or node, structure means that the description of certain network elements is practically impossible. For example, the highway half-interchange[24] obviously exists and yet it is impossible to represent it by a node under the current state of the theory.

Therefore, it is necessary to improve network modeling, to call into question this assumption of the neutrality of nodes and to consider their structures, the possibility of describing them while respecting the axioms and fundamental properties of graph theory.

We put forward another working hypothesis: the development of drawing rules, of representation as well as a reflection on the duality must make it possible to pass from the traditional graph to percolation and the use of MSA (multi-agent systems) in a cellular network, with all the modifications of hypotheses with respect to application that this enables and implies. Indeed, to describe and to simulate, even to optimize, a transport system as a whole is a type of problem that implies certain assumptions that we will clarify; on the other hand, describing the individual behavior of an agent in a network is another problem and does not necessarily require the same assumptions, since sometimes it renders them particularly illusory. Similarly, even thought graph theory indeed makes it possible to represent a network with facility and flexibility and with a certain degree of generality in order to solve some of the problems mentioned above, it enables very little with respect to the graph area which is enclosed between its various arcs.

That is obvious because graph theory is only preoccupied with the relations between nodes. However, the property of duality makes it possible to take this space into account. Nonetheless, this property is closely related to the planar representation of the graph and therefore requires a certain number of additional hypotheses and the contribution of additional precise details in order to answer the questions posed above. This problem will be tackled later in the book.

Lastly, one of the problems that dynamic network modeling comes across is the development of networks, the extension of the graph and of its grid, either simultaneously or subsequently.

We will see that the possibility of describing simple fractals as graphs makes it possible to find a solution for these problems.

If we wish, as is the case, to be able to achieve comparable and repetitive drawing and, thus, the layout of maps in Waldo Tobler's sense, we will note that additional assumptions are necessary to specify the type of possible layout. Similarly, we observe that with some additional assumptions and constraints we are able to draw graphs that have all the properties of fractals.

---

24 See Part 3: a half-interchange does not enable the exit and entry to the highway in both directions.

An explicit assumption must be stressed due to its consequences on later applications: in graph theory we adopt the hypothesis of perfect rationality and total knowledge of the network. Indeed, let there be a graph corresponding to the definition above, where all its elements are known to us via a complete and finite list of nodes and that of the relations between nodes, which is described by the list or the adjacency matrix. Such a definition is absolutely necessary to be able to deal with the basic problems of graph theory and, in particular, the problems of optimization of paths and maximum flow. The demonstration is obvious: if a single arc was missing, it would be impossible to obtain the matrix of the minimal path, since at least one of them might not exist if an arc was added. *A fortiori* if a node is missing.

This significant hypothesis is no longer necessary for the simulation of agents in a "cellular graph": in order to find a path, the local knowledge such as, for example, the number of arcs starting at the node or the next node and their orientations with respect to a reference direction may suffice. The set of necessary assumptions will thus have to be specified for each application and each representation of the graphs.

The plan

The plan of the book follows the development of the questions that we have touched upon above and comprises three parts: the first is entitled *Graph Theory and Network Modeling*; the second, *Graph Theory and Network Representation* and the third, *Towards Multilevel Graph Theory*.

Bibliography

[ALD 00] ALDOUS J.M., WILSON R.J., *Graphs and Applications, an Introductory Approach*, Springer-Verlag, London, Berlin, Heidelberg, 2000.

[BEA 92] BEAUQUIER D., BERSTEL J., CHRETIENNE P., *Eléments d'algorithmique*, Masson, 1992.

[BER 70] BERGE C., *Graphes*, Dunod, Paris, 1970.

[BER 87] BERGE C., *Hyper graphes*, Bordas, Paris, 1987.

[BRU 87] BRUNET R., *La carte mode d'emploi*, Fayard/Reclus, Paris, 1987.

[CAU 86] CAUCHY A.L., Recherches sur les polyèdres, Premier mémoire J. Ecole Polytechnique 9 (Cah.16) (1813), p. 68-86., in N.L. Biggs, E.K. Lloyd and R.J. Wilson, *Graph Theory 1736-1936*, Oxford University Press, 1976, 1986.

[CHA 97] CHAPELON L., L'HOSTIS A., MATHIS P., *Les logiciels MAP et NOD*, 1997.

[CHO 64] CHOW Y., CASSIGNOL E., *Théorie et application des graphes de transfert*, Dunod, 1964.

[CHR 33] CHRISTALLER W., *Die zentralen Orte in Sûddeutschland*, G. Fischer, Paris, 1933.

[COR 94] CORMEN T., LEISERSON C., RIVEST R., *Introduction à l'algorithmique*, Dunod, Paris, 1994.

[EUL 1736] EULER L., "Solutio problemats ad geometriam situs pertitnentis", *Commemtarii Academia Scientiarum Imperialis Petropolitanae*, 8, p. 128-140, *Opera Omnia*, vol. 7, p. 1-10, 1736.

[EUL 1758a] EULER L. "Demonstreatio nonnullarum insignium proprietatum quibus solidra hedris planis inclusa sunt praedita", *Novi Comm. Acad. Sci. Imp. Petropl. 4* (1752-1753, published in 1758), 140-160, *Opera Omnia* (1), vol. 26, 94-108, 1758.

[EUL 1758b] EULER L., *Elementa doctrina solidorum. Novi Comm Acad. SCI. Imp. Petropol. 4* (1752-1753, published in 1758), 109-140, *Opera Omnia* (1) vol. 26, 72-93, 1758.

[FAU 75] FAURE R., *Eléments de recherche opérationnelle*, Collection "Programmation", Gauthier-Villars, Paris, 1975.

[FIN 01] FINKE G., *Recherche opérationnelle et réseaux*, Hermès, Paris, 2001.

[FOR 68] FORD L.R. Jr., FULKERSON D.R., *Les flots dans les graphes trad*, J.C. Arinal (ed.) Gauthier-Villars, Paris, 1968, *Flows in Networks*, RAND Corporation, 1962, Princeton University Press.

[GEN 00] GENRE GRANDPIERRE C., Forme et fonctionnement des réseaux de transport : approche fractale et réflexions sur l'aménagement des villes, PhD Thesis, Besançon, 2000.

[HEA 86] HEADWOOD P.J., "Map Colour Theorem", *Quarterly Journal of Pure and Applied Mathematics* 24 (1890), 332-338, *Graph Theory* 1736-1936 Biggs N.L., Lloyd E.K., Wilson R.J., Oxford University Press, 1976, 1986.

[KAU 64] KAUFMANN A., *Méthode et modèles de la recherche opérationnelle*, Vol. 1 and 2, Dunod, Paris, 1964.

[LHU 86] LHUILLIER S.6A.6J., "Mémoire sur la Polyédrométrie", *Annales de Mathématiques 3* (1812-3), p. 169-189, in Biggs N.L., Lloyd E.K., Wilson R.J., *Graph Theory* 1736-1936 Oxford University Press, 1976, 1986.

[MIN 86] MINOUX M., BARTNIK G., *Graphes, Algorithmes, Logiciels,* Dunod, Paris, 1986.

[POI 86] POINSOT L., "Sur les polygones et les polyèdres", *J. Ecole Polytechnique* 4 (Cah.10) (1810), p. 16-48., in Biggs N.L., Lloyd E.K., Wilson R.J., *Graph Theory* 1736-1936, Oxford University Press, 1976, 1986.

[PUM 97] PUMAIN D., SAINT JULIEN T., *Analyse spatiale 1 Localisations dans l'espace*, Armand Colin, 1997.

[ROS 91] ROSEN K.H., *Mathématiques discrètes*, Chénelière/McGraw-Hill, 1991.

[SER 00] SERRHINI K., Evaluation spatiale de la covisibilité d'un aménagement, sémiologie graphique expérimentale et modélisation quantitative, PhD Thesis, Tours, 2000.

[XUO 92] XUONG N.H., *Mathématiques discrètes et informatique*, Masson, Paris, 1992.

PART 1

Graph Theory and Network Modeling

Chapter 1

The Space-time Variability of Road Base Accessibility: Application to London

This chapter presents an attempt to integrate traffic conditions into an accessibility calculation model associated with road networks at the regional scale. For identical networks, congestion, which is closely linked to the use of infrastructure, is responsible for variations of traffic speeds. In order to optimize accessibility calculations it is of primary importance to obtain the "real" speed section by section by taking into account the level of traffic. To that end we endeavored to systemize the empirical approach initiated by the *Transport Research Board* and developed in its *Highway Capacity Manual* [TRB 98]. This approach, which is associated with a specific road network modeling method in the form of valuated graphs, can indeed account for space and time variability of motor vehicle traffic speed on roads.

1.1. Bases and principles of modeling

1.1.1. *Modeling of the regional road network*

Graph theory is one of the most effective means to model transport networks. The vertices of a graph are associated with network nodes and the arcs are associated with links. Along with the principle of mechanical graph description, this type of modeling makes it possible to benefit directly from the calculating power offered by computers.

Chapter written by Manuel APPERT and Laurent CHAPELON.

1.1.1.1. *From network to graph*

A priori no constraints restrict the choice of the graph vertices. The size of nodes in terms of population, of economic or administrative functions, for example, may be given prevalence. However, several factors must be taken into account in order to improve the precision of the results.

First of all, the strategic role that certain nodes may play in the organization and operation of the road network (crossroads, interchanges) makes it necessary to compute them. Then, when the design features of the infrastructure or their environment (urban/rural) change, it is advisable to create a new arc. The vertices are placed consequently. Lastly, the graph can also adapt itself to the available structure of traffic measurement. The limits separating two measurements series are therefore likely to be regarded as vertices of the graph.

Optimally, an arc corresponds to a section of the road presenting relatively homogenous design features and intensity of use over its entire length.

The arcs of the graph must render with precision the design features of the road infrastructure, insofar as they directly influence the traffic speed. That relates simultaneously to the structure and the quality of the roads. Among the elements that may be noted let us cite the number, width, pattern, slope, sinuosity of lanes, existence of a central reservation on dual carriageways, width of lateral clearances, etc.

Within the framework of our applications these elements made it possible to define 9 classes of infrastructure (Table 1.1). If the number of classes is not restrictive, we postulate that the discrimination applied here is adequate to model road networks on a regional scale with precision (Figure 1.3). Free speed is an important input for the model. For a given vehicle type (here a car) this is the maximum traffic speed technically allowed by each class of infrastructure within the limits authorized by the Highway Code and without congestion. The variations observed between French and British free speeds are due to differences in maximum speeds authorized by the Highway Code and to differences in design features present for the same infrastructure class. The geometry of roads in the French network often has better quality than that of the British roads. Infrastructures are assigned to each class by the operator on the basis of the technical information provided, in particular, by roadmaps and road agencies.

Class	Design features of the infrastructure	Free speed	
		France	Great Britain
R1	Highway.	130 km/h	113 km/h
R2	Intercity expressway: dual carriageway of the highway type and principal or regional connection road with a separate roadway.	110 km/h	113 km/h
R3	Urban expressway: highway or dual carriageway of the highway type with a reduced shoulder (< 60 cm).	80 km/h	80 km/h
R4	Trunk or regional connection road with 3 or 4 lanes or 2 broad lanes, width: [7 m-9 m].	90 km/h	96 km/h
R5	Trunk or regional connection road with 2 "standard" lanes, width: [5 m-7 m].	80 km/h	80 km/h
R6	Trunk or regional connection road and local service road with 1 or 2 narrow lanes, width: ≤ 5 m.	70 km/h	49 km/h
R7	Urban artery: penetrating or structuring by-pass of the urban road network.	50 km/h	49 km/h
R8	Urban boulevard: principal urban access road.	35 km/h	35 km/h
R9	Street: local urban access road.	35 km/h	35 km/h

Table 1.1. *Infrastructure classes*

1.1.1.2. *From graph to machine*

Two types of files enable the electronic alphanumeric description of graphs. Node files contain the name, code, population and geographical coordinates of vertices. Arc files contain the codes of the vertices of origin, destination, class of infrastructures to which the arc is attached, the length of the arc in kilometers and the occupancy rate of the road lane or the hourly (or working day) rate needed when taking congestion into account.

The principle of modeling makes it possible to work with orientated and two-ways arcs simultaneously. The orientation applies to infrastructure classes. Since the classes have at least one orientated arc (single direction of traffic, traffic data differentiated by direction) all of their orientated arcs are in the database. In this case a one-way road is described by an arc (i, j) and a two-way road by two distinct arcs (i, j) and (j, i). The arcs of other classes are automatically regarded as orientated in both directions and therefore they do not have to be computed twice in the database.

1.1.2. *Congestion or suboptimal accessibility*

In the case of small scale road modeling, the functioning of the system, which is an essential factor in determining the accessibility levels, is often undifferentiated in time as well as in space.

In this case traffic speeds are fixed regardless of the geographical location and the time of day; we do not consider the effective operation of the infrastructure and, thus, the possible local failures of the system to satisfy the mobility of people and goods in the modeled zone. Taking into account the operating conditions of the road network, i.e. congestion phenomena, appears less relevant. The use of average speeds applied to the length of arcs in kilometers is an alternative that can be adapted for the calculation of the durations of journeys, since for each pair "vehicle type/infrastructure class" they are gauged with precision [CHA 97, CHA 98, CHA 00].

On the other hand, as the geographical scale increases, the omission of traffic conditions becomes more problematic when road modeling is sought. For regional applications, it is indeed necessary to evaluate real traffic speeds with more precision, arc by arc, in close relation with localized system operation failures. Finally, on a very large scale the analysis of vehicles movements at intersections uses another category of methods known as microscopic.

Aiming our observations at a regional scale we have to measure the accessibility of sites according to their geographical position, structure and quality of the network, traffic codes and rules and vehicle types, as well as the congestion phenomena, which, all things being equal, may imply a significant decrease in accessibility.

Thus, the gains in accessibility that could be achieved by developing a fast regional network may be partly cancelled out by congestion on its links or at its intersections. We will therefore highlight situations of suboptimal accessibility, for which taking into account real speeds by road segment modifies the accessibility, which would be calculated on the basis of fixed average speeds.

If network failures were only accidental, taking them into account would no longer be essential, since these congestion phenomena would not correspond to network structural logics that could be subjected to modeling.

On the other hand, if the regional area under research manifests recurring traffic problems with respect to space and time, we may seek to find out their impact on accessibility. It would then be a question of modeling a progressive phenomenon, a function of daily and hourly variations of network use (holiday periods, start of the working week, rush hour during working days, etc.). We will thus admit the relevance of a congestion function.

A short-term constant, i.e. the infrastructure capacity, is subjected to modulated demand according to the stretch of road and time slots. The supply and demand ratio measures the intensity of use of the infrastructure and, thus, its level of congestion. To this end we have two types of demand variables: one measured in time (flow) and the other in space (occupancy rate) [ABO 85], whose theoretical characteristics and possible applications are different (Figure 1.1).

We will see how to introduce each type of variable into a specific congestion modeling method adapted to accessibility calculations in a valuated graph.

Flow q is defined as the number of vehicles passing through the counting point per unit of time. By extension, the obtained results are assigned to the road segment delimited by two intersections.

To formalize: "let us consider a point x on a road and count the vehicles that circulate there in one or both directions during the interval $[t, t + \tau\Delta]$. Let us note by $n[t, t + \tau\Delta]$ the number obtained. The flow at point x during the period $[t, t + \tau\Delta]$ is by definition the number of vehicles passing by per unit of time, that is [LEU 93]:

$$q[t, t+\Delta t] = n[t, t+\Delta t] / \Delta t\text{"} \qquad [1.1]$$

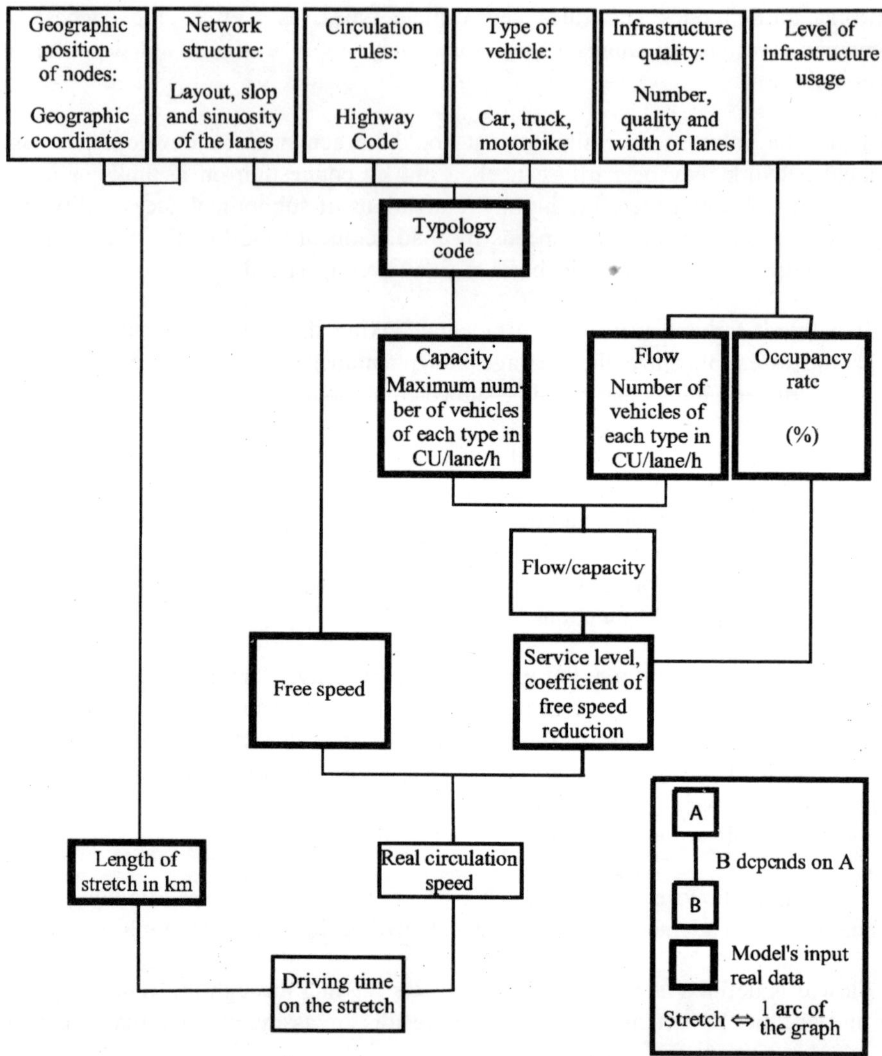

Manuel Appert, Laurent Chapelon, 2002, UMR ESPACE 6012, Montpellier

Figure 1.1. *Calculation of driving times*

This type of data, which is currently used, poses a theoretical problem limiting its use. The flow-speed curve (Figure 1.2) allows the same flow for two different traffic speeds. A low flow may be interpreted either as a low traffic level (high speed), or over saturation characterized by an extremely reduced flow (low speed).

Figure 1.2 shows in this respect that speed is a decreasing function of flow to the critical point where capacity is reached. From this point the two variables decrease simultaneously [COH 90]. Considered up to the critical point indicating infrastructure saturation, the flow-speed ratio can be used both for affected and observed flows.

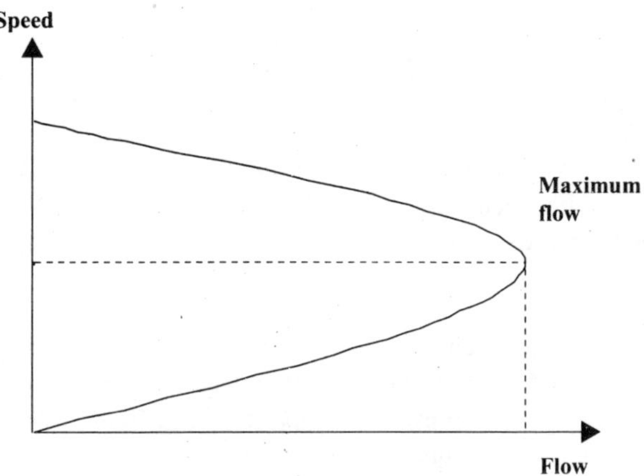

Figure 1.2. *Flow-speed curve*

Once we have flows affected by a "traditional" traffic model [CER 90] as $emme_{/2}$[1] on a very complex graph, over saturated traffic states are avoided. Indeed, in these sequential models with four stages (generation, distribution, modal split and application) the last phase that consists of applying modeled traffic demand to the routes proceeds by iterations of applications to the quickest paths between the starting point and the destination. Consequently, once axis saturation is attained, the ways not yet affected are used as alternative routes. The traffic conditions will thus be saturated at most and the use of flows remains relevant.

On the other hand, when we have effective (observed) flows for a given time slot it is impossible, as emphasized previously, to distinguish low traffic levels from over saturation. In this case, good knowledge of traffic states of the analyzed network makes it possible to locate the sections where a risk of over saturation exists and to consequently adapt the flows. The ideal is, obviously, to get as much information related to concentration (occupancy rate) as possible.

The occupancy rate of a road is also a specific measure, which can however be easily linked to concentration, which is a variable of space. It is the part of the

1 Sequential traffic modeling software used, in particular, by highways agencies.

measurement period during which a vehicle occupies the measurement point. More formally: "at a moment t let us consider a stretch [x–Δx, x] of a road and count the vehicles that are there. Let us note by n[x–Δx, x, t] the number obtained. The concentration k[x–Δx, x, t] at the moment t on the segment [x–Δx, x] is by definition the number of vehicles per unit of length [LEU 93], that is:

$$k[x-\Delta x, x, t] = n[x-\Delta x, x, t]/\Delta x\text{"} \qquad [1.2]$$

"Experiments with homogenous traffics show (…) that the speed of the flow decreases as concentration grows. The assumption of the fundamental diagram (statistical relation between flow and speed) is the postulate that there exists a functional dependence u = u(k) between the flow speed and concentration" [LEU 93].

In this sense speed, which is the opposite function of occupancy rate, can be calculated regardless of the present level of traffic, as opposed to what the use of observed flow enables us. The use of occupancy rate is more relevant theoretically and makes it possible to measure with precision road congestion and, in particular, accessibility in cases of over saturation.

However, the method that we will endeavor to clarify hereafter allows the use of two types of input variables, i.e. flow or occupancy rate, since the latter is seldom available. This method is inspired by the *Highway Capacity Manual*[2].

1.2. Integration of road congestion into accessibility calculations

1.2.1. *Time slots*

The amount of traffic in a given axis varies according to time. To evaluate the real performance of a network we generally look at the period of its most intensive use. Observable traffic points my arise by accident when they are the result of particular occurrences or can be recurring when they are allotted to regular mass movements, such as daily commutes (to or from employment centers), mass weekly departures (during the weekend at the outskirts of large agglomerations) and mass seasonal travel (in the outskirts of coastal towns and on intercity corridors).

2 The *Highway Capacity Manual* presents a method of road infrastructure capacity and service level estimation. It does not propose legal standards, but rather represents the commonly accepted American points of view summarized by the *Transport Research Board* of Washington.

In urban or outlying environment it is common to carry out observations during rush hour in the morning or in the evening, that is between 8 and 9 am and 5 and 6 pm respectively, even though variations may be recorded depending on the areas studied [EPD 97]. Therefore, the occupancy rates and flows observed depend on the choice of a particular time slot.

1.2.2. *Evaluation of demand by occupancy rate*

The estimation of traffic demand must take into account the various vehicle types. When the occupancy rate is used this is automatically enabled "by an electromagnetic loop placed under the road surface which makes it possible to measure the occupancy rate, by recording the duration of electromagnetic contacts, from which the duration of the real contacts is deduced, by taking into account the influence of the circuit length" [LEU 93]. The number and length of vehicles are thus associated to provide a reliable picture of spacing between vehicles responsible for variation in traffic speeds.

Occupancy rates are measured over short time periods: from a few seconds for networks equipped with dynamic regulation modes (real-time traffic management, such as the Gertrude[3] and Petrarque systems) to several minutes for networks controlled by static devices. They can then be aggregated over an hour.

We then obtain the average percentage of the road occupied by the vehicles for a given section of the road and a given time. 0% corresponds to zero traffic and 100% to maximum occupation, literally bumper to bumper. Thus, occupancy rates accurately account for the real level of traffic stretch by stretch and make it possible to better take into account the localized problems.

Expressed as rates, they free the users from having to evaluate road capacity, since they represent a supply and demand ratio. The measurement techniques used provide information for each traffic direction, implying the use of an orientated graph.

However, it is advisable to ensure an appropriate location of electromagnetic sensors under the road. Indeed, placed near a vehicles stoppage area (ie: controlled or uncontrolled intersection), these sensors may provide information unsuited to the method used here. It is generally preferable to use several measurement points on the stretch and to calculate an average or, failing that, a single judiciously positioned measurement point [ABO 85, LES 92].

---

3 http://www.gertrude.fr.

1.2.3. *Evaluation of demand by flows*

1.2.3.1. *Flows*

Flows measured or approached using a "traditional" traffic model are generally expressed in vehicles per unit of time. However, finer models can differentiate between cars, motorcycles and trucks. Moreover, if we consider their technical differences (size, flexibility), the flow measurement would have to be homogenized.

Thus, the larger trucks occupy a greater area of the road. Moreover, their length makes overtaking them more difficult. Taking into account their weight and the speed limits imposed on them, these vehicles circulate at a lower speed, causing blockages in the traffic flow. These become stronger if the road is sloped and overtaking possibilities are limited. Even when a lane is available for overtaking, the speed differential between the overtaking vehicle and the one being overtaken causes a loss of space that could have been occupied in front of the vehicle being overtaken. Finally, in case of jams[4], trucks react more slowly to mode changes. The same applies to buses, which may, moreover, slow down traffic during stops.

Since cars constitute the majority in the distribution of traffic volume, they are often selected as standard. Consequently, the volume of trucks is represented in automobile equivalent, such as 1 truck equals 2 cars, which is an average. Variations are observable in reality, in particular, according to the gradient of the slope[5]. Contrarily, motorcycles are considered less important than other vehicles. It is generally retained that 1 motorcycle corresponds to 0.4 cars.

When these two standardized vehicles classes (trucks and motorcycles) are added to the volume of cars we obtain for each section the total volume expressed in car units (CU), which is then divided by the number of lanes and a unit of time (CU/lane/h).

Depending on the availability of information, the flow may be differentiated according to traffic directions, implying that the graph is directed and the value is divided by the number of lanes in the direction considered. If it is not differentiated (unorientation graph) we divide the value by the total number of lanes (both directions together).

If the data consists of average annual daily flows (in CU/day), it is advisable to apply the percentage of daily traffic circulating during the time period that we wish to study to the value. This percentage may vary depending on the class of infrastructure

4 At traffic lights, roundabouts, stop or give way signs.

5 In this respect, a finer method is presented in the *Transport Research Board, Highway Capacity Manual*.

to which the stretch belongs, which, in this particular case, is in fact an additional input into the model. This calculation makes it possible to obtain flows in CU/h.

1.2.3.2. *Capacity determining factors*

First of all, it is necessary to isolate exogenous factors that are not taken into account as they are not particularly explanatory of structural congestion problems. For example, weather conditions can be responsible for variations [MAY 89]. Among the endogenous factors we may cite the frequency of bus stops and intersections with and without traffic lights.

If the intersections are controlled by traffic lights, the fact of allocating a fraction of time per movement phase (passage) involves considerable capacity variations, especially as a safety lag must be observed between two phases and the reaction of users is not instantaneous.

Moreover, certain endogenous factors may prove to be not easily quantifiable, on the one hand, because they are very subjective, such as the differences in behavior between commuters and occasional road users and on the other hand because they refer to incidental occurrences, such as traffic accidents or roadworks.

Two major factors determine capacity. Firstly, the maximum speed authorized by the Highway Code in the sense that for identical roads a higher authorized speed can enable greater traffic flow. In addition, the number of lanes has the same effect because, all things being equal, it can be estimated that the more the number of lanes increases, the more the infrastructure capacity increases. Indeed, the multiplication of lanes makes it possible for several vehicles to overtake simultaneously at the same point.

In addition to legal speed limits and the number of lanes, technical features of the infrastructure are major elements for an efficient capacity evaluation. Indeed, the capacity of two roads with the same number of lanes may differ, since they exhibit different cross-sections (width of lanes, existence of a central dual carriageway separator, distances between the lanes and obstacles on the sides, etc.) Let us consider, for example, a reduction of lateral space. It is generally noted that users tend to compensate for this reduction by increasing the distance with the car in front, which contributes to reducing supply capacity [CHAU 97].

Taking these design features into account in the evaluation of theoretical capacities can be done in two manners: either by applying reduction coefficients specific to the optimal theoretical capacity of the infrastructure category considered [TRB 98], which are coefficients that take into account technical, weather or behavioral factors, or, as we described previously, by adapting the principles of graph modeling to a fine description of the road network in order to distinguish the various types of roads and the various traffic conditions that can be encountered.

At this time in the development of the method it is advisable to take into account the scale, on which the analysis is carried out. Indeed, on a local scale, the most exhaustive possible consideration of microgeographical elements that have just been mentioned proves essential but within the framework of an analysis carried out on a regional scale it is different. In the latter case only the most important elements have to be retained for the obvious reasons of cost-efficiency-precision of modeling.

1.2.3.3. *Capacity*

Capacity may be interpreted as the maximum observed flow. We will discuss theoretical capacity, which is intrinsic to supply (number and width of lanes, linear or singular installations, access conditions), corresponding to optimum usage conditions. We will also suppose that flows indicate "downstream fluid flow, a sufficient upstream demand to approach capacity [and] vehicles sharing space homogenously" [LEU 93].

Infrastructure class	**Capacity in CU/lane/h**
Highway (R1) Intercity expressway (R2) Urban expressway (R3)	1,800
Trunk or regional connection road with 3 or 4 lanes or 2 broad lanes (R4) Urban artery (R7) and urban boulevard (R8)	1,250
Principal or regional connection road with 2 standard lanes (R5) Principal or regional connection road and local service road with 1 or 2 narrow lanes (R6) Street (R9)	1,000

Table 1.2. *Capacity of the road infrastructures*

Thus, each of the 9 road classes used for the road modeling network is associated with a capacity provided by the infrastructure engineering departments (Table 1.2). In cities the capacity may be evaluated in two ways: either by using an "intersection by intersection" modeling logic, which requires integrating average waiting times at each crossroad into the calculation of movement time, or by considering the most important road sections (from 1 to 3 km) in order to obtain an average capacity, by taking into account crossings of intersections controlled or uncontrolled by traffic lights.

The temporal constraint of the latter will thus be expressed by a loss of capacity leading to reduced speeds and therefore driving times. The meso-geographical approach that we choose encourages us to adopt the second approach, the first being preferentially applied on a microgeographical scale.

1.2.3.4. *The flow/capacity ratio*

Having established the flow and the capacity (in CU/lane/h) for each section of the road network for a given time slot, it is then advisable to calculate the flow/capacity ratio necessary for the section by section evaluation of traffic conditions.

1.2.4. *Calculation of driving times*

To evaluate traffic conditions we use the classification by level of service (LOS) developed in the *Highway Capacity Manual* on the basis of flow or occupancy rate criteria (Table 1.3) [TRB 98, LEU 93]. For each stretch the level of service is given by the occupancy rate value or by the flow/capacity ratio value (Tables 1.4 to 1.6). The correspondence between the level of service and occupancy rate or flow/capacity ratio varies according to the infrastructure class to which the stretch belongs. A speed reduction coefficient defined as the ratio between the sought real traffic speed and free speed (Tables 1.4 to 1.6) corresponds to each LOS and each infrastructure class. Thus, for each stretch the real speed is obtained by multiplying this coefficient by the free speed of the infrastructure class to which the stretch belongs:

$$\text{Real speed} = \text{free speed} \times \text{speed reduction coefficient}$$

Then, by dividing the length of the stretch in kilometers by the real speed we obtain the time needed to traverse it:

$$\text{Driving time} = \text{length of the section/real speed}$$

The components of the method presented above were integrated into the NOD[6] software which was conceived to evaluate the projects aiming to modify transport supply.

LOS	Traffic	Description of traffic conditions
LOS A	Fluid	Traffic conditions are ideal. Users tend to move at optimal speed without inconveniencing each other.
LOS B		Traffic is fluid. Overtaking, however, may prove difficult and the observed speed is close to the authorized limit.
LOS C	Busy	Traffic is busy, maneuvers become difficult. Speed starts to drop.
LOS D	Dense	Traffic becomes denser. Lane changes are very difficult and cause considerable deceleration.
LOS E		Traffic is strongly slowed down, flow becomes saturated. Traffic conditions are very unstable, even a minor incident leads to LOS F
LOS F	Saturated	Capacity is insufficient, backlog phenomena occur. Traffic is slowed down or even stationary. The flow is no longer continuous.
LOS G	Over saturated	

Table 1.3. *Description of the level of service*

6 Chapelon L., L'Hostis A., Mathis P., 1993-2002.

Levels of service for highways (R1), intercity expressways (R2) and urban expressways (R3)			
	Occupancy rate (%)	Flow/capacity	Coefficient of free speed reduction
LOS A	[0 – 4]	[0 – 0.318]	0.86
LOS B	[4 – 7]	[0.318 – 0.509]	0.81
LOS C	[7 – 11]	[0.509 – 0.747]	0.77
LOS D	[11 – 16]	[0.747 – 0.916]	0.66
LOS E	[16 – 25]	[0.916 – 1]	0.43
LOS F	[25 – 40]		0.33
LOS G	> 40		0.21

Table 1.4. *Table of the level of service for highways and expressways*

Levels of service for roads (R4), (R5) and (R6)			
	Occupancy rate (%)	Flow/capacity	Coefficient of free speed reduction
LOS A	[0 – 4]	[0 – 0.25]	0.92
LOS B	[4 – 7]	[0.25 – 0.4]	0.83
LOS C	[7 – 11]	[0.4 – 0.6]	0.75
LOS D	[11 – 16]	[0.6 – 0.8]	0.67
LOS E	[16 – 27]	[0.8 – 1]	0.54
LOS F	[27 – 40]		0.33
LOS G	> 40		0.21

Table 1.5. *Table of the level of service for interurban roads*

Levels of service for urban arteries (R7), urban boulevards (R8) and streets (R9)			
	Occupancy rate (%)	Flow/capacity	Coefficient of free speed reduction
LOS A	[0 – 4]	[0 – 0.25]	0.9
LOS B	[4 – 7]	[0.25 – 0.49]	0.7
LOS C	[7 – 12]	[0.49 – 0.6]	0.5
LOS D	[12 – 20]	[0.6 – 0.8]	0.4
LOS E	[20 – 33]	[0.8 – 1]	0.3
LOS F	[33 – 50]		0.24
LOS G	> 50		0.15

Table 1.6. *Table of the level of service for streets*

Examples of applications

Use of the occupancy rates

Let us take the example of a 5 km long stretch of an interurban highway with 3 lanes each way. This stretch belongs to the R1 class (highway). In France the free speed is 130 km/h. What is the real speed between 5 and 6 pm?

– determination of the free speed reduction coefficient: the information contained in the data file used in model input provides us with the average occupancy rate of each stretch by time period, distinguishing between traffic directions (directed graph). For the stretch that is of interests, i.e. between 5 and 6 pm, this rate is 15% in the north-south direction and 10% in the south-north direction. With an occupancy rate of 15% the north-south direction forms part of the [11 – 16] class of the table of level of service for highway infrastructure (R1). The stretch exhibits dense traffic conditions (LOS D), with which a free speed reduction coefficient of 0.66 is associated. The occupancy rate of the south-north direction corresponds to the [7 – 11] class in this same table, that is, busy traffic (LOS C). In this direction the free speed reduction coefficient equals 0.77;

– calculation of the real traffic speed: the real traffic speed is obtained by multiplying the free speed by the reduction coefficient calculated above. The real peed is thus 86 km/h in the north-south direction (130 × 0.66) and 100 km/h in the south-north direction (130 × 0.77);

– calculation of driving time for the stretch: the time needed to drive the entire length of the stretch is about 3 and a half minutes in the north-south direction (5 × 60/86) and 3 minutes in the south-north direction (5 × 60/100).

Use of flows

Let us take the example of a 4 km long highway stretch with 3 lanes in both directions located in a residential area with a lateral clearance less than 60 cm. This stretch belongs to the R3 class (urban expressway). In France, as in Great Britain, free speed is 80 km/h and the average annual daily traffic is of 100,000 vehicles (including 10% of trucks) in both traffic directions. What is the time needed to traverse the stretch between 5 and 6 pm?

– differentiation and homogenization of traffic: (100,000 × 10)/100 = 10,000 trucks circulate. Moreover, we consider 1 truck ⇔ 2 cars. Upon standardization this represents 20,000 trucks (in equivalent cars or CU) and 90,000 cars (100,000 – 10,000). In total 110,000 CU circulate, that is, 18,333 CU/lane/day (110,000/6);

– calculation of hourly rate: with respect to urban expressways we could observe that 9.1% of the total daily volume circulate during the height of the evening rush hour, namely 5-6 pm. The theoretical flow per lane for this time slot is thus 1,668 CU (18,333 × 9.1/100);

– determination of the free speed reduction coefficient: a theoretical capacity of 1,800 CU/lane/h is associated with "urban expressway" infrastructure class. Consequently, the flow/capacity ratio is equal to 0.93 (1,668/1,800). The expressway is nearly saturated and therefore tends to operate in LOS E. From Table 1.4 we deduce that the free speed reduction coefficient is equal to 0.43;

– determination of the real traffic speed: the real traffic speed is thus 34 km/h (80 × 0.43). This speed applies to both traffic directions, since the initial data does not make it possible to make a distinction;

– calculation of the driving time for the stretch: the time necessary to drive the entire length of the stretch is about 7 minutes (4 × 60/34) in each direction.

1.3. Accessibility in the Thames estuary

London is the centre of the most populated urban area in Europe, i.e. the South-East of England. The geographical area studied here relates more particularly to the enlarged East Thames Corridor which is the eastern half of this area. The latter includes the zones bordering the Thames estuary, namely, Essex, Kent, Greater

London, but also Cambridgeshire, Hertfordshire, Sussex and Surrey. These counties have been affected by profound spatial changes that have upset the distribution of settlements during the last 50 years.

The processes of suburbanization and counterurbanization, which are at work throughout the entire East Thames Corridor since the interwar period, have indeed resulted in strong space dilution of activities and residences beyond the green belt surrounding London in a radius of 30 to 40 km from the city centre [HALL 89]. The large number of commuters and the lengthening of daily travel distances lead to heavy traffic regardless of the distance from the centre of London. An analysis of the distribution of average annual daily flows shows peak traffic on the M25 (200,000 vehicles/day), which is the ring road that encircles London at a distance of 25 km from the centre.

The road network, although improved over the last 20 years (M20, M25), is not always able to support such amounts of traffic. This brings about high levels of congestion. In addition to intensity, congestion is characterized by its large extent, since recurrent traffic difficulties are observable within distances of more than 50 km from central London [IAU 98].

The graph of East London conceived by using MAP©[7] software comprises 891 vertices and 1,321 arcs (Figure 1.3 in the color plates section). It corresponds to the state of the road network in 1998 as described on the Michelin map number 404. The graph contains the first eight infrastructure classes described in Table 1.1. The free speeds used are presented in the same figure.

The arcs of the graph are not orientated since traffic measurements provided by the Department of the Environment, Transport and the Regions (DETR) correspond to average annual daily flows from 1997.

Hourly rates are obtained by using load diagrams of DETR [DEP 97]. The proportions of daily traffic circulating during the two time periods used in this study are summarized (Table 1.7).

---

7 Alain L'Hostis, Philippe Mathis, 1993-2002.

Infrastructure classes/time slots	1-2 pm	5-6 pm
Motorway (R1) Intercity expressway (R2)	5.6%	8.9%
Urban expressway (R3) Urban artery (R7) Urban boulevard (R8)	5.7%	8.5%
Trunk or regional road with 3 or 4 lanes or 2 broad lanes (R4) Trunk or regional road with 2 standard lanes (R5) Trunk or regional road and local service road with 1 or 2 narrow lanes (R6)	5.8%	8.6%

Table 1.7. *Proportion of daily traffic during the 1-2 pm and 5-6 pm intervals*

The flows in CU/h obtained using these percentages are then divided by the total number of lanes in the section (both directions together) in order to obtain traffic data in CU/lane/h.

The Thames ferry crossing between Woolwich and Silvertown (Greater London) is added to the description of the road network. The crossing's modeling used different principles than those described previously. Indeed, congestion does not directly influence the crossing time (four minutes). On the other hand, the time spent in waiting areas on both banks can vary strongly depending on the time of day. We estimated this time at 15 minutes for the 1-2 pm period and 30 minutes for the 5-6 pm period. The duration of the journey (wait + crossing) is directly associated with the corresponding arc.

1.3.1. *Overall accessibility during the evening rush hour (5-6 pm)*

Among accessibility indicators the sum of minimal driving times between each node and all the others is one of the most synthetic:

$$t_r(i)=\sum_{j=1}^{n} t_r(i,j) \qquad [1.3]$$

where:

– i is the index of the node of origin;

– j is the index of the node of destination;

– $t_r(i, j)$ is the minimal driving time through the network between i and j;

– $t_r(i)$ is the sum of minimal driving times through the network between i and every other city.

The interpretation of results must take into account the four principal explanatory factors that affect this indicator simultaneously, namely: geographical position of the nodes, road network structure, infrastructure quality and congestion.

This indicator highlights the accessibility structure for the selected time slot (5-6 pm). The central part of the estuary and notably the right bank benefit from the best accessibility (Figure 1.4 in the color plates section). The centre-periphery ratio is high, ever so slightly higher than 2.5. The A2/M25 interchange is the central point of the structure and on the contrary Frinton, in Essex, is the least accessible place.

The class structure of the most accessible nodes, around the A2/M25 interchange, is characterized by a diffusion extended to the east. On the other hand, it is strongly limited to the west by the congestion in the densely built-up area. The relatively good functioning of the fast network on the right bank makes it possible to maintain the same structure as far as central Kent. Conversely, the traffic difficulties in London and on the M25 (Dartford) tend to sharply reduce accessibility to the west and, to a lesser extent, in the left bank corridors. More generally, the progressive character of the diffusion structures in the east tends to be replaced in the west, at least in the agglomeration, by a juxtaposition of highly spatially restricted classes. Moreover, accessibility is diffused more homogenously in the semi-rural territories of Essex than in Sussex, since the latter has more irregular topography (inside of the North and South Downs cuestas), which often leads to a more sinuous infrastructure.

The weight of the geographical position of the nodes is the dominant explanatory factor of the accessibility levels observed in Figure 1.4 (see color plates section). The choice of the space studied is not neutral. This is particularly clear for London and the coastal cities. This edge effect precludes a good legibility of the structure, of the quality and operation of the axes connecting the peripheral nodes. It is possible to eliminate it and to isolate the sole network performance by calculating a specific accessibility indicator.

1.3.2. *Performance of the road network between 1 and 2 pm and 5 and 6 pm*

The network performance indicator for each network node is the ratio between the sum of the direct distances to all of the other nodes and the sum of minimal driving times to these nodes:

$$v_e(i) = \frac{\sum_{j=1}^{n} l_e(i,j)}{\sum_{j=1}^{n} t_r(i,j)} \times 60 \qquad [1.4]$$

where:

– i is the index of the node of origin;

– j is the index of the node of destination;

– $l_e(i, j)$ is the Euclidean length separating i from j;

– $t_r(i, j)$ is the minimal driving time through the network between i and j;

– $v_e(i)$ expressed in km/h is an indicator representing the influence of network performance on the accessibility of node i.

The value thus obtained places all the nodes on an equal footing from the point of view of their geographical localization. The minimal driving time present in the denominator is a function of the geographical position of the cities, structure, quality and operation of the network. As a result, the Euclidean length in the numerator erases the influence of the geographical position without deteriorating the impact of network performance.

1.3.3. *Network performance between 1 and 2 pm*

The general structure of network performance during off-peak hours is clearly of the "corridor" type. The best performances follow the layout of the expressways (Figure 1.5 in the color plates section). The variations in accessibility exhibit a ratio of 1.7 between the two extremes. The M11 interchanges show the best performances, whereas the nodes of central London, such as Hyde Park Corner, suffer from the worst performances in the network.

The London agglomeration encircled by the M25 ring-road presents the most heterogenous picture. The strong proportion of urban boulevards, which are generally saturated, reduces average velocities. This is not only true for central London, but also for the south of the agglomeration.

The borough of Croydon is thus exceptionally inaccessible by comparison with others boroughs of Outer London. The A23 that connects it to London and the urban roads (A232, A212) that connect it to other urban centers in the southern suburbs are of poor quality and present a saturated operation.

The western end of the estuary, in turn, benefits from a much better accessibility. This is even better to the south of the Thames where the A2 and A20 expressways that are lightly congested during the off-peak hours maintain a "corridor" effect. The Blackwall tunnel, even when congested, makes it possible to maintain relatively good accessibility to the Greenwich and Stratford boroughs. The A406 inner ring also makes it possible to diffuse the general network performances in a narrow corridor in the north-east of the agglomeration.

Finally, the M25 does not seem to play a major part in diffusing accessibility in the agglomeration. Its good performances are limited by the congestion of the urban road connectors leading to it. On the other hand, it is on the external side where its performances and its relative good functioning are truly diffused through the radials and thus benefit a larger number of cities [LIN 91]. The nodes located at both ends of the M25 should be more affected mechanically than those positioned halfway through its course. This is verified for interchanges 6 and 8. On the other hand, the interchanges 25, 26 and 27 suffer from the effects of congestion of the north-eastern quadrant of the M25.

In the rest of the metropolitan area, "rural" East Sussex and the estuary peninsulas (Essex) remain disadvantaged by the lack of fast infrastructures. For Southend-on-Sea, which in a "dead-end" position in the network, the problem is exacerbated by the congestion of the axes connecting it to the main network.

This indicator also proves relevant to demonstrate the strengthening of "tunnel effects". While the Eastern Kent network is performing well and is only lightly congested maintaining good general accessibility, the towns of Rochester and Maidstone suffer from the congestion of the axes connecting them to the main network (terminal connections) that is still doing well. Finally, the interchanges of the axes connecting the exterior of the agglomeration maintain a relatively good performance, in particular for those furthest away from London and the M25.

1.3.4. *Network performance between 5 and 6 pm*

The overall accessibility structure for this time slot is more difficult to detect than for the previous time slot (Figure 1.6 in the color plates section). The highways and expressways that are, generally, the principal structuring elements no longer exhibit the distribution of the accessibility levels so clearly.

The distribution of the latter yields a ratio of 1.9 between the most advantaged node (Dover) and the worst off node (Hyde Park Corner). The London agglomeration brings together the strongest contrasts. The saturation of the urban boulevards which prevails in the city centre is responsible for the poor performances detected of the central London nodes. This is in no way exceptional, nor preoccupying, since, actually, only 14% of journeys in this area are taken by cars. However, it is rare to find these low accessibility levels 18 km away from the city centre. Thus, Croydon and Brixton have a score identical to Westminster. The saturation of the South London axes, urban arteries and urban boulevards implies the constitution of a corridor with extremely poor accessibility.

This is reinforced by the saturation of the M25, which cannot offer a proper level of functionality, in particular, for routes undertaken from the south of the agglomeration, as opposed to what is observed during off-peak hours. It is also the case for nodes located in North London. Indeed, the congestion of the M25 and the urban sections of the M11 deteriorates the performances of the Northern corridor, which is, however, less congested. Nonetheless, since only the eastern part of the London area has been adopted as the subject of this study, we have disadvantaged the accessibility of this Northern corridor. Should we consider that the M25 is saturated over its entire length, then the longer the stretch driven by the motorist, the poorer the calculated performance will be and vice versa. We may thus explain the relatively good performances of the nodes located along the M25 from Dartford to the M25/M20 interchange. Indeed, within the selected area the nodes in question are in the middle of the course of the M25. Its use will thus not be as systematic as for the routes stemming from the north or south corridors.

The nodes located near the A12 and A10 owe their relatively poor accessibility to the combination of delays on these two axes, as well as on the M25. Moreover, the poor performance of the nodes located on the estuary peninsulas is explained by the congestion of the Essex axes and by the sinuosity of the infrastructure used to reach it. Southend-on-Sea is the one town, aside from London, which benefits the least from network performance. The rural areas of inland Sussex are also at a disadvantage due to the sinuosity and the spatial patterns of the routes, which lead to long detours of the optimal paths, taking them away from the path as the crow flies.

The eastern end of Kent appears to be an exception. Not only accessibility is greater there, but it is still structured in corridors. The highest values follow the course of the A299 (Faversham-Margate), A2 (Canterbury-Dover) and A20/M20 (Ashford-Folkestone). The remote location and the isolated nature of the county also limit the traffic volumes in this part of England.

By comparing the two time slots (1-2 pm and 5-6 pm), we realize that the most significant losses are found in, or in the vicinity, of the axes of the intercity London suburb. This is verified inside as well as outside of the M25. The latter, whose interchanges suffer from heavy losses, are once again poised to play a major part. Nodes that are most likely to be used are the main victims. This is all the more true as the axes that connect the M25 are already congested. It is the case, in particular, in the North-East and the South. Conversely, the nodes of the agglomeration are definitely less concerned. Since traffic Speeds are already low during off-peak hours, the fall is somewhat attenuated during peak hours by comparison with the rest of the metropolitan area. Similarly, the rural zones, which are often more remote, are barely affected. It is particularly clear for South Sussex, East Kent and the cities and villages at the "dead-end" of the estuary.

Taking into account the flows during peak hours highlights real accessibility levels when network operations are the least effective. We note, by proceeding in a similar manner, that it is primarily the inter-urban roads that suffers most from accessibility losses, which is not exceptional. On the other hand, the scale and the location of these losses are particular to London. Indeed, not all cities can complain about accessibility reductions of more than 30% in a band ranging from 25 to 65 km from the main urban centre!

1.3.5. *Evolution of network performances related to the Lower Thames Crossing (LTC) project*

Figure 1.7 (see color plates section) highlights the importance and location of the performance gains related to the construction of the *Lower Thames Crossing*. The graph comprises 892 vertices and 1,327 arcs. The gains are obtained by absolute difference between the nodal results with respect to two distinct states of transport supply. On the one hand, the initial network dated in 2005, when the recalibrated A13 and A2 will be operational and, on the other hand, the final network including the *Lower Thames Crossing* (LTC) envisaged by 2015 [GLA 01].

This new Thames crossing comprises a 1.5 km bridge connected to the A13 to the North and the A2/M2 to the South by expressway stretches (R2) with 2 × 2 lanes for a total of 14 kilometers. Two arcs are added to the initial graph: one between the M2 J1 interchange and Shorne and the other between Shorne and the A13/A1044 interchange. The traffic at the two sections should reach 1,500 CU/h in both directions, that is, 375 CU/lane/h [DEP 93]. For the remainder of the network we preserve the 1997 flows for lack of available reliable data of prospective nature. The bridge has a triple objective. First of all, it is intended to improve connections between the two economic centers of the estuary, namely, the Southend-Basildon-Wickford conurbation in the North and the Medway Towns in the South. It is also an issue of reducing the isolation of the peninsulas in the North of the estuary and seeking an alternative route to the congested M25 motorway especially around the Dartford area. According to the simulation, the gains observed overall are high and diffused. This is explained by the relative position of the bridge in the estuary. Built 10 km away from to the East of Dartford it would be the first fixed link bridging the two banks of the Thames. Moreover, the performance of the fast networks in Essex (A12, A127, A13) and Kent (A2/m2, M20, A229) would make it possible to reach a vast hinterland. Lastly, these gains are also the result of a decrease in detours. The obligatory passage by Dartford, which is imposed by the configuration of the network, would thus be cancelled for the North-East to South-East routes.

The distribution of gains is paradoxically asymmetric. The left bank benefits more from the new infrastructure. The Basildon area, in particular, benefits from particularly improved performances, whereas there is no equivalent on the right bank. This is explained by the difference in the development of routes between the two banks. The northern bank, and more particularly the area of Southend, has suffered more from detours imposed by the network structure, since the axes connecting it and thus the minimal paths following them currently tend to veer away more from straight lines. This is in particular the case of the routes bound for the South of the estuary. On the other hand, by comparison, on the southern bank the orientation pattern of the A2/M2 and the M20 limits the detours for the routes toward the northern half and thus, the impact is less strong.

The routes between Kent and Essex could thus be partly taken over by the *Lower Thames Crossing*. Similarly, the relatively high gains recorded in the north-eastern quarter of the London agglomeration show that the A406 and A13 corridors are likely to drain optimal routes to or from Kent via the *Lower Thames Crossing* instead of using the M25 from Dartford. The stretch of the A13 between the M25 and the LTC, which is less congested than its counterparts on the right bank, the A2/M2, could perhaps support this concentration. More certainly, the M25 would be relieved at the Dartford passage. Another consequence of the LTC is the improvement of the link between the two main urban centers located on either side of the bridge; Rochester and Southend would benefit from an improvement of 13 and 25% of their accessibility.

The space-time evaluation of networks benefit from a new dimension when the real infrastructure operation is taken into account. Speed losses are not distributed homogenously and congestion seems to differentiate network space. We could thus represent the temporal dilution of urban and intercity areas of London. In this context even a minor modification of road supply may result in a proportionally more significant impact on the network performance. Thus, a simple infrastructure modification may involve an upheaval of the optimal routes because of the heterogenous distribution of congestion levels. However, planners need to know the consequences of the investments which they prepare to carry out, because the functioning and the balance of the whole network are concerned. The principles of modeling and the tools presented in this chapter clearly fit into this perspective of assisting decision-making in regional planning.

1.4. Bibliography

[ABO 85] ABOURS S., *Estimation des temps de parcours sur un axe urbain à partir de taux d'occupation*, no. 76, Arcueil INRETS, 1985.

[APP 99] APPERT M., Intégration de la congestion dans l'évaluation spatio-temporelle d'un réseau routier régional: le cas de l'estuaire de la Tamise, Thesis, Montpellier 3, 1999.

[APP 02] APPERT M., CHAPELON L., Planification des transports régionaux en Languedoc-Roussillon et Nord-Pas-de-Calais: évaluation de la concurrence rail-route, Research Program INRETS/GRRT, Maison de la Géographie, Montpellier 2002.

[AUR 94] AURAY J.-P., MATHIS P., "Analyse spatiale et théorie des graphes", *Encyclopédie d'économie spatiale*, Economica, Paris, 1994.

[BER 83] BERGE C., *Graphes*, Gauthier-Villars, Paris, 1983.

[CER 90] CERTU, *Les études de prévision de trafic en milieu urbain: guide technique*, CERTU, Lyon, 1990.

[CHA 97] CHAPELON L., Offre de transport et aménagement du territoire: évaluation spatio-temporelle des projets de modification de l'offre par modélisation multi-échelles des systèmes de transport, PhD Thesis, Tours University, 1997.

[CHA 98] CHAPELON L., "Evaluation des projets autoroutiers: vers une plus grande complémentarité des indicateurs d'accessibilité", *Les cahiers scientifiques du transport*, no. 33, p. 11-40, 1998.

[CHA 00a] CHAPELON L., "Accessibilité routière et périphéricité des villes atlantiques", *Les cités atlantiques*, Publisud, Paris, p. 139-170, 2000.

[CHA 00b] CHAPELON L., *Transports et énergie, Montpellier – Paris*, RECLUS, La Documentation Française, Atlas de France Collection, 2000.

[CHAU 97] CHAUVIN *et al.*, "Une expérimentation de réduction de profil en travers sur autoroute urbaine", *RGRA*, no. 756, 1997.

[COH 90] COHEN S., *Ingénierie du trafic routier*, Presses de l'ENPC, Paris, 1990.

[DEP 93] DEPARTMENT OF THE ENVIRONMENT, Transport and the Regions, River crossings to the east of Tower Bridge, Report of the British Ministry of the Environment, Transport and the Regions, 1993.

[DEP 97] DEPARTMENT OF THE ENVIRONMENT, Transport and the Regions, London traffic monitoring report, Report of the British Ministry of the Environment, Transport and the Regions, 1997.

[DUP 96] DUPUY G., *Les territoires de l'automobile*, Economica, Paris, 1996.

[DUP 99] DUPUY G., *La dépendance automobile*, Economica, Paris, 1999.

[FAY 96] FAYARD A., "La congestion du réseau routier de rase campagne", *Revue d'économie régionale et urbaine*, no. 1, p. 41-52, 1996.

[GLA 01] GREATER LONDON AUTHORITY, Transport strategy (transport development plan for the city of London), 2001.

[HAL 89] HALL P., *London 2001*, Unwin Hayman, London, 1989.

[IAU 98] IAURIF, *Paris, London, New York, Tokyo: comparaison des systèmes de transport de quatre métropoles*, IAURIF, Paris, 1998.

[LES 92] LESORT J.-B., Étude d'indicateurs de circulation en milieu urbain, INRETS report, no. 142, INRETS, Arcueil 1992.

[LES 93] LESORT J. B., "Théorie simplifiée des flots de traffic", *Modélisation du traffic*, INRETS, Arcueil, p. 141-163, 1993.

[LEU 93] LEURENT F., Proposition d'indicateurs macroéconomiques des conditions de circulation sur un réseau routier, INRETS report, Arcueil, 1993.

[LIN 91] LINNEKER B.J., SPENCE N.A., "An accessibility analysis of the impact of the M25 London orbital motorway on Britain", *Regional studies*, vol. 26, p. 31-47, 1991.

[MAT 93] MATHIS P., *L'effet paradoxal des transports à grande vitesse*, DATAR/Editions de l'Aube, Paris, 1993.

[MAY 89] MAY A.D., BONSALL P.W., MARLER N.W., Travel time variability of a group of car commuters in North London, Working papers, 1989.

[MER 84] MERLIN P., *La planification des transports urbains: enjeux et méthodes*, Masson, Paris, 1984.

[MOG 90] MOGRIDGE M., *Travel in towns: jam yesterday, jam today and jam tomorrow?*, MacMillan, London 1990.

[TRB 98] TRANSPORT RESEARCH BOARD, *Highway capacity manual*, 2nd ed., TRB, Washington, 1998.

Chapter 2

Journey Simulation of a Movement on a Double Scale

An increasing number of urban dwellers choose to spend a few hours of relaxation in the countryside. Over the past years we could observe an increase in leisure and recreation activities around cities [DEW 90, DEW 92, DEW 93]. This phenomenon of visiting natural spaces is scarcely studied, in the sense that there are certain difficulties in determining the movement conditions of these visitors [TOU 98]. Indeed, if certain sites are frequented more than others, then necessarily they are more attractive and able to captivate more visitors than nearby sites. This phenomenon thus poses questions to the researcher on the nature of the attractiveness of a site. These questions are central to our work: should it be attributed to the activities on offer, to its generally remote location in a region, to its natural environment or its fame?

If there is movement, there is a goal. This goal is to take a walk in "nature", to relax in the spare time. By recreational visits to a natural environment we mean a process comprising two distinct displacements: access to the site and pedestrian diffusion in the environment with the aim to enjoy the entertaining nature of the activities that the visitor seeks at the site. Consequently, there arises the central question of taking into account the space and the choices of the visiting individuals related to the network, i.e. the network of sites structuring the area is likely to determine induced forms of distribution and diffusion in the natural environments which are open to public visitation.

Chapter written by Fabrice DECOUPIGNY.

2.1. Visitors and natural environments: multiscale movement

2.1.1. *Leisure and consumption of natural environments*

For many places of interest a large part of the visits arises from local entertainment usage [CUV 98, IFE 00]. So-called tourist infrastructures are increasingly used by the local population and in particular by urban dwellers [PL 98]. The study of visits carried out in the Hautes-Vosges [DEC 97] has revealed large numbers of local or regional visitors[1] with rates that can attain 75% on some Sundays. In short, we noted that visits double on Sundays over the entire massif[2] and that they correspond to rush hour phenomena during the day, with the peak of visits occurring between 4 and 5 pm. Therefore, throughout this chapter, we prefer the term visitor to this of tourist[3] or excursionist. As a more neutral concept, first of all, it enables us not to presuppose a segmentation of the visitor population that could introduce a bias into the analysis by taking into account that a tourist *a priori* exhibits different on-site behavior than an excursionist. We will therefore assume that there is a visitor population that is homogenous not in its social practices (choice of types of natural environments – sea, mountains, countryside – and recreational activities), but in its behavior in discovering a natural environment (type of movement and diffusion over the area), moving away from a place of residence towards one or more sites offering natural environments.

Before carrying on, it is necessary to identify the two constitutive stages of movement: choice and visit. The former involves socio-cultural conditions: why do people choose the mountain over the sea, one activity rather than another? Why do people leave for a period between one and four weeks, a weekend or just one Sunday [CAZ 92]? The practices related to leisure involve a decision taken individually. Although it is not our intention to analyze social practices, we believe that it is important to determine from them the observable behaviors (the manner in which the visitor populations are diffused). The very first observable behavior manifests itself through the use of cars as movement vectors between the place of residence and natural environments [CHA 99].

---

1 Bordering departments and/or regions.

2 These observations are also verified in other areas (Ecrins, Pyrenees, Pont du Gard, PNR Brotonne).

3 For the OMT a tourist is a person who travels for leisure and spends at least a night at the site.

2.1.2. *Double movement on two distinct scales*

Behaviors may thus appear varied in the choice of visited natural sites, but they are always made according to three principal characteristics of the natural environment: protected nature that offers relaxation, managed nature that has been made safe and pedestrian accessibility to natural attractions from the parking spots [ENV 92]. The main difficulty in evaluating visitor displacements consists of integrating a double scale of movements according to the behaviors linked to the three factors of site visiting (safety, accessibility, protection). This double scale depends on two movement processes: one or more displacements by car through a network followed by one or more pedestrian displacements at the natural sites.

2.1.3. *Movement by car*

The first movement scale corresponds to the first process – movement by car – that takes place in space and calls upon the morphology of networks and the accessibility of sites at the entrance points, which we will refer to as "gateway cities"[4].

Natural recreational areas consist mainly of natural sites that do not all have the same value in the eyes of the visitors. Visitors are likely to come to several of them during a displacement and they make choices to visit certain sites on the basis of the latter's better or worse infrastructure, beauty, accessibility from a car park or proximity to places of residence, etc. It is then possible to model this area by using a network of sites (a reception network) which is characterized by a hierarchical structure based around one or several main sites that condenses the flows and redistributes them to secondary sites during the second displacement. One of the recurring characteristics on these networks is the existence of more important sites which will act as "attractors" for the nearby sites. The visitor who comes by car will be partly attracted to one or more sites that he wishes to discover. Once the visit is over, it may be that he decides to see another site in the vicinity, thus carrying out a tourist tour. The very structure of the network, which is generally hierarchical, induces forms of recreational consumption of natural environments. It is important to know if the very structure of a reception network or, more precisely, its morphology and hierarchy, which comprises various sites connected to each other by a road network, is likely to direct or to transform certain types of movements.

4 These terms are generally used by regional natural reserves for the principal cities with direct access to the parks and we will use them in turn to describe the cities emitting the main incoming visitor flows.

2.1.4. *Pedestrian movement*

The second movement scale is related to the way in which individuals use the area once they leave their car – pedestrian movement, i.e. the way in which visitors diffuse through the area. The study of visits to the Hautes-Vosges[5] highlighted a *radio-concentric* diffusion of pedestrian flows[6] around access points, which are the car parks. Whether a German tourist, a Parisian or a local resident of a higher socio-professional category or not, all these individuals engage in pedestrian movement around the parking, by doing either a loop circuit or an outward and return journey. We could determine three types of visitors according to the pedestrian diffusion areas that the visitors were likely to attend: "simple visitors", walkers and hikers (Table 2.1). Based on concepts of demographic and ecological load capacity[7] of a natural environment this classification makes it possible to quantify the human pressures and to obtain information that is essential in developing the sites' potential to accommodate visitors.

Type of visitors	Diffusion area	Length of displacement
"Simple visitors"	R < 500 meters	< 1 km in an outward and return journey
Walkers – "simple visitors"	R < 500 meters	2 to 4 km in a loop
Walkers – "simple visitors"	R < 1,000 meters	2 to 4 km in an outward and return journey
Walkers	R < 1,000 meters	6 and 8 km in a loop
Walkers	R < 3,000 meters	6 to 8 km in an outward and return journey
Hikers	R < 3,000 meters	10 to 20 km in a loop
Hikers	R > 3,000 meters	20 km and more

Table 2.1. *Typology of visitors and radio-concentric areas of pedestrian diffusion*

---

5 Observations that have been verified by several studies: Pont du Gard in 1997 by the Nimes CCI, Pyrenees national parks [DOC 94] and Ecrins [DOC 92].

6 During the study of the visits to the Hautes-Vosges it was extremely interesting to note that some paths were used only for descending or climbing.

7 The demographic load capacity is the maximum quantity of visitors that a site can accommodate, the ecological load capacity is the maximum quantity of visitors that a natural environment can accommodate without degradation.

The classification of visitor populations according to the types of diffusion areas in natural environments enabled us to obtain groups of individuals segmented according to behaviors that appear common to all visitor types: pedestrian movement. This typology makes it possible to regard visitor groups, when formalizing our assumptions, as spatial consumption agents of recreational natural environments.

Interactions between network morphology and path behaviors manifest themselves in variations linked to visit times and the number of sites attended in one day. These variations induce a fluctuation of visitor distribution through natural environments, by creating a greater or smaller number of visits of the attended sites in time and quantity. Low movement times between sites may multiply the number of visited sites and decrease the attendance times at each site concentrating the visitors and the impacts on an area of 500 meters, whereas a sparse sites network minimizes the number of visited sites but increases visiting times, which imply visits and impacts that are more diffused but have a greater impact on the natural environments.

2.2. The FRED model

2.2.1. *Problems*

On the basis of the main observations made about the extent of the tourist infrastructure of an area and the types of recreational consumption of the natural sites, the aim of constructing the FRED model [DEC 00] was to resolve a scale problem in order to integrate the two movement forms. Let us recall that one of the main difficulties is to define the variables and to determine the parameters in order to model a multiscale process: kilometric scale for car movement and hectometric scale for pedestrian movement.

According to the assumption that space predetermines or directs a certain number of space consumption behaviors we suppose that there exists an area with flows of visitors that are homogenous not in their behaviors but in the type of activities that they have come to engage in (indeed, it seems obvious that people do not come to Ecrins to practice water sports!). Thus, we suppose that there are areas offering a set of sites that are homogenous with respect to the panel of proposed activities ("spatial offer"). That makes these spaces more structured and attractive by forming geographical entities for the discovery of natural environments.

The aim of formalizing is to obtain a relative value of the area according to the type of recreational consumption represented by a distribution behavior. Thus, we consider that the value of the natural environment differs according to the activity that the visitors engage in. Whether the visitor is a "simple visitor", a walker or a hiker, the attractiveness to the site is determined by the types of recreational activities for the visitors, a "practical value" of the site. That is, the attractiveness of a site depends not only on the structure of the area, but also of the type of visitor's recreational behavior – a relative attractiveness for a visitor type. All the assumptions exposed below and the originality of the model are based on the fact that there is no absolute value of space. Space is subjected to filters that are different according to the types of practices of recreational space consumption. An area is not seen the same way by "simple visitors", walkers or hikers.

2.2.2. *Structure of the FRED model*

To this end, we exploit two calculation modules. The FRED model is broken up into two parts and meets precise aims corresponding to specific modeling methods.

2.2.2.1. *The calculation module of visitor access to the sites*

The first models processes of spatial interactions linked to our three previously defined spatial constraints: spatial offer, absolute attractions and accessibility of sites. It models the distribution of visitors at the sites following car movement initiated at the place of residence (or gateway cities) and ending at the car park with access to natural sites. The simulation model is based on the architecture of a gravitation model [HUF 62], which defines for each visitor type the attractions of the various car parks with access to natural sites. These attractions are calculated according to three spatial constraints.

On the basis of the various observed elements we have constructed a simulation model which consists of calculating the probability that a car park has to fixate tourist flows at certain entrance points determined by taking into account:

– the accessibility of the places of interest: "distance time" between the entrance and the various car parks of the tourist area;

– the absolute attractions of each site in the network according to a recreational practice of space consumption of the tourists (diffusion area);

– the existing spatial offer around a car park (level of infrastructure and accessibility to natural attractions: mountain tops, panoramic viewpoints, lakes).

On the basis of accessibilities to a movement starting point (a "gateway city") the FRED model then calculates the probability that a car park can fixate differentiated visitor quantities ("simple visitors", walkers and hikers). By using the graph theory and the former work by the CESA laboratory on the use of this theory with respect to transport [AUR 94, HO 97] we have modeled a tourist reception network in a graph which is defined by arcs that symbolize the road connections in a regional area and by nodes, connected by the arcs, that represent car parks with access to natural environments. The modeling principle consists of calculating the distribution probabilities for each visitor type – “simple visitors”, walkers and hikers – on the graph.

2.2.2.2. *Diffusion calculation module*

The second module models visitors’ routes through the natural environments. In this case it is a calculation of pedestrian movements of the individuals during their walks through the natural sites on the basis of cellular automata programming. The aim of this programming is to provide a tool enabling us to evaluate the potential pressures of visitors’ movement by simulating the routes in the natural environments.

On the basis of a maximum movement time corresponding to the distances of 500, 1,000 and 3,000 meters diffusion areas we calculate the number of times that an arc of the path graph is used during a displacement with a car park (or point of entry) as origin and a “stop” point (either a natural attraction or a natural site) as destination. Movement is taken into account, if and only if the site can be reached within the allotted time. The assumption here again is that movement in a diffusion area necessarily implies a minimum time of presence at the natural site.

2.3. Part played by the network structure

Our prior work [DEC 00] clearly highlighted that the three spatial constraints – accessibility, vicinity and spatial offer of the site – determined the distribution of visitors on car parks with access to natural environments. We could therefore identify several movement types subjected to threshold phenomena.

A first access movement oscillates between 45 minutes and an hour, then a second movement (from 10 to 15 minutes) of redistribution on the reception network (or circulation between sites). The results are interesting, since they made it possible to determine spatial movement indicators which demonstrate very selective recreational practices in the natural environments.

These results give us implicit information about the attendance time of the site visitors. Indeed, if the visitors move in a network with poorly accessible car parks, they visit few sites, perhaps only one. We may then suppose that walking time is spent at one site only. In this case pedestrian diffusion is likely to be more penetrative of the natural environment, i.e. movements tend more to be within the 1,000 and 3,000 meters diffusion areas around the car parks. Otherwise, a network which makes it possible to stop at several sites in one day implies that walking time is divided between several sites. Pedestrian diffusion then tends to be only within the 500 meters zone from the car park and leads to a concentration of impacts around the car parks.

As mentioned in the beginning with regard to territorial management it is interesting to know the factors of visitor flow distribution in a site network. We assume that network structure has an influence on the distribution forms. If the accessibility of sites to "gateway cities" appears to be a factor of visitor distribution, the density and the connexity of the relations between car parks with access to natural sites seem determinant in visitor redistribution to the nearby sites.

Road network coverage, the distribution of sites through the mountain and spatial characteristics associated with the graph (access distances of the sites from the gateway cities and distances of mutual site interaction) reveal a tendency to homogenize the distributions throughout the territory by assigning visiting probabilities to all sites. However, heterogenous space distribution of the natural environment offer of the sites (taking into account site hierarchy) would then polarize the visits to clearly defined areas. The various sites of these sectors will have an attraction effect by concentrating flows in their area, like first order places of interest that redistribute flows to the surrounding sites. This phenomenon is all the more amplified by the road network morphology.

Visitor distribution probabilities through the Hautes-Vosges car parks

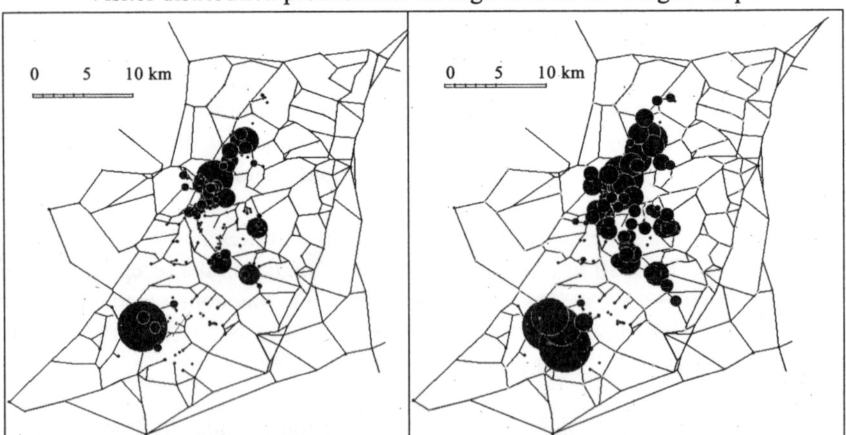

Distribution of car parks starting at the gateway cities after a 45-minute movement

Cumulative distribution of visitors on car parks after a three-stop tour

Cartography using the FRED model. Fabrice Decoupigny, CESA Laboratory, 2000

Figure 2.1. *Distribution of visitor flows throughout the Hautes-Vosges massif*

The proximity effect would thus reveal geographical aggregates of very attractive sites (Figure 2.1). Indeed, these sites are then able to mutually exchange visitors, thus increasing their attractiveness by comparison to sites that are generally isolated in the network. It then becomes interesting to evaluate the influence of the network structure (location of car parks and the road graph morphology) in the forms of visitor distribution throughout the reception area.

2.4. Effects of the network on pedestrian diffusion

2.4.1. *Determination of the potential path graph: a model of cellular automata*

First of all, we calculate the practicable area, i.e. the area where it is possible for an individual to move normally. This practicable area is calculated by introducing the constraint of a limit slope. Any arc with a slope over 50% cannot be used as a path arc and therefore cannot be taken by an individual. In a way the program determines a potential area of practicable diffusion. This operation consists of tracing on the digital terrain model (DTM) graph from any remarkable point to every other all the outward and return routes that can exist between the specific natural environment nodes (car parks and natural attractions). The path laying process is conducted node by node following an 8-connected diffusion. Once the diffusion is over we create a file of the nodes and arcs of the graph of the potential routes obtained so as to simulate the routes and impacts according to a calculation of optimization of the minimal paths in the second phase of the simulation.

The aim of the diffusion is to simulate movement from the departure point to an arrival point passing by various DTM nodes supposing that the individuals have no knowledge of the environment but see the node that has to be reached. The assumption is very different from that of minimal path optimization. First of all, the automaton defines all the potential routes in an area and only afterwards the course through the paths is optimized. Moreover, this assumption also uses another element that comes into play during the choice of routes: visibility. The cellular automaton representing the walker describes routes without anticipating the total difficulty of the path, the only information that it has is the location of the arrival point. It optimizes movement only on sections with connected nodes. In this manner we can take into account a very common type of visit for walkers: movement at sight. Indeed, the assumption, as it is defined, supposes that the arrival point site is known or seen; we thus suppose that the visitor knows the general direction of the node to be reached.

2.4.2. *Two constraints of diffusion*

The observed impacts on natural environments do not make it possible, *a priori*, to evaluate movement types because the multiplicity of the impact types (shortcuts through sharp bends, spontaneous routes, widening of the footpaths) in the natural environments do not enable us to decide on the type of movements that generate the impacts. We wish to test two types of movements in the environment (Figure 2.2 in the color plates section). The question that we pose is to find out if the visitors move in natural environments using the shortest distance path (called *Shortest*) or the path that requires the least effort (called *Easiest*).

In the first case, the assumption of the model concerning the choices of paths includes the general hypothesis of "the law of least effort". It consists of stating that people try to reach a point with a minimum of effort. A person will choose the connected node according to the variation in altitudes making it possible to associate minimal difficulty to the easiest path. In the second case, the choice of the related node is made by minimizing the Euclidean distance without involving difficulty. However, the program stores in memory the difficulty of the traversed arcs. Since, if the visitor chooses the shortest path, he remains subjected to the constraints generated by slopes and will not spend the least time to reach the target, it may even be that this type of path is the longest in time.

Thus, it becomes possible for us to differentiate between outward and return routes. Indeed, various observations have demonstrated that the routes between two points may be different depending on the direction of walking. The outward and return routes are not symmetrical.

Taking into consideration the results obtained by the cellular automaton, we have a potential path laying area with no preferential routes. It is thus necessary to reprocess the information obtained and to establish among these routes or portions of routes those which are more likely to be used than others.

Indeed, we may admit that people have a certain knowledge of the areas either because they have already visited, or because they use maps, or quite simply follow a footpath and then retrace their steps. The objective of the model is to find the factors governing movements in a natural environment and the induced impacts on the basis of a simulation of the genesis of a network of paths.

Therefore, contrary to the assumption put forward previously regarding the forms of movement, we suppose that people move from one point to another optimizing the path in a potential network (*Easiest* or *Shortest*). Thus, we introduce the assumption of preferential choice of path made according to an optimization of the minimal path under the movement type constraint. This assumption of the optimization of the minimal paths is admissible insofar as we can suppose that when the visitor leaves his vehicle he wishes to go and reach the natural site as soon as possible. The path graph will be used as a basis to calculate the impacts of the movements of the various types of visitors on the natural sites. For the two path graphs (*Shortest* and *Easiest*) we will apply "Floyd's" algorithm in order to search the minimal paths. The fact of having differentiated outward and return movements and the difficulty of the arcs makes it possible to have a dissymmetrical return path layout compared to the outward one. The goal of the calculation of the optimal paths on the path graphs consists of calculating origin-destination movements between the various points of interest (car parks, natural sites) and distinguishing the arcs which are likely to be the most used to get from one node to another under the constraint of minimization of the distance covered according to the two types of movement behavior in the natural environment, either the *Shortest* or the *Easiest* path.

Three types of routes are considered: access movements from car parks to the natural sites (or outward movements); walking movements (or circulation) between the natural sites, return movements from the natural sites to the car parks. The two movement assumptions (*Easiest* and *Shortest*) associated with the movement thresholds in the diffusion areas enable us to put forth assumptions about the types of visitor movement through the natural environments. We will thus be able to see whether the impacts generated by movement correspond to there and back walks or loops. The initial conditions of the simulations observe movement times corresponding to the diffusion areas of the various visitor types (Table 2.2).

For a walking speed of 3 km/h			
Types of visitor	"Simple visitors"	"Simple visitors" and walkers	Walkers and hikers
Walking time less than	10 minutes	20 minutes	60 minutes
Maximum distance covered	500 m	1,000 m	3,000 m

Table 2.2. *Typology of movement spatial behavior*

The impacts of the routes are yielded by the frequencies of use of the graph arcs each time there is movement from one point to another. If the probabilities of fixating on car parks are integrated in the form of the weight of pedestrian flows entering through one, we do not proceed to the calculation of probability of path usage. The results give only potential impacts on the area regardless of the choices of direction that may take place in the natural environment (Figure 2.3). That is, starting at a car park we calculate only the number of times that an arc is used once all the itineraries have been calculated.

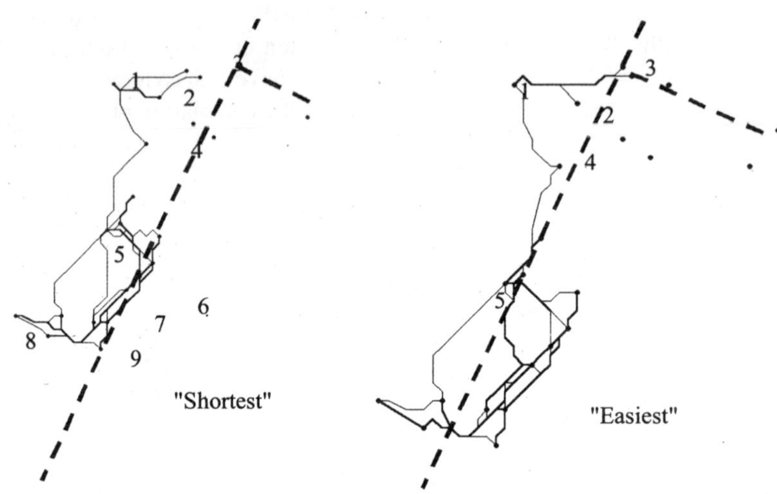

Point 1: Car park at the nature reserve of Tanet - Gazon du Faing
Point 2: Panoramic table
Point 3: Soultzereneck
Point 4: Gazon du Faing
Point 5: Dreieck car park (or Collet du Lac Vert)
Point 6: Lac Vert
Point 7: Lac Vert car park
Point 8: Tanet car park

Figure 2.3. *Comparison of a simulation of routes for a 20-minute movement in a 1,000-meter diffusion area (nature reserve of Tanet and Gazon de Faing – the Hautes-Vosges massif)*

2.4.3. *Verification of the model in a theoretical area*

We have simulated movements in a theoretical space under the same conditions as at the Vosges. The aim of this comparison is to verify if the results obtained for the Tanet area reveal specific characteristics due to local conditions (geomorphology, location of the car parks, etc.) or to conditions of a more general nature. To this end we have created a hilly theoretical area modeling the mountain relief with altitudes oscillating between 600 and 1,400 meters and two potential path graphs: *Shortest* and *Easiest* (Figure 2.4 in the color plates section).

We can note that movement simulations yield different results than those obtained at the Vosges (Figure 2.5 in the color plates section). The results are less pronounced in the theoretical area. According to the type of movement constraint (*Easiest* and *Shortest*) the impacts are of a different nature: either linear pressure or a multiplication of the paths (routes generating impacts such as shortcuts, path branching, etc.) The main difference stems from simulations of the movement hypothesis corresponding to the *Easiest* constraint. At the Vosges we have movements concentrating on routes along footpaths, whereas in the theoretical area we see extensive movements multiplying the paths (see Figure 2.5). The differences between the two movement assumptions are much more marked in the theoretical area. Contrary to the Vosges the induced impacts are reversed: for the "Easiest" movements we have impact types that would tend to multiply the footpaths spontaneously, whereas for the "Shortest" movements the impact types would tend to widen the footpath.

What do these two contradictory results mean? At first we may think that the geomorphology of the site leads to different movements. However, simulations in a theoretical area should have yielded results similar to these obtained for the Vosges, since we took the same physical characteristics. It is therefore necessary to seek the answer elsewhere. If it is not just the relief that induces impact types, we may consider that the reception facilities may play a part in the diffusion forms in natural environments.

In order to verify this assumption, we will again address the specificities of the Tanet area (Figure 2.6 in the color plates section) in a theoretical area. To that end we have kept only three car parks at mid-slope and two at the bottom of the slope to obtain a structured area similar to the Tanet nature reserve (with the same number of natural sites and car parks). We then obtain completely different results compared to the first simulations (Figure 2.6) corresponding to those obtained for the Vosges. For the latter we note that the number of routes is denser when there is a certain number of natural sites and network entrance points.

Although theoretical space is denser at natural sites we observe the same types of routes as these at the Vosges, with a linear concentration of routes on the *Easiest* path graph and a multiplication of the routes accompanied by a north-south axis of pedestrian circulation on the slopes of the *Shortest* path graph. This axis also appears at the Vosges. These simulations show that areas are differentiated not only according to the types of spatial usage and site but also according to the reception facilities and accessibility of the massif.

These results are fundamental. We recall that one of the observations made during the study of visits to the Hautes-Vosges [DEC 97] tended to say that visits to a massif depended on the number of car parks and not on the number of places. That is, the pressure of visitor diffusion on natural environments depended more on the number of access points than on the load capacity of the car parks.

The model shows that the impacts to a large extent depend on the reception facilities in natural environments. The number of car parks and their location can play a determinant part in the types of impacts of anthropogenic pressure. Depending on the network structure, movement type and geomorphology of the area, anthropogenic pressures can take different forms. Not only does the network structure play a dominating part in car movements and the choice of the site to visit, but it also seems to be the case for pedestrian movements within the natural environments.

As regards the planning and protection of natural environments, network management may differ from one area to another if we wish to concentrate or diffuse the visits to certain places of interest. To this end tourist visits must be evaluated by taking into account three interacting elements: the site, the network and the visitor. This amounts to saying that the management of visitor flows and their impacts on natural environments are apprehended simultaneously on two distinct scales: car movement associated with a type of pedestrian movement.

2.5. Bibliography

[AUR 94] AURAY J-P., MATHIS P., “Analyse spatiale et théorie des graphes”, in J.P. Auray, A. Bailly, P.H. Derycke, J.M. Huriot (ed.), *Encyclopédie d’économie spatiale: concepts, comportements, organisations*, Economica, Paris, p. 81-88, 1994.

[CAZ 92] CAZES G., *Fondements pour une géographie du tourisme et des loisirs*, Bréal, Paris, 1992.

[CHA 99] CHARDONNEL S., Emplois du temps et de l’espace. Pratiques des populations d’une station touristique de montagne, PhD Thesis, Grenoble Joseph Fourier University, 1999.

[CUV 98] CUVELIER P., *Anciennes et nouvelles formes de tourisme: une approche socio-économique*, L’Harmattan, Paris, 1998.

[DEC 00] DECOUPIGNY F., Accès et diffusion des visiteurs sur les espaces naturels. Modélisation et simulations prospectives, PhD Thesis, Tours University, 2000.

[DEC 97] DECOUPIGNY F., Etude fréquentation Hautes-Vosges, Tours, 1997.

[DEW 90] DEWAILLY J.-M., *Tourisme et aménagement en Europe du Nord*, Masson, Paris, 1990.

[DEW 92] DEWAILLY J.-M., “Les citadins et les loisirs de nature”, in *Tourisme et environnement*, La Documentation Française, Paris, p. 30-32, 1992.

[DEW 93] DEWAILLY J.-M., FLAMENT E., *Géographie du tourisme et des loisirs*, SEDES, Paris, 1993.

[DOC 92] "La fréquentation touristique", Documents scientifiques du Parc National des Ecrins, no. 4, 1992.

[DOC 94] "La fréquentation touristique", Documents scientifiques du Parc National des Pyrénées, no. 28, 1994.

[ENV 92] MINISTERE DE L'ENVIRONNEMENT, MINISTERE DU TOURISME, *Tourisme et environnement*, La Documentation Française, Paris, 1992.

[FER 95] FERBER J., *Les systemes multi-agents, vers une intelligence collective*, Paris, InterEdition, 1995.

[HUF 62] HUFF D.L., *Determination of intra-urban retail trade areas*, University of California, Los Angeles, 1962.

[IFE 00] INSTITUT FRANÇAIS DE L'ENVIRONNEMENT, *Les indicateurs: tourisme, environnement, territoires*, Orléans, 2000.

[LHO 97] L'HOSTIS A., Image de synthèse pour l'aménagement du territoire: la déformation de l'espace par les réseaux de transport rapide, PhD Thesis, Tours, 1997.

[PLA 98] Premier Ministre, Commissariat Général au Plan, Secrétariat au Tourisme, Direction du Tourisme, *Réinventer les vacances. La nouvelle galaxie du tourisme*, La Documentation Française, Paris, 1998.

[TOU 98] SECRETARIAT D'ETAT AU TOURISME OBSERVATOIRE NATIONAL DU TOURISME, L'évaluation par les flux: outil d'analyse touristique territoriale, Observatoire national du tourisme, Paris, 1998.

Chapter 3

Determination of Optimal Paths in a Time-delay Graph

3.1. Introduction

In urban planning, the applications of graph theory have demonstrated their effectiveness in modeling transport networks and evaluating their impacts on topological space through the contraction of space-time and its representation in the form of Chronomaps [LHO 97], on accessibility using NOD software [CHA 97], on the operation of the network, i.e. the organization of potential movement flows, or on its evolution considering a dynamic network evolving over time [BAP 99].

First of all, these approaches distinguish the "containing" part (linear and nodal transport infrastructure) from the "content" part (means of transport); both of these are nevertheless joined together in the form of functional binomials associating a certain infrastructure with a certain means of transport. Each one is distinguished from the other according to the announced aim: analysis and evaluation of supply modifications, taking demand into account, integrating the temporal factor. Using the latter objective we can classify the types of graph necessary for simulations into three categories.

The graph referred to as "traditional" distinguishes between arcs according to their physical characteristics (rail, road, number and width of lanes, driven by electricity or not, etc.) refined by topography and sinuosity constraints, the valuation of these arcs in duration or cost resulting from all the above criteria. The means of

Chapter written by Hervé BAPTISTE.

transport are considered jointly leading to the development of binomials with permanent functionality. Anyone in a particular vehicle can use a road at any moment, unconstrained by time, if we agree to neglect climatic risks (shutting-down of snow-covered sections, for example), works on roadways or structures. This type of graph paves the way for diachronic analyses by comparing the performances of two graphs according to a distinct transport supply.

The adaptive graph is, in fact, a graph of the same type (described by permanent functional binomials, a valuation of the arcs of the same nature), but considered from a temporally dynamic, long-term point of view. On the basis of an initial graph, the improvement of a transportation axis frequently and durably exceeding its maximum capacity leads, following a cyclic process, to the plotting of a series of new graphs, which are described by arcs with different characteristics (widening of a road followed by doubling of the number of lanes or pure and simple creation of a new highway arc, for example).

On the other hand, the time-delay graph adds another constraint that is not only infrastructural, but rather more functional, thus by definition discontinuous in time for a mass transport network. This constraint is necessary when infrastructure evaluation stems from supply quality analysis. Indeed, the existence of a TGV train station in a town is only of limited interest for an individual if the number of effective stops is restricted. The main requirement was thus to consider binomials with temporary functionality via a coupling of permanent infrastructures (rails, roads even air corridors, etc.) with means of transport functioning punctually throughout the duration (trains, buses, planes, etc.), hence the name of a time-delay graph with functional temporization.

To this end an algorithm, based on Floyd's algorithm, was developed and aimed to determine the optimal paths between all the nodes of a time-delay graph, thus integrating departure and arrival timetable constraints of trains and other mass transport vehicles, leading to simulations of the type: which cities can I reach in less than an hour, departing from my origin at a certain time? From which city can I leave in order to arrive at the destination before a certain time? Or more generally, what timetables do I have at my disposal to go from any station to another?

We will not develop the founding model, i.e. Floyd's algorithm [BAR 86], nor its modifications directed more specifically towards urban planning simulations and more precisely towards access problems in the transport domain, as it has already clarified in [MAT 78]. We will only point out its main characteristics, namely, the determination of the optimal path (in time, in cost, etc.) from any point to any other point by enumerating the series of arcs and we will use an example in order to describe its operation.

On the other hand, the evaluation of the quality of service being the principle basis for the chosen modeling, we will develop the concepts of "door to door" movement and interconnection constraints (time needed for a journey on foot, using public transport, etc., number of changes beyond which the path is abandoned, whether in favor of a possible alternative or not, etc.).

The "traditional" and "timetable" versions of the developed Floyd's algorithm are distinguished above all by the structure of the descriptive matrix of the arcs, by introducing the third dimension needed to account for temporal discretization of the relations (a train that left five minutes ago has lost its functionality for the user who arrived late at the platform, although the infrastructure remains). In the matrix, all nodes taken two by two yield a number of relations, defined by departure and arrival times, existing for a given period (a day, a week, etc.), to which relations with permanent functionality can be added.

Initially neglecting, in order not to confuse matters, the connection constraints specific to towns of different sizes or with several train stations and induced by the time wasted to get off a train and get on another one, the set of paths is determined by successive loops. For a given origin and a destination, for each departure during the initially chosen period and regardless of the intermediate destination, the algorithm will construct, if it exists, the series of arcs that respond to the constraint of the time of arrival at an intermediate node before the time of departure from this same node for the rest of the journey.

Formalization itself will be clarified by choosing an example of several nodes connected by several arcs described in the form of a timetable, which will be progressively detailed in frames as the calculation is developed.

3.2. Floyd's algorithm for arcs with permanent functionality

Before describing Floyd's algorithm that is specifically developed to integrate a transport timetable offer we begin by briefly pointing out the operating mode of the primitive algorithm, with complexity of the 3rd order which is based on triple looping through n nodes (n^3). For didactic reasons we take an example limited to four nodes A, B, C and D and three arcs connecting A and B, B and C then C and D (Figure 3.1) by chronologically exposing the computing process in a formal manner and referring the reader for the algorithmic description to [BAP 99, CHA 97].

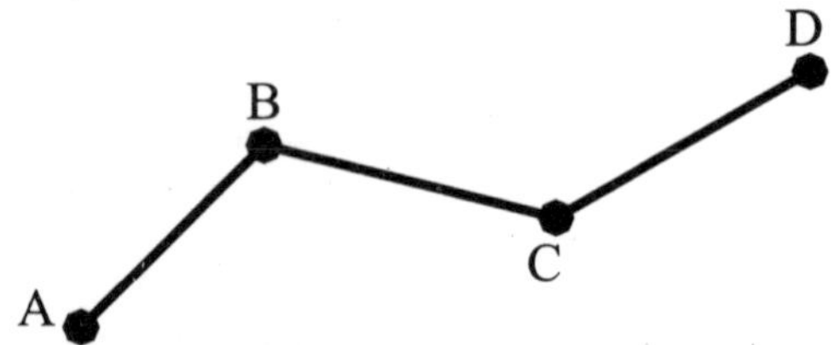

Figure 3.1. *Example of four stations connected by a railway*

For each pair of nodes AC, AD, BD (by considering undirected relations and therefore neglecting symmetric relations) which are connected by paths[1] (in the opposite case, i.e. a pair of nodes connected by an arc, the calculation continues without processing the relation for this pair) each node is regarded as an intermediate transit node. If the graph were directed, 64 relations would be possible including loops and counting outward and return journeys along an arc or a path separately. Relations AB, BC and CD are also processed on the assumption, which is invalid in this particular example, of a shorter duration when passing by one or more intermediate nodes.

Thus, after processing the intermediate node A, which is done quickly since it is located at the end of the network, the intermediate node B is taken for the pair AC by connection of the AB and BC arcs. For the AD pair, node B is an intermediary but no path is determined because the path between B and D has not been constructed yet. Then starting with the intermediate node C, relation AD is determined as associating the AC path (built previously) and the CD arc, as well as the BD path associating the BC and CD arcs. Finally, since node D is in the same situation as node A, the calculation is completed by having indexed all of the possible relations (paths and arcs) between each pair of nodes. We have purposefully omitted symmetric relations, which are of course determined during double looping at the origin and then at the destination for each intermediate node (that is, the CA, DA and DB paths).

The brief presentation of this algorithm shows its interesting adaptation to determining the paths with permanent functionality, where the selection of the arcs and the shortest paths (in time, in cost, etc.) constitutes a first extension of the primitive algorithm. For this latter aspect a preliminary selection of the optimal arcs between each pair of nodes is carried out when it is connected by several arcs of different nature (traditional in the transportation field where a local or regional road is doubled by a trunk road with greater throughput). This procedure initiates the calculation of minimum paths by being positioned upstream of the main algorithm.

1 Understood as a juxtaposition of at least two arcs.

In addition, the functionality of the primitive algorithm that enables determining "antecedents", i.e. for each node the immediately preceding node in a route with multiple transit nodes, is preserved. In particular, it makes it possible to determine the principal corridors which are identified when a significant number of paths follow the same arcs.

The description that has just been made deserves a comment on its possible application to a network that is completely or even partially comprised of public transport connections. The permanent functionality of road transport networks allows interconnection without load rupture, in other words, immediate passing from one arc to another (by neglecting the time spent at a node, which in addition is solved by multiscale modeling and is applicable to Floyd's algorithm by simply considering "interurban" and "urban" functional binomials, the latter connecting complementary intra-urban nodes [CHA 97]).

3.3. Floyd's algorithm for arcs with permanent and temporary functionality

3.3.1. *Principle*

The particular case of a network consisting of binomials with temporary functionality, to which a user is constrained by the times of departure and arrival, has imposed a major modification of Floyd's algorithm, which requires to take into account the load rupture times and an adequate distribution of the timetables in time during connections[2].

For this reason several versions were developed according to the needs of studies and research [COL 96, BAP 02]. On the basis of an algorithm making it possible to determine the set of "timetable" paths we have successively developed new functionalities, such as the need for arriving at a destination before a certain schedule (*arrival time constraint*), or for leaving an origin only after a certain time (*departure time constraint*). Moreover, according to the space scale considered and the mass transport means (rail, air, mixed), specific connection constraints have been worked out particularly in individualized processing of paths requiring a connection with a change of station in the same city (Paris, etc.), which is described by only one node but by several stations, as well as constraints on terminal stops in order to consider door to door path laying.

2 This transformation of the algorithm was imposed when it became necessary to pass from a simple logic of transport supply (infrastructural) to a quality of service logic, within the framework of research for the DTT and the OEST (ex-DAEI).

We will develop hereafter the most general version determining the *set of paths within a time range* defined by the user.

3.3.2. *Description*

Similarly as for the traditional version of Floyd's algorithm we will expose its operation by using an example limited to four nodes, which are connected successively by three links: each proposing two departures during a certain considered period. By neglecting the extremities of the line we have certainly chosen a less probable example because the return is impossible (a case that is, nevertheless, not useful here, since its study concerns the same method (Figure 3.2)). We will follow the algorithmic steps chronologically by presenting the concerned procedures successively.

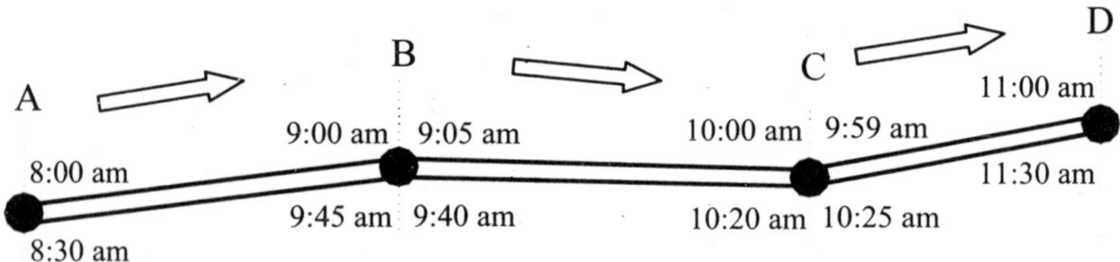

Figure 3.2. *Four stations connected by rail with timetables*

3.3.2.1. *Creation of the minimum arcs matrix*

In the first instance, the arcs matrix is processed in order to check the potential existence of arcs with permanent functionality in addition to the timetable arcs (Algorithm 3.1). According to this hypothesis, if two relations with permanent functionality exist for the same arç, the fastest is preserved (example of a relation by walking and a faster one by car). Similarly, if there are also relations with fixed timetables, they will only be preserved if they are faster than a relation with permanent functionality. Flexibility is retained here to privilege an unconstrained timetable relation if it performs better with respect to duration (a relation of the same duration will also lead to privileging that without timetable constrains). At this level nothing would prevent implementing the opposite choice resulting from taking into account unmotorized travelers and privileging in the long term the mass transport route constrained by timetables.

Moreover, the maximum number of timetables for a given arc (in other words the maximum matrix thickness) is determined; the timetable arcs are placed starting from the second layer, whereas the best performing permanent arc with functionality, if it exists, is placed in the first layer.

```
CREATEMATARCSMINTIMETABLE
Knowing the set of arcs referenced by the iOri indices of the origin node and the iDes indices of the destination nodes, their means of transportations,
traveling time taken from the corresponding input file,
    For each arc k = 1, 2,..., n (of iOri origin and iDes destination indices):
        If the arc k has permanent functionality:
            If the minimum arc (iOri, iDes) does not yet exist (value equals zero),
            or if the arc k is shorter than the minimum arc, (iOri, iDes) already stored:
              * The journey duration and means of transportation are stored in the "ArcsMin" matrix of the minimum arcs with indices iOri, iDes.
              * The new minimum arc (iOri, iDes) is affected to the depth of 1
              (layer allocated to the permanent functionality relation that is the most performing in terms of duration).
            End If
        Otherwise, if the arc k has temporary functionality:
            If the minimum arc (iOri, iDes) does not yet exist (value equals zero):
                If its depth is zero (no arc with permanent functionality):
                  * A depth of 2 is allocated (to position the timetables starting from the second layer)
                Otherwise:
                  * An additional layer is allocated.
                End If

                If the depth obtained is greater than the maximum depth until now:
                  * Maximum depth is adjusted to this new value.
                End If
            Otherwise, if the arc k is faster or has the same duration as the minimum arc (iOri, iDes):
                An additional layer is allocated,
                If the depth obtained is greater than the maximum depth until now:
                  * Maximum depth is adjusted to this new value.
                End If
            End If
        End If
    Next arc k
```

Algorithm 3.1. *Creation of the minimum arcs matrix*

3.3.2.2. *Determination of the range of timetable processing*

Secondly, the user defines the time range within which the search for the minimum path must be carried out, which is limited by processing start and end times. The start time is immediately assigned to a "reference time", which will progressively increase during processing to determine the valid paths chronologically.

3.3.2.3. *Partial and provisional filling-in of the minimum paths matrix*

As a precondition to the main calculation to determine the paths, a matrix of precedents must be partially filled (Algorithm 3.2): this matrix makes it possible, if necessary, to recompose the path by using intermediate nodes, as was underlined previously.

```
FLOYDPREPARATION
For each node i = 1, 2,..., n:
    For each node j=1, 2,..., n:
        If the minimum arc (i, j) exists:
            * The cell (i, j) of the matrix of precedents recovers the index i and the value of the minimum arc (i, j).
        End If
    Next node j
Next node i
```

Algorithm 3.2. *Preparation of the calculation of the minimum paths*

3.3.2.4. *Processing of the minimum arcs*

The following procedure still concerns the preparation of the calculation of the optimal paths themselves, specifying, nonetheless, elements depending on input data (Algorithm 3.3). In fact, the specific case of trains leaving one day and arriving the following day requires particular processing in order to have an arc in the matrix that is coherent with respect to the operation during a single day. Indeed, if the aim is to determine the set of routes and timetables possible during one working day, 24 hours are added to the time of arrival, thus meaning arrival is later than departure[3].

Moreover, if the true paths, i.e. cumulating at least two minimum arcs, are only created afterwards, the minimum paths matrix during this procedure will be partially filled using the already stored minimum arcs, with neither connection, nor stop required.

In order to optimize the timetable search procedure adapted to work out the shortest possible path, the timetable permitting, for a given arc, to arrive at the earliest is placed in the first layer[4].

---

3 A specific version of Floyd's algorithm was developed to determine minimum paths in the form of timetables and is identical to that presented here, but considering days as well as hours. The addition of 24 hours is no longer necessary, since it is substituted by checking the day.

4 If arcs with permanent functionality are mixed with "timetable" arcs, the latter are placed in the first layer, because they are always adapted to supplement a path.

```
PREPCALCPPATHTIMETABLE
For each node i = 1, 2,..., n:
    For each node j=1, 2,..., n:
        If the thickness value of the minimum arc (i, j) exceeds the maximum thickness for all arcs:
            # Maximum thickness is assigned to this value.
        End If
    Next node j
Next node i

For each node i = 1, 2,..., n:
    For each node j = 1, 2,..., n:
        If i=j:
            # The (i, j) path is assigned the value of 32,000 ('infinite')
        Otherwise:
            If the value of the minimum arc (i, j) is zero (no permanent functionality means):
                If the depth of the minimum arcs matrix for (i, j) is zero (no timetable either):
                    # The (i, j) path is assigned a value of 32,000.
                Otherwise (the arc is only described by timetables):
                    # The (i, j) path is set to 32,000.
                    For each layer k = 2, 3,..., n' of the minimum arc (i, j)
                      (sweeping the entire depth of the arc apart from the first):
                        If the (i, j) arc in layer k has an arrival time earlier than the departure time
                          (change of day):
                            # The arrival time is modified by adding 24 hours.
                        End If

                        If the departure time of the (i, j) arc in the layer k is before the reference time:
                            # We add 24 hours to both departure and arrival times.
                        End If
                        # Filling in the paths matrix (value, departure and arrival times, means of transportation,
                          number of the start and finish line, index of the nodes i and j).
                        * If the (i, j) arc in layer k also makes it possible to arrive before the time is stored in
                          the first layer (reserved for the arc that arrives the first for a given relation):
                            # It is placed in the paths matrix in the first layer with all the characteristics
                              (value, departure and arrival times, means of transportation, number of the start
                              and finish line, index of the nodes i and j).
                        End If
                    Next layer k
                End If
            Otherwise (the arc is described by a permanent functionality relation):
                ** The minimal arc (i, j) is placed in the paths matrix storing the departure time as reference time,
                  the arrival time as this same time plus the duration, as well as the means, its value and the indices i and j of the nodes.
                If the arc is also described by timetables:
                    For each layer k = 2, 3,..., n' of the minimum arc (i, j):
                        If the (i, j) arc in layer k has an arrival time before the departure time (change of day):
                            # The arrival time is modified by adding 24 hours.
                        End If

                        If the departure time of the (i, j) arc in layer k is later than the reference time:
                            # It is placed in the paths matrix with all its characteristics (value, departure and arrival times,
                              means of transportation, number of the start and finish line, index of the nodes i and j ).
                        End If
                    Next layer k
                End If
            End If
        End If
    Next node j
Next node i
```

Algorithm 3.3. *Processing of minimum arcs*

3.3.2.5. *Determination of paths*

Once these first initialization procedures have been implemented the calculation itself to determine all the "timetable" path can begin. Triple looping on k, i, j implemented for the "traditional" version of the algorithm is kept because the innovations relate to taking on board the potential multiplicity of timetables for a given arc.

As in the traditional version of Floyd's preparation, we will not consider the cases where A and D are intermediate nodes, which are not of interest here since they do not enable the development of a path.

For the intermediary B, the first origin A is initially tested to verify the existence of an arc or path between A and B, in which case the processing with the following destination point C is useful. In the opposite case, the following origin is considered, since AB does not exist, the interest to calculate BC is zero and unnecessarily expensive in terms of time to find a path AC by B.

For the BC timetable, a first comparison of the arrival time is carried out with the arrival time of the AC path (if there is any) in order to preserve the shortest timetable of arrival at C and at the same time minimize the number of connections.

In our example there is no AC path and a new test is carried out to verify the adequacy of the time of arrival at B from A and the departure time of B to C. The former must logically be earlier than the latter in order to enable load unloading at B[5]. Information over the connection time at B is thus found using the difference in the two timetables above.

In practice, for the relation AC with B as intermediate point, the two timetables (8:00-9:00 am) between A and B, and (9:05-10:00 am) between B and C thus constitutes a first path, which did not exist until now. The next timetable (9:40-10:20 am) between B and C does not replace the previous one since the arrival time is later.

Still with B as intermediate point, the relation AD cannot be determined, since the BD path has not yet been found. It will be determined now by considering the intermediate point C. Let us immediately observe that the second timetable (8:30-9:45 am) between A and B is considered only later on when the reference timetable supplants the current one (that is, at the first passage, the initial processing timetable).

---

5 A connection is accepted by default which is a minimum time of two minutes, or when it indicates a simple stop at a station (which is identified when the line number does not change).

According to point C the AD relation will be on the basis of the already existing path AC with the timetable (8:00-10:00 am) and passing by point B (with five minutes connection time) and the arc CD. The possibility (9:59-11:00 am) is quickly discarded, since it does not enable the connection at C from A to D. It is thus the latter which will be retained, leaving at 10:25 am from C and arriving at 11:30 am. The complete path thus includes two connections at B at C, leaving A at 8:00 am and arriving at D at 11:30 am.

From B to D the path found starts at 9:05 am, arrives for the connection at C at 10:00 am and again necessarily follows the timetable (10:25-11:30 am) between C and D. This result is stored at this stage or only written down if the following reference hour is later than the departure of BD.

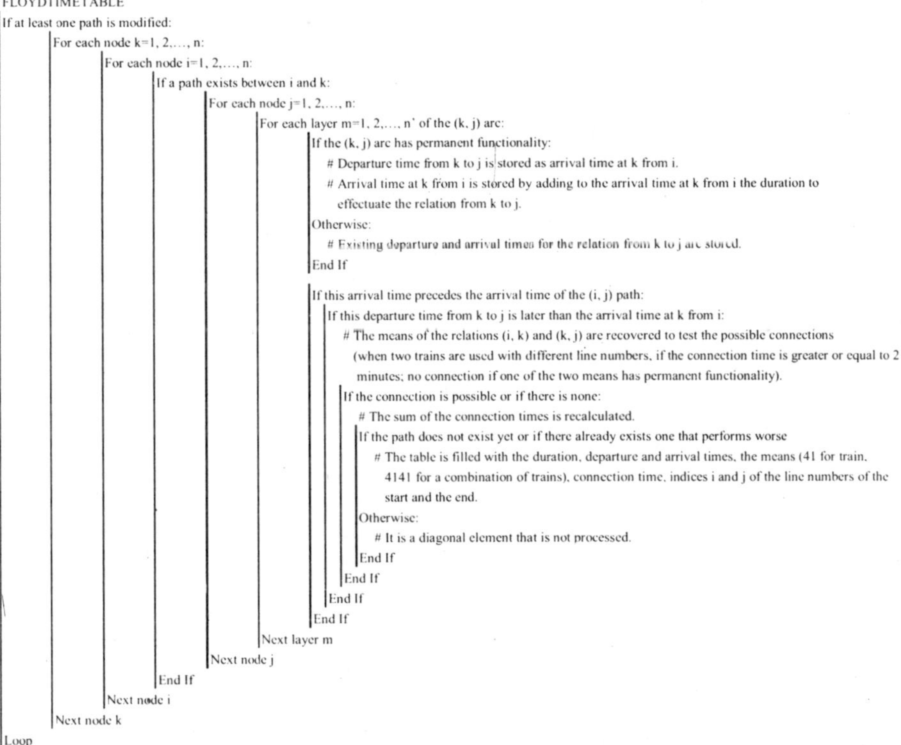

```
FLOYDTIMETABLE
If at least one path is modified:
    For each node k=1, 2,..., n:
        For each node i=1, 2,..., n:
            If a path exists between i and k:
                For each node j=1, 2,..., n:
                    For each layer m=1, 2,..., n' of the (k, j) arc:
                        If the (k, j) arc has permanent functionality:
                            # Departure time from k to j is stored as arrival time at k from i.
                            # Arrival time at k from i is stored by adding to the arrival time at k from i the duration to
                              effectuate the relation from k to j.
                        Otherwise:
                            # Existing departure and arrival times for the relation from k to j are stored.
                        End If

                        If this arrival time precedes the arrival time of the (i, j) path:
                            If this departure time from k to j is later than the arrival time at k from i:
                                # The means of the relations (i, k) and (k, j) are recovered to test the possible connections
                                  (when two trains are used with different line numbers, if the connection time is greater or equal to 2
                                  minutes; no connection if one of the two means has permanent functionality).
                                If the connection is possible or if there is none:
                                    # The sum of the connection times is recalculated.
                                    If the path does not exist yet or if there already exists one that performs worse
                                        # The table is filled with the duration, departure and arrival times, the means (41 for train,
                                          4141 for a combination of trains), connection time, indices i and j of the line numbers of the
                                          start and the end.
                                    Otherwise:
                                        # It is a diagonal element that is not processed.
                                    End If
                                End If
                            End If
                        End If
                    Next layer m
                Next node j
            End If
        Next node i
    Next node k
Loop
```

Algorithm 3.4. *Main calculation to determine the minimum paths*

Processing continues in this manner with looping until no timetable change can be obtained. This assumption can sometimes be verified, since the creation of a new path *n* may make it possible to find at the following iteration, for another relation, a better performing path *p* which nevertheless takes the path *n*. Algorithm 3.4 describes in formal language the main procedure to determine optimal paths.

3.3.2.6. *Determination of a new reference time*

Following the previous main processing the new reference time is calculated by considering all the departure times of all the relations, after the previous time of the start of processing (Algorithm 3.5). The nearest time (by at least 1 minute) serves as the new basis for the Floyd calculation. According to the following example, 8:30 am is the new reference time.

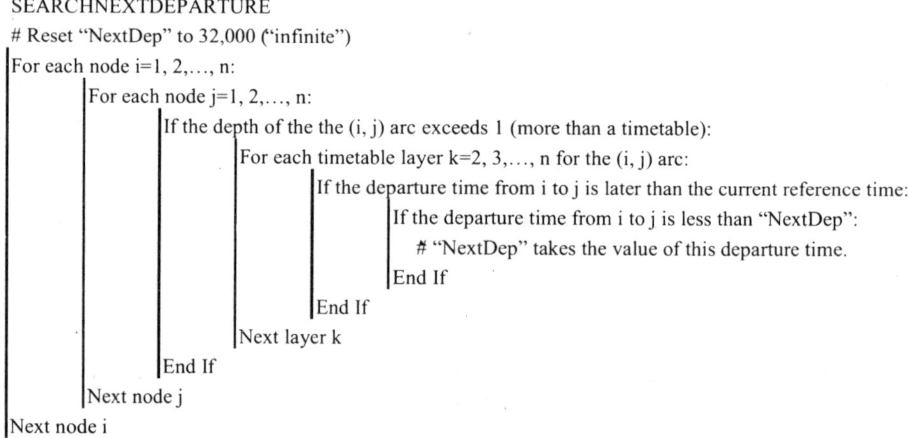

```
SEARCHNEXTDEPARTURE
# Reset "NextDep" to 32,000 ("infinite")
For each node i=1, 2,…, n:
    For each node j=1, 2,…, n:
        If the depth of the the (i, j) arc exceeds 1 (more than a timetable):
            For each timetable layer k=2, 3,…, n for the (i, j) arc:
                If the departure time from i to j is later than the current reference time:
                    If the departure time from i to j is less than "NextDep":
                        # "NextDep" takes the value of this departure time.
                    End If
                End If
            Next layer k
        End If
    Next node j
Next node i
# The new reference time is from now on equal to "NextDep", which is at least one minute later than the
  previous reference time.
```

Algorithm 3.5. *Search for the next departure time*

3.3.2.7. *Validation of the minimum paths matrix*

It is only at this stage that the paths which are included between the previous reference time and the new time are really written (Algorithm 3.6). The paths already found in this range are stored in order to avoid heavy repetitive calculations in calculation time and written later when they are included in the range of reference times.

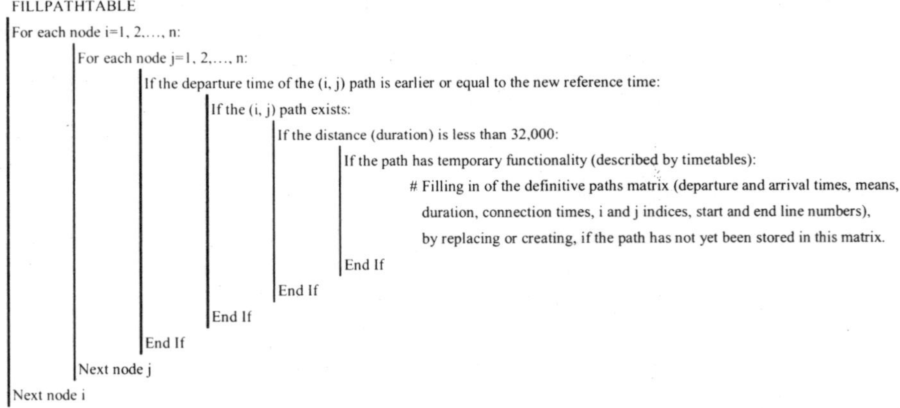

```
FILLPATHTABLE
For each node i=1, 2,..., n:
    For each node j=1, 2,..., n:
        If the departure time of the (i, j) path is earlier or equal to the new reference time:
            If the (i, j) path exists:
                If the distance (duration) is less than 32,000:
                    If the path has temporary functionality (described by timetables):
                        # Filling in of the definitive paths matrix (departure and arrival times, means,
                          duration, connection times, i and j indices, start and end line numbers),
                          by replacing or creating, if the path has not yet been stored in this matrix.
                    End If
                End If
            End If
        End If
    Next node j
Next node i
```

Algorithm 3.6. *Filling in of the minimum paths matrix*

By recalling the example with the new reference time equal to 8:30 am and *using the intermediate point B* no new relation AC is stored, insofar as the timetables (8:30-9:45 am) from A to B and (9:40-10:20) from B to C are not compatible for a connection.

Now by considering C, the AD link is not possible, insofar as AC is incompatible as shown above. On the other hand, BD can be constructed with the timetables (9:40-10:20 am) then (10:25-11:30).

By following the process of minimum path determination we have voluntarily refrained the symmetric relations from D to A from the demonstration, since this is incompatible with the creation of new paths. However, since the algorithm loops all the points, these symmetric relations are also analyzed.

3.3.2.8. *Process looping*

The process then continues as long as the reference time calculated afresh precedes the processing end time given by the user at the start.

The procedure of preparation of the minimum paths matrix (Algorithm 3.2) is carried out again, then Floyd's algorithm itself (Algorithm 3.4) and, finally, the search for the new reference time and the filling in of the final paths matrix (Algorithms 3.5 and 3.6).

However, an additional procedure is inserted between the preparation of the paths matrix and Floyd's algorithm, thus substituting the preparation procedure of the calculation of the timetable paths (Algorithm 3.3). This is done in order to avoid recalculating the already determined paths that have not been stored in the final paths matrix.

3.3.2.9. *Creation of standardized files*

At the end of the process a file of standard format is produced, in which each column of fixed width refers to an element of the path description: names of the stations of origin and destination, departure and arrival times, means of transportation (single train or series of several trains), length of trip, connection time, numbers of the start and end lines, codes of departure and arrival stations specific for each one.

3.4. Conclusion: other developments of Floyd's timetable algorithm

3.4.1. *Determination of the complete movement chain*

Before describing the timetable development of Floyd's algorithm we have raised the specific case of connection constraints as well as the necessary addition of pre- and post-routing durations to the timetable matrix in order to refine the quality of service concept up to a door to door processing. We will now describe these initial developments, which have been operational since 1996.

To describe the algorithmic developments specific to connection constraints we will recall a study carried out in 1996 on behalf of the DTT and the OEST, which was tasked with defining a level of service grid for the SNCF and interior aerial transportation system within the framework of evaluating the governing principles of the transport infrastructure. In the general algorithm no constraints are applied to the number of connections and the maximum waiting time; only the minimum duration must be longer than two minutes. At national level it was necessary to substitute these considerations by stricter constraints. For interior aerial transport only one connection was accepted, with a maximum connection time of two hours when there was no change of airport, or ranging between one and three hours on the assumption of a connection with change of airport.

For rail transport two connections were accepted with methods exploiting a variable duration depending on whether a city like Paris or a provincial town was considered. In Table 3.1 we indicate the assumptions envisaged and introduced into Floyd's algorithm regarding the allowed connection times.

In addition, the individualized knowledge for each graph node of the average duration observed to reach a public transport station from home, or conversely to reach home after leaving the station would yield a first response to evaluating the quality of service over the entire movement chain. The lack of this type of data for all the nodes, nevertheless, led us to consider approximate times which, however, corresponded to a certain observable reality and were variable according to the size of the cities and the means of public transport used for the main journey.

Connection 1	**Time (in hours)**	**Connection 2**	**Time (in hours)**	**Total time (in hours)**
G[6]	≤ 2	–	–	–
Gp1–Gp1[7]	≤ 2	–	–	–
Gp1–Gp2[8]	1 ≤ ≤ 3	–	–	–
G	–	G	–	≤ 2
Gp1–Gp1	–	G	–	≤ 2
Gp1–Gp2	1 ≤ ≤ 3	G	≤ 1	–
G	–	Gp1–Gp1	–	≤ 2
G	≤ 1	Gp1–Gp2	1 ≤ ≤ 3	–

Table 3.1. *Connection constraints*

Therefore, the initial travel times to reach an airport are traditionally longer than those to reach a railway station. According to this principle and with the help of a matrix of the nodes valuated by duration of the two final routes, the optimal paths matrix is modified by bringing forward the departure times to integrate the duration of preliminary routing and moving back the arrival times to take into account the final leg of the journey. In addition, a time known as safety departure is added to account for the "anxious behavior" of a user not wishing to miss his train or plane. Two other developments stem from the above considerations.

6 G: connection at a provincial station.
7 Gp1-Gp1: connection in Paris without change of station.
8 Gp1-Gp2: connection in Paris with change of station.

3.4.2. *Overview of all the means of mass transport*

First of all, Floyd's algorithm, which has been specifically obtained to provide an account of a means of transport with temporary functionality, is also adapted for railways as for other means such as aerial, maritime, river transport or on another spatial scale for transport in a city bus, tram, etc. The availability of timetables and constitution of the corresponding matrices are the only constraints, which when exceeded become open to possible comparative analyses of the efficiency of each means of individual or public transport. This method of sector by sector analysis of the means of transport does not at all overshadow the possibility of evaluating a multimodal service incorporating all these means, in particular, to account for the *hub and spokes* operation of aerial transport, frequently requiring pre-routing by TGV, which in turn implies an initial journey on foot or by taxi from home.

3.4.3. *Combination of means with permanent and temporary functionality*

Secondly, the combined processing of means with temporary and permanent functionality, which is possible by using the algorithm presented, can solve the problem of approximating the lengths of the final journey. According to the nodal zoom method [CHA 97] the creation of new intra-urban nodes (stations, airports, sites attracting employment, housing, etc.) and of new arcs connecting these stations or airports to workplaces or homes by using such means as walking, bus, urban methods of transport, etc., considerably refines the movement chain by individualizing each path determined by the algorithm. To take but one example, the departure time from home to an SNCF railway station can thus be calculated on the basis of the departure time of the main route, which is brought forward by a length of time determined for an initial pedestrian arc. Other urban means of transport are recognized in Floyd's algorithm. Associated with various types of infrastructure (generally extensive public roads system, at a site of generally high urban density, at a usual site or a site with exclusive right of way), this prospect leads to distinguishing three functional binomials for the bus (8, 15 and 25 km/h depending on traffic density or the width of lanes), two binomials for the tram (18 and 45 km/h depending on the type of material), one for the subway (24 km/h), or two more for walking (3 and 5 km/h). The list is naturally dependent on the needs that have arisen up until now, but remains perfectly open to other means (roller skates, etc.) or finer differentiations (pedestrian on a conveyor belt, on a narrow or a broad sidewalk, etc.).

Graph theory and Floyd's algorithm offer a significant development potential, which is limited only by the model creator's preoccupation with precision. Thus, why not imagine the processing of the performance of river and maritime transport networks by considering the constraints of current, regulation speeds, starting with directed arcs with permanent functionality, or even more exactly using a series of arcs with temporary functionality required by the effect of tides and closings associated with locks and able to repeatedly prevent a ship from moving upstream?

3.4.4. *The evaluation of a timetable offer under the constraint of departure or arrival times*

Another extension related more to an algorithmic modification than to a graph problem still has to be specified here. In the introduction we have already posed the problem of simulations using an additional constraint of departure or arrival times; i.e. which means of transport should I choose if I wish to travel from Lyon to Marseilles and be there before 8:00 am? The performances of the line assigned to TGV at first glance appear as the most adequate alternative. Nonetheless, introducing a constraint on the time of arrival at destination may lead to prefer the use of an individual vehicle for the flexibility that this means of transport allows while being free from departure time constraints.

The general algorithm already enables the determination of the set of paths for a given time range. The remaining developments stem from these particular problems of arbitration between several means of transport, depending on the imposed arrival or departure timetables. They were implemented by Alain Hostis and Christophe Decoupigny [LHO 00].

If a chronological presentation of the algorithmic development for a user is preserved, the procedure of creating the minimum arcs matrix (Algorithm 3.1) remains unchanged.

Then, depending on the imposed constraint of arrival time, the user may input the reference arrival time, after which no path will be determined, since they become of no interest for a user who does not wish to be late for a meeting or work, for example. Let us recall that in the general case the user delimited the processing time range using both departure and arrival times. Except for this modification, the partial and provisional filling of the minimum path matrix (Algorithm 3.2) remains unchanged.

The true change is observed in the main processing of the minimum arcs (Algorithm 3.3).

If there is no arc valuated by a means of transport with permanent functionality, it is the IF/End If loop indicated with an asterisk in the general procedure that is modified. For a given arc (origin, destination, time) this is stored in the first layer of the paths matrix only if the arrival time is earlier or the same as the reference time and if the departure time is later than that already found (meaning the possibility of leaving the origin later, while minimizing the total duration of the journey).

If there is an arc valuated by a mode with permanent functionality, the modification of the algorithm is found at the level of the double asterisk in Algorithm 3.3. The duration of the arc with permanent functionality added to the reference time in order to provide the new arrival time is now deduced from the reference time to provide the departure time. This minor change has little importance insofar as it is the durations that are really processed thereafter. Nevertheless, in the actual calculation to determine the paths (Algorithm 3.4), it makes it possible not to generate paths that can arrive after the arrival time constraint when a permanent arc is added to a time arc.

Next, the determination of the minimum paths is carried out using a procedure similar to Algorithm 3.4 except for the fact that for the general algorithm we have looked for the optimal paths between *i* and *j*, for a given relation *ik* (k being the intermediate node) by sweeping all of the relations *kj*. With the constraint on the time of arrival for each given relation *kj* we work out the paths by sweeping the relations *ik*, so as not to exceed the arrival time at *j* from *k*, which has been imposed at the start.

In the following procedure (Algorithm 3.6), which is specific to the arrival constraint, we gradually fill in the matrix when the arrival time of a path precedes the reference time (whereas in the general case the departure time of a path had to be earlier than the reference time, but the arrival time could be later). Moreover, the path being processed is preserved only if, at the time of arrival equal to another path for the same origin/destination, it leaves later, thus minimizing the total duration of the path, while respecting the arrival constraint.

Lastly, we find the same looping already clarified for the general algorithm which makes it possible to look for the reference time of the next departures, and so on until no longer entering new minimum paths.

On the contrary, the constraint on the departure time is a simple alternative to the determination of the minimum paths within a time range which is limited by processing start and end times. The procedures of partial and provisional filling of the minimum paths matrix (Algorithm 3.2), of processing the minimum arcs (Algorithm 3.3) and of determining the paths (Algorithm 3.4) are carried out without modification. However, the looping leading to the search for the next departure within the time range and so on for all the ranges is not carried out. Indeed, at the end of the first iteration all the paths that leave starting from the time indicated as departure constraint are given.

The possible applications for arrival and departure constraints are most frequently based on the comparison between, for example, individual transport and the railway transport. The arrival time being known, it is enough to compare the possible departure time by car (by deducting from the imposed arrival time the minimum transport duration according to its type) with that determined for public transport (i.e. the latest departure time to arrive at the nearest possible time to the one imposed).

3.4.5. *Application of Floyd's algorithm to graph properties*

The description of the developments of Floyd's algorithm or, more generally, of the methods implemented to specify multimodal routing of the door to door type would be incomplete without introducing the latest publications relating to the use of graph properties to evaluate a transport timetable offer.

The quality of service of public transport is traditionally evaluated, as we have demonstrated, by explicitly characterizing the relations between one or several origins and one or several destinations. This approach is naturally relevant with a view to timetable optimization, better spatial distribution of the relations, or measuring the possibility of effectively and quickly linking the possible destinations with a given origin, and thus evaluating the spatial diversity of the offer.

However, the study of graph properties, such as external and internal half-degrees, enables to provide additional information by evaluating the importance of a station, an airport, etc., according to the number of time relations carried out by train, by plane, etc., arriving or departing from a city. Such calculations are implemented on the basis of the matrix of minimum paths which are determined beforehand using Floyd's algorithm. This application is thus strictly speaking not a development of the algorithm, but rather a new analysis of the results emanating from it by partially answering to the deficiencies of graph theory to describe and characterize the nodes.

In practice, for a rail transport network, for example, the sum of the relations with a given city as destination (thus the number of arrival times) provides an interesting indication about the convergent traffic (the internal half-degree) and from that about the importance of the hierarchical position of a city in an urban network[9].

---

9 Despite everything the results must take into account a certain observation, in the sense that it is the number of arriving times from all the origins at a given destination that is measured, and not the number of trains or planes really arriving at a station or at the airport. By way of illustration, a train may very well leave Bordeaux, stop at Poitiers, then arrive at Tours, thus accounting for two relations on arrival at the last station, but for only one train. Consequently,

This indicator may be supplemented by opposite information, the external half-degree, by summing the number of departure times from the same city, the observation for its interpretation being in the same veins as for the internal half-degree.

3.5. Bibliography

[BAP 99] BAPTISTE H., Interactions entre le système de transport et les systèmes de villes: perspective historique pour une modélisation dynamique spatialisée, Thesis, Tours, 1999.

[BAP 02] BAPTISTE H., L'HOSTIS A., Analyse de la qualité de service du réseau de transport collectif en région Nord-Pas-de-Calais et Languedoc-Roussillon, recherche menée pour le compte du Conseil Régional de la région Nord-Pas-de-Calais, CESA, Tours/INRETS, Villeneuve d'Ascq, 2002.

[BAR 86] BARTNIK G., MINOUX M., *Graphe, algorithmes, logiciels*, Bordas, Paris, 1986.

[CHA 97] CHAPELON L., Offre de transport et aménagement du territoire: évaluation spatio-temporelle des projets de modification de l'offre par modélisation multi-échelles des systèmes de transport, Thesis, Tours, 1997.

[COL 96] BAPTISTE H., LARRIBE S., MATHIS P. (ed.), Grille de niveau de service, recherche menée pour le compte de la DTT/OEST, Rapport de recherche, Tours, 1996.

[LHO 97] L'HOSTIS A., Images de synthèse pour l'Aménagement du territoire: la déformation de l'espace par les réseaux de transport rapide, Thesis, Tours, 1997.

[LHO 00] L'HOSTIS A., DECOUPIGNY C., MENERAULT P., MORICE N., *Cadencement et intermodalité de l'offre en transport collectif en Nord-Pas-de-Calais: analyse et propositions d'amélioration*, INRETS, Villeneuve d'Ascq, 2000.

[MAT 78] MATHIS P., Economie urbaine et Théorie des systèmes, Thesis, Tours, 1978.

---

the result must be interpreted as a measurement of the richness of the material offer coupled with the diversity of possible origins.

Chapter 4

Modeling the Evolution of a Transport System and its Impacts on a French Urban System

4.1. Introduction

A quick analysis of the present French urban network shows a rather good adequacy between the size of cities and the quality of transport infrastructure servicing them. Taking into account the advanced state of the implementation of various road and railway projects, the largest agglomerations will all be connected by TGV connections and/or highway type links within a few decades [REY 93].

This development, whose foundations were laid around the 15th century with the centralization policy implemented by Louis XI, accelerated appreciably since the 18th century with the development of postal roads [ARB 92, STU 95]. The improvement of transport networks to the benefit of the largest cities made it possible to set up an urban network of a higher level, to some extent providing the latter with an advantage compared to their respective countryside, thus forming a network among networks. These mechanisms initially found their origin in the strong political will to control the territories which led to imposing an "invisible hand" stretching from Paris towards the main French ports through ease of communication [BRA 90b].

Today the various laws on regional planning still encourage greater accessibility of the worst connected territories, still and always to the benefit of the capital or the largest agglomerations [JOU 95]. However, a dominant evaluation method intended to estimate the needs and choices with respect to the construction of transport

---

Chapter written by Hervé BAPTISTE.

infrastructure was added to the political aspect. Indeed, it is shown that costs/benefits analysis leads to a pattern according to which the demand for transport generates the resulting supply and tends to reinforce the existing tendencies with respect to movements, as well as increase the polarization of territories (see [BAP 99], p. 56). Nonetheless, this mechanism supports the concentration of population on a regional or national scale, which also results in the opposite in more restricted areas where the increasingly lengthening intra-urbanization follows the improvement of transport networks and the development of mobility.[1]

The main objective here is to estimate the long-term consequences of the evolution of departmental transport networks on the spatial distribution of alternating migrations and then on definitive migrations, but also its opposite, whose processes tend to connect the best performing transport networks with the most important cities. In other words, it is a matter of specifying, of dynamically measuring the interactions between a departmental system of cities and the transport system.

Modeling by using graph theory is an effective response to this type of question. With respect to the multiplicity of interrelationships between several hundreds of cities which are, in particular, manifested through a significant number of transport infrastructure installations, the description of area by nodes and arcs offer invaluable assistance in the understanding of such mechanisms. While by necessity simplifying reality, the data processing tool in this context remains the most appropriate means of analysis by simultaneously making it possible to take into account tens of thousands of spatial and quantified data inputs.

4.2. Methodology: RES and RES-DYNAM models

4.2.1. *Modeling of the interactions: procedure and hypotheses*

Two types of models are used for this purpose: the first (RES) aims to create theoretical areas from any part of a graph responding to many constraints in order to reveal similarities with the areas observed. The second (RES-DYNAM) is a calculation model intended to exploit the previously created graphs by implementing processes of transport network modification, generation of commuting and definitive migration.[2]

1 See [PUI 94] for a development of the polarization concept, its definition, manifestations and causes.

2 These two models are more broadly clarified in [BAP 99].

Modeling by using RES-DYNAM is based on two functionally and temporally dependent processes (see Figure 4.1): on the basis of any graph (of theoretical or real area) the optimal routes are determined following a temporal measurement of the transport network (1)[3]. A gravity model[4] is used at this level (2) to determine the mutual influences exerted by the cities and the potential movement flows (in particular the home-work commute), the latter being finally assigned to the various arcs of the graph (3). The spatial configuration of the transport network, the quality of the infrastructure play a major part in the distribution of movement, since space acts as a brake for various flows.

This module qualified as static supposes an immediate determination of the attraction of each city and an immediate adaptation of movement flows to the transport supply. This philosophy of simulation, restrictive as it is, is, however, widespread in this field because it makes it possible to model the observed flows with unquestionable efficiency.

A dynamic module is then exploited leading to a redistribution of a fraction of the population by definitive migration, according to a global mobility rate selected for the area used (4). The decision to migrate definitively and especially the destinations of each individual are determined according to the generally attractive nature of each city, which is itself derived from the results of the gravity model.

However, a factor that proves attractive to one person may be repulsive for another. For example, the attraction of a level of services, employment or important infrastructure can lead to a migration towards a city of considerable size (the correlation between size and the functional level of a city being significant), land costs or quality of life are then not perceived as determinant. However, on the contrary, a stronger attachment to the latter criteria can lead to a migration towards smaller towns, which are close to a large employment pool, and reveal a diagram of migration by centrifugal redistribution from large cities. In other words, these two "residential strategies" are just one example among all the modeled motivations integrated into the RES-DYNAM model. At this level, a looping is carried out (5): the new settlement pattern of each city is placed at the origin of a new gravity calculation which determines the new influences exerted by cities. Loop after loop the hierarchy of city sizes changes, each one of these iterations being comparable to a year when the selected global definitive mobility rate is coherent with real life observations.

---

3 For a clarification of the model used based on Floyd's algorithm see [CHA 97].

4 The model used here is derived from the model of Mathis [MAT 78] which is based on the population and temporal distance measured in the network, respecting the hypotheses of mutual influences between cities (attraction of cities taken two by two and being then relativized with respect to all other cities) and of dissymmetry of flows.

Until now, the hypothesis of temporal modification of the transport network characteristics has not been used. Its integration is based on the existence of saturation thresholds of the transport axes: after gravity calculation, determination of potential movement flows and new partial population redistribution, the crossing of the threshold would cause an infrastructure modification (6) using methods clarified later on[5]. Once the transport network configuration has been modified the optimal routes have to be calculated afresh during the following iteration (7), since the new journey times are integrated in the gravity calculation. These durations of journey, as well as the population, will necessarily intervene in distributing the influence between cities and, in the long term, new population localizations. Thus, there is clearly a dynamic interaction between the system of cities and the transport system.

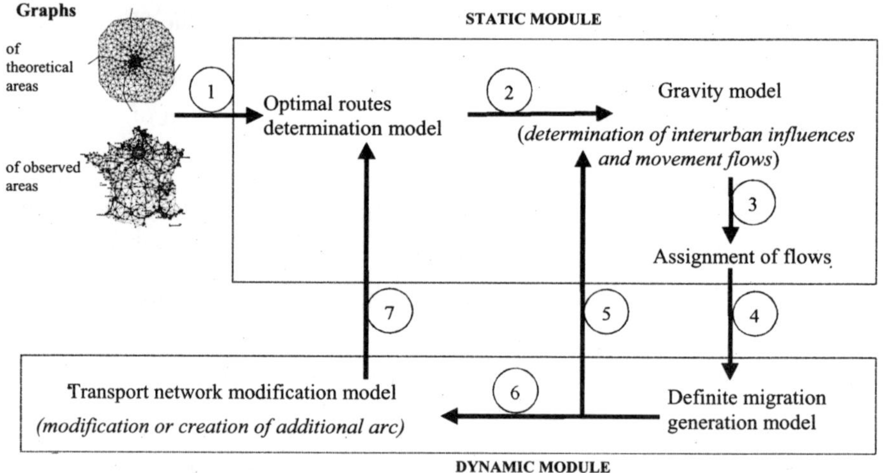

Figure 4.1. *Schematic modeling diagram*

The results are interpreted using a comparative method: two simulations are carried out, one using a temporally invariable transport network, the other considering the transport network evolution through the improvement of performance along certain axes. The analysis by differences leads to a conclusion regarding the role of transport networks in the processes of commuting and definitive migration, the dynamic process of population redistribution being in turn rigorously identical, just as the parameter values and the calculation hypotheses.

5 If the saturation threshold is not reached, the iterative process continues without modification of the transport network (5).

4.2.2. *The area of reference*

The graph of theoretical area used below, which is generated by the RES model, is constructed on the basis of simple assumptions to observe a certain similarity with the area of the Indre-et-Loire department.

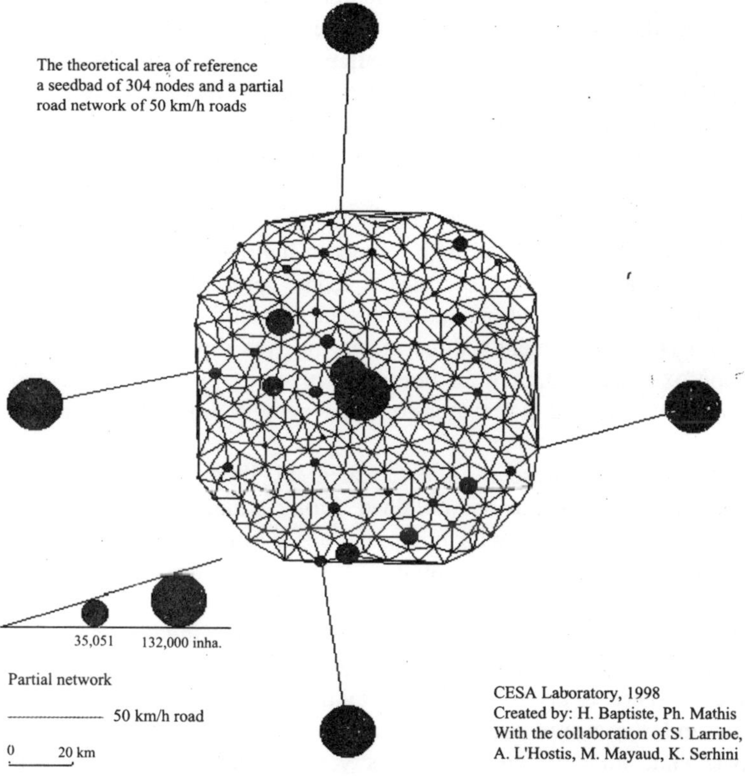

Figure 4.2. *Initial theoretical area*

The theoretical area extends over 100 km from East to West and 110 km from North to South and is described on the basis of 300 towns, which are located randomly but, nevertheless, respect a minimal spacing of 5 km, which is the average observed distance (Figure 4.2). Four towns were added afterwards in order to benefit from a certain openness of the system of cities located between 80 and 100 km from the center of the theoretical area, which is a distance compatible with the spacing observed between two prefectures of bordering departments[6].

The populations were distributed according to the rank-size law, following a form that makes it possible to preserve common points with the Indre-et-Loire department: the total population of the theoretical area reaches 520,000 inhabitants (not including the four peripheral towns), the main town – approximated to the department prefecture – is assigned a population of 130,000 inhabitants, as well as the four peripheral towns, thus making it the equivalent of the prefectures of the bordering departments. The location of the second town by size is selected near the first city, similarly to Joue-le-Tours spatially which is contiguous with Tours.

All the towns are connected by a partial infrastructure network[7] of "50 km/h roads", which can be assimilated with respect to the maps published by Michelin to roads captioned as: "regional or release roads" presenting strong sinuosity, or "surfaced or unsurfaced road or road with bad viability". These are two-lane axes, whose width is most frequently three meters in each direction, where the average speed is approximately 50 km/h. In addition, a planarity principle is adopted, such as can be generally observed in real areas of this type with this kind of road network.

The interest of the presented theoretical graph then lies in enabling a certain generalization of the conclusions obtained during simulations, while preserving a degree of reality compatible with interpretations that could be made regarding the departmental area of Indre-et-Loire alone.

Moreover, the process of transport network evolution involves additional partial networks: the "70 km/h road" is captioned in roadmaps as a "main road" since the speed selected here is an average value taking into account, in particular, stops, crossroads and successive decelerations. The "60 km/h road" refers to the same roadmap legend, but with stronger sinuosity and inducing a more limited average circulation speed. Finally, the expressways and the highways are considered separately according to the respective average speeds of 90 and 110 km/h.

6 Let us note, however, that the RES model enables the representation of more atypical situations, like that of Alençon in the Orne department, which is located at the periphery of the department, at a distance of about 50 kilometers from Le Mans.

7 A partial infrastructure network can be defined as the entirety of the infrastructures of a given means of transport and presenting a certain homogenity particularly in terms of average circulation speed (see [LHO 97], p. 60).

4.2.3. *Initial parameters*

4.2.3.1. *Static modeling*

According to the description of the transport network given above, each arc and each path from any node to every other node is assigned a minimum time value measuring the optimal duration to reach all the nodes in the graph.

Specifically for average movement durations within the same town we choose a time equal to 15 minutes, according to the results obtained beforehand during the model calibration phase using alternating commutes in Indre-et-Loire in 1990 (see [BAP 99], p. 306). The movements during simulations retained hereafter will thus relate only to the home-work commutes. Only the active population with an employment will be considered and determined on the basis of the population of each town according to an average ratio of a number of people for each employee.

4.2.3.2. *Dynamic modeling of population evolution*

The dynamic process of evolution of populations in town follows the characteristics observed between 1968 and 1990 in the Indre-et-Loire department only for definitive internal migrations. However, the values of attraction indicators were modified slightly to simulate an evolution towards a more marked area polarization, in order to make the final results more perceptible. In other words, we accentuated the attraction of the towns in terms of employment.

The total rate of final mobility is chosen to be equal to 5% per iteration[8] for all simulations. Each one of these iterations is thus – by the selected parameter values – about one year long and the simulations relate to a period of 100 years.

4.2.3.3. *Modeling of transport network evolution*

The dynamic process of transport network improvement implemented for one of the two following simulations works according to a "following" transport policy, in the sense that it meets existing needs, represented by saturation of road axes, but with a chronological delay between the initial circulation difficulties and the implementation of correction measures.

In fact, we considered a period between 10 and 20 years depending on the case, during which the infrastructure capacity was regularly exceeded before an improvement could be offered to the users (in a network that he could use, the

8 That is, a probable value signifying a sufficiently strong mobility compared to the average annual rate of 3.6% which was observed on an inter-commune scale in France from 1982 to 1990.

construction period was included). The very basis of transport network evolution thus has to be sought among the capacity values of each transport axis.

The capacity of road infrastructure corresponds to the maximum flow of vehicles that it can support. On the one hand, it depends on the geometrical characteristics: the number and width of lanes, the number of lines of traffic circulating in the same direction, slope, visibility range and curvature radius. In addition, it is dependent on the circulating vehicles, with respect to their types (cars, trucks with and without trailers, etc.) and speed of circulation.

The practical capacity of an infrastructure is a measurement of the maximum flow that still preserves good circulation conditions, in terms of speed, safety or overtaking maneuvers. At this level of traffic the circulation starts to be disturbed, the average speed of the vehicles decreases, the first indicators of obstruction appear, but still under conditions which are considered bearable (see [COQ 72], pp. 140 and 146). This value thus corresponds to the flow limit, up to which the infrastructure is sufficient. It corresponds to the saturation threshold, beyond which stops in the traffic flow are more and more prolonged; the difficulties of circulation increase requiring frequent braking and acceleration, which leads to reducing the average speed of the vehicles.

The "possible capacity", also called physical capacity or saturation flow, is the most intense theoretical flow that the infrastructure can cope with, where the vehicles are in a situation of "driving in a procession" ([COQ 72], p. 146): they follow each other at intervals equal to the minimum safety distance, at reduced speed. At this level of traffic any circulation incident (breakdown, accident, crossroads, etc.) results in the full halt of the entire line of vehicles concerned.

The capacity is generally expressed as the number of vehicles per hour observed at a point, the result then being extrapolated to a portion of the infrastructure presenting equivalent characteristics. From this unit the capacity can then be converted into the average number of vehicles circulating daily, by considering that the normal rate during rush hour (repeated periodically during a short period of time, such as during commuting migrations, for example) is equivalent to a sixth of the average daily flow calculated over a year [COQ 72, p. 36]. The values of infrastructure capacity mentioned further on are therefore not determined on the basis of the largest flow values, if these are specific for a time of year (in particular, holiday departures). The traffic observed during these exceptional rushes cannot indeed be retained as a reference value, since economic conditions cannot reasonably lead to over-develop the road networks to absorb certain exceptional traffic jams.

In addition, if the concept of average is recurrent in the determination of infrastructure capacities, it is because the latter are implanted in geographical space. Thus, topographic constraints (slopes, sinuosity minimizing visibility, etc.) make it is extremely difficult to adopt the same value of capacity for the entire infrastructures of the same partial infrastructure network. Moreover, many other factors intervene to modify the capacity of the same road: the existence or absence of obstacles at points of access (low walls, bridge railings), the percentage of trucks or periodic traffic fluctuations.

For reasons of feasibility we, nevertheless, chose a single value for each of these partial networks, each one of those having been determined to group infrastructures with similar characteristics (in terms of width, number of lanes, sinuosity and slope). Consequently, the bias is relatively weak.

Finally, the concept of economic capacity is used to determine a maximum economically sustainable flow. Thus, a higher value in an urban environment is acceptable, in the sense that the distances covered by users are small, urban constraints are stronger that the maximum traffic is observed during a short period of time during a day, all this leading to a raised saturation threshold. On the other hand, the threshold is fixed at a lower value for interurban environments where temporally durable saturation is no longer acceptable for long distances. Since the arcs of the graphs connect different towns or cities, we then consider values related to circulation in an interurban environment.

In any case, hereafter we have retained values relating to saturation thresholds and not the "possible" capacity, insofar as we saw that the traffic flow using the latter concept is extremely unstable, and unsustainable in an interurban network, where the inconvenience caused to vehicles is no longer acceptable for longer distances.

Finally, the criteria used to determine the saturation threshold of each partial infrastructure network are:

– interurban circulation only;

– total value by type of partial network;

– value expressed as an average daily number of vehicles, since the traffic figures assigned to arcs are not segmented by time;

– balanced circulation in both directions for infrastructures with undivided highways, i.e. symmetric distribution of traffic in each direction. However, the traffic seldom obeys this situation (in particular for the alternating commutes), which leads us to overestimate the true capacity of these types of roads;

– saturation thresholds determined for circulation include 10% of commercial vehicles, which once again leads us to overestimate the real capacity of the infrastructure, insofar as the distribution of road traffic in France by type of vehicle shows that trucks, coaches and light commercial vehicles represented approximately 20% of the total traffic in 1996. However, we minimize the share of these categories of vehicles because light commercial vehicles have a less considerable impact on capacity reduction than trucks or coaches;

– value provided without crossroads, or any other device involving difficulties for the traffic flow. Once again the indicated capacity will therefore tend to overestimate the real figures.

Partial network	**Number of vehicles without commercial vehicles**	**Number of vehicles with 10% of commercial vehicles**	**Observations**
70 km/h road	2,700	2,400	Good geometrical characteristics, visibility more than 450 meters at any point
60 km/h road	2,400	2,100	Higher sinuosity (visibility lower than 450 meters for approximately 30% of the length of the road)
50 km/h road	2,100	1,800	Reduction of the width of the lanes to 3 meters
Expressway	11,125	10,000	Twice two ways
Highway	14,400	13,000	Twice two ways

Table 4.1. *Saturation thresholds for one direction*[9], *average number of vehicles per day*

Overestimating the real capacities is justified by the fact that the improvement of transport networks or the construction of a new axis generally takes a relatively long period of time, during which the true saturation threshold is exceeded. A higher value then makes it possible to take this inertia into account in the overall response

9 Assumption of balanced circulation in interurban environment without taking crossroads into account.

time. On this basis the capacities retained for each partial network integrated into the process of transport network evolution are determined (Table 4.1).

For the "70 km/h road" partial network the lanes are, with rare exceptions, broad (3.5 meters) and come in twos (one in each direction). Roads with three lanes exist in a minority and we have adopted a saturation threshold valid for a seven meters wide road with good geometrical characteristics (visibility over 450 meters at any point of the infrastructure, flat ground and weak slopes).

The capacity of such infrastructure is 900 vehicles/hour in both directions if we consider that only cars are in circulation, in the absence of crossroads and in interurban environment. By considering 10% of commercial vehicles the decrease of the above capacity by extrapolation also amounts to approximately 10% of an interurban road (see [COQ 72], p. 145). It is then equivalent to approximately 800 vehicles/hour in both directions, that is, an average of 4,800 vehicles per day and, thus, 2,400 vehicles in each direction (considering balanced circulation).

For the "60 km/h road" partial network the increased sinuosity compared to the previous partial network cuts down visibility values. Roger Coquand estimates that when 20% of the stretch of road studied has a visibility below 450 meters, its capacity decreases from 900 to 840 v/h; from 900 to 760 v/h for 40% of the length of this road (see [COQ 72], p. 150). In view of these observations we choose a capacity of 800 vehicles/hour (both directions, without commercial vehicles), i.e. approximately 2,100 vehicles per day on average (with 10% of commercial vehicles in each direction).

Roger Coquand determines the reduction in capacity between a type "70 km/h road" infrastructure and a three meters per lane road, that is, a "50 km/h road", at 23% (see [COQ 72], p. 149), which yields an approximate capacity of 1,800 vehicles per day and direction on average.

Infrastructures with a central separator have a higher capacity by lane in the sense that flow limitation no longer depends on increased or decreased difficulty of overtaking maneuvers. Nonetheless, traditional crossroads (left turns, etc.) may still exist on expressways, whereas only interchanges exist on highways. However, if transverse circulation is considerable, the traffic flow on the main roads may be affected by it, thus decreasing the saturation threshold (see [COQ 72], p. 272).

The saturation threshold for a highway lane is fixed at 1,200 v/h. A two-lanes highway track thus has an approximate capacity of 14,400 vehicles per day and direction on average. Considering that expressways are an intermediate type of infrastructure between highways and divided highways with four lanes without a central reservation (capacity of 7,500 vehicles per day and direction [COQ 72, p.

159]), we arbitrarily adopt for them a saturation threshold equal to the average value, that is, 11,125 vehicles per day and direction.

The values indicated above apply to circulation made up exclusively by cars. By extrapolation, according to methods presented previously, the capacities obtained with 10% of commercial vehicles are decreased to obtain the following figures (expressed as number of vehicles per day and direction on average, for four-lane divided highways):

– expressway: 10,000;

– highway: 13,000.

The simulation of the transport network evolution during 100 iterations for a theoretical area is obtained on the basis of capacity figures integrating 10% of commercial vehicles.

The transformation itself of partial networks road (from 50 km/h to expressways assigned an average speed of 90 km/h) is carried out by widening existing lanes, thus by substitution of the successive networks, whereas a highway is added to the existing networks. We have, nonetheless, rendered the mechanism more complex according to the following assumptions:

- The successive transformation of a "50 km/h" partial network into an expressway is carried out in 10 year periods upon exceeding capacity; on the other hand, the construction of a highway requires an interval of 20 years before it is implemented. Then, by supposing an already stated construction schedule, the later sections are placed at the disposal of users after 10 years.

This expresses quite well the idea, according to which "in general, a generation needs to pass between the moment when the idea to implement new transport facilities for a given service emerges and when this idea becomes popularized and is ready to become operational" (see [MER 94], p. 6).

The durations taken into account here are, however, lower, insofar as the times chosen for this application stem from a transport policy which is based more on permanent adjustment to the movement demand – but shifted temporally – than on real long-term planning. However, we roughly respected the financial constraints related to the differentials of capital costs according to infrastructure types, leading to higher time intervals before the installation of a highway stretch[10].

10 Merlin thus evaluates approximately the capital costs (in francs, in 1990) at *1.5 million francs per kilometer for a road in town* (the equivalent of the "50 km/h road" partial network), *3 million for a secondary road with 2 lanes* ("60 km/h road"), 8 million for a trunk road with three lanes, 18 million for an expressway with 2×2 lanes, and above all *25 million*

Table 4.2 shows in detail and in chronological order the various terms, during which infrastructure capacity is exceeded, before any improvement or construction.

Initial network	Resulting network	Necessary duration (years)	Type of modification
50 km/h	60 km/h	10	improvement
60 km/h	70 km/h	10	improvement
70 km/h	90 km/h (expressway)	10	improvement
90 km/h (expressway)	110 km/h (highway)	20	construction
then 90 km/h (expressway)	110 km/h (highway)	10	construction

Table 4.2. *Transport network modification assumptions*

4.3. Analysis and interpretation of the results

4.3.1. *Demographic impacts*

First of all, we show the final configurations obtained for the system of cities under consideration before clarifying its generating mechanisms.

Demographic trends which were only attributed to the inequalities in spatial distribution of definitive migration at the end of 100 years show an increase in urban concentration regardless of the hypothesis, but following different patterns (Figure 4.3):

– by densification of the main town following the hypothesis of a transport network invariable in time (this town increasing from 132,000 inhabitants to 181,986 inhabitants[11]);

for a highway (see [SEA 94], p. 227). Since in the application the expressway substitutes existing roads by simple improvement, we considered a 10-year modification just as for networks of lower hierarchical levels despite the higher construction costs.

11 If we continue the simulation for an additional 50 iterations, the population of the main town stagnates, to the benefit of its primary crown, which is explained by a saturation threshold caused by an elasticity lower than 1 of one of the attraction indicators; nevertheless, peripheral diffusion remains very limited beyond the primary crown.

– by peripheral diffusion of housing in the second simulation (see right-hand side of Figure 4.3), with decreasing intensity as the distance from the city-center increases, which is hereafter compared to the prefecture of the theoretical area (its population decreasing from 132,000 to 129,632 inhabitants).

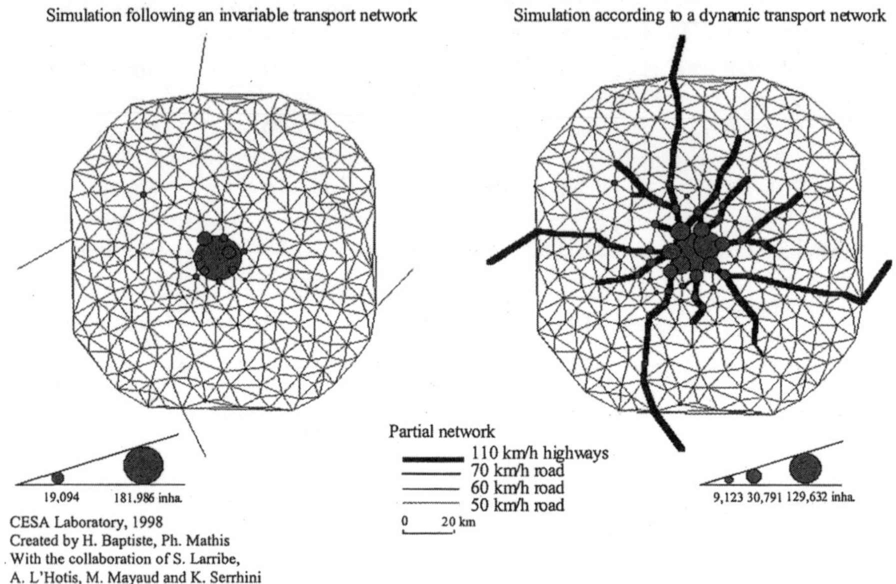

Figure 4.3. *Populations after 100 iterations*

Nevertheless, for both hypotheses the peripheral zones of the area suffer from demographic loss due to the effect of definitive migration in the direction of the large employment pool. on the other hand, on the basis of Figures 4.4 and 4.5, we note that regardless of the chosen simulation hypothesis the hierarchy of the sizes for the 300 towns (not including the four bordering prefectures), according to the rank-size law representation, is generally respected over time.

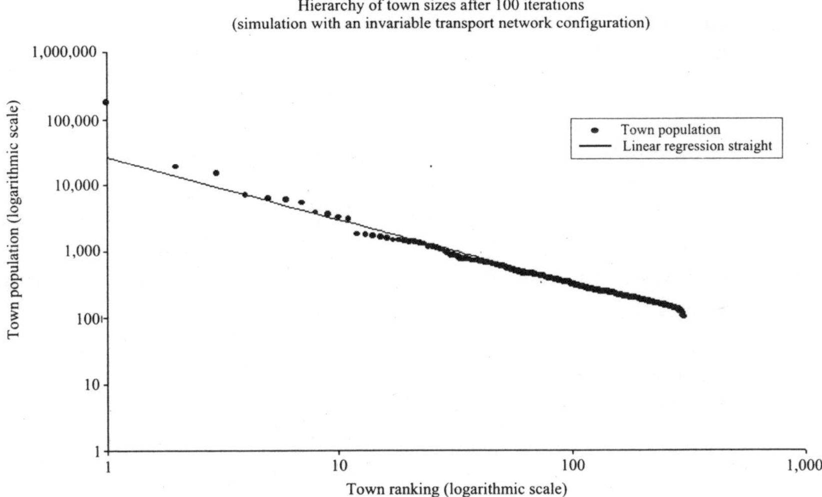

Figure 4.4. *Rank-size distribution of towns with an invariable network*

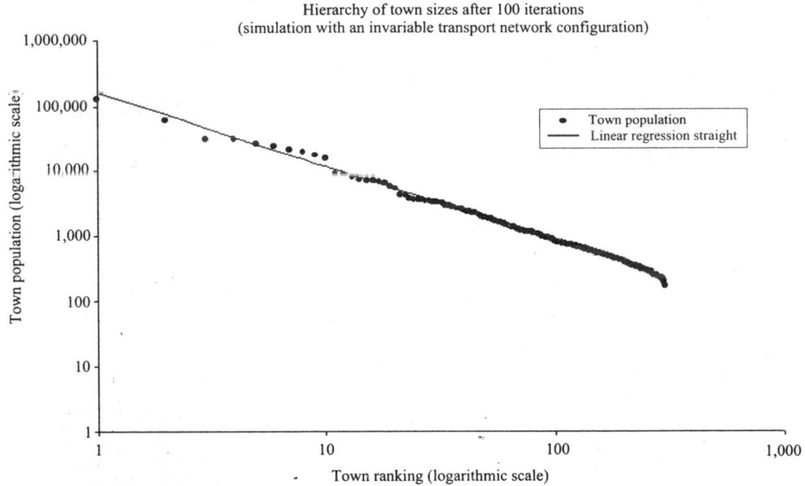

Figure 4.5. *Rank-size distribution of towns with a dynamic network*

From a negative slope initially equal to 1.2, the slopes are respectively reduced to 0.95 for an invariable configuration of the transport network and 1.15 by considering the dynamics of the transport networks. The primacy of the main city is especially clear in the first case, whereas the intermediate levels are much better represented in the second.

4.3.2. *Alternating migrations revealing demographic trends*

In order to clarify the mechanisms leading to the trends above, at this stage we detail the orientations of alternating morning migration flows, i.e. the proportion of active population with employment moving from one town to another following the indicated arcs, and the proportion of active population remaining in their town of origin, according to the color of the nodes (Figure 4.6). The initial theoretical area shows a strong polarization of the main town and its primary crown, which is more marked than for the Indre-et-Loire department. Some other towns remain attractive, but flows between peripheral towns are very weak. The towns sending more than 40% of their active citizens outside their territory are few in number; only the prefecture and a town located in the east of the area fall into this profile[12].

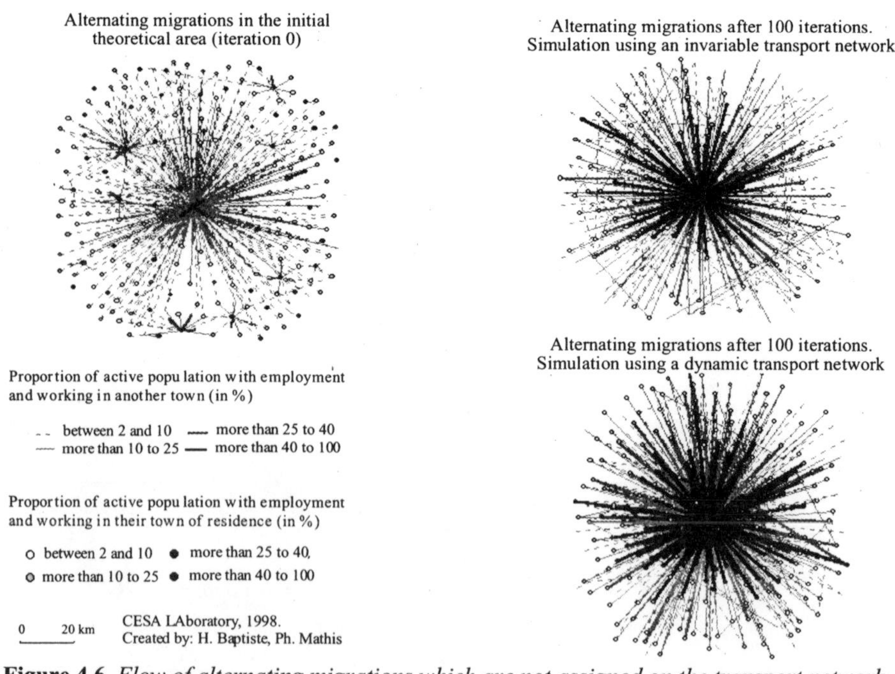

Figure 4.6. *Flow of alternating migrations which are not assigned on the transport network*

12 However, we did not represent the directions of morning flows for map legibility purposes.

After 100 iterations the structures of attraction are appreciably different according to the two hypotheses. The polarization increases in both cases, which is visible through the strong reduction in the number of towns at the periphery of the area offering employment to more than 25% of their active population (black circles), and the increase in the number of towns sending more than 40% of their active population to the center of area (black thick arcs) thus indicating a major reduction in their own attraction. However, the distance at which this polarization is felt varies quite clearly depending on the hypothesis.

For an invariable transport network configuration, for which the single average speed of circulation is 50 km/h, the towns sending more than 40% of their active population to the prefecture or its close periphery are located within a perimeter of approximately 10 to 20 km around the center, as opposed to 20 to 30 km for a transport network undergoing successive improvements.

In addition, despite the greater area polarization and regardless of the hypothesis, after 100 iterations we observe the conservation of certain spatial diversity of employment for a dynamic transport network in towns located at the periphery of the area: those offering employment to 25% of their active population, or even up to 40%, are still rather numerous compared to the initial area. On the other hand, for an invariable transport network almost all of these towns provide a little less than 10% (a significant number preserving even less than 2%, which is visible on the map by the absence of a circle). This is explained by a smaller number of definitive migrations originating at these small peripheral towns for a dynamic transport network and could mean the conservation of a certain number of day-to-day services, this being a point which we will reconsider later on.

On the other hand, the decreased influence of peripheral towns during the first simulation with invariable transport network is explained as follows: since the distance resistance remains strong in the first area where the spatial concentration of employment increases, the residents of towns far away from the employment pool migrate definitively to the close periphery in order to go to work there. In addition, exchanges between the towns of this area and the four other bordering prefectures are negligible (less than 2% of active population sent by each of the 300 towns, including the prefecture, and most often less than 0.5%).

The closest towns (located at about 50 km) are mainly attracted towards the prefecture of the studied area, which is better described and takes into account a wider employment pool. On the other hand, the inter-departmental exchanges with the main city of the area require approximately and 1.5 hour-long journey in the morning, which plays here a strongly discouraging role; with the improvement of the transport network under the second hypothesis – decreasing this journey time to less than 45 minutes – this type of exchanges becomes largely favored, as we will see in the map of flows assigned to transport networks[13].

4.3.3. *Evolution of the transport network configuration*

Up to now we have limited the analysis to the clarification of modifications with respect to spatial orientation of alternating migration flows between cities within the area of reference and those obtained after 100 iterations following the two assumptions regarding transport networks. The results reflect the way in which the definitive migrations take place, being strongly dependent on this scale on the sites of employment, which, finally, contribute only "to emptying" the peripheral zones of the theoretical areas of their inhabitants to a larger or smaller extent.

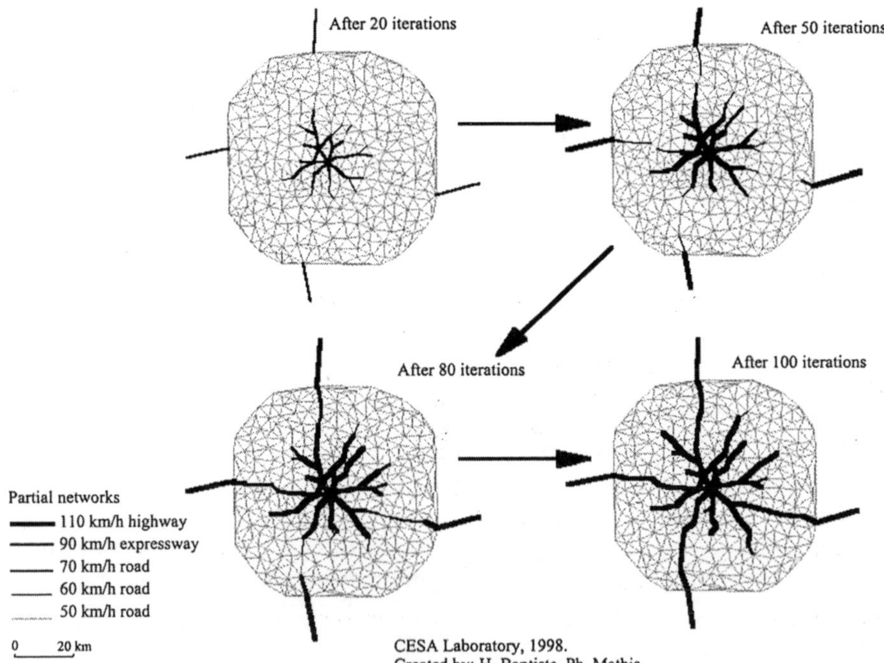

Figure 4.7. *Evolution of the transport network*

13 Indeed, no commune of the area taken individually sends 2% or more of its active populace to the bordering prefectures, regardless of the hypothesis upon 100 iterations. On the other hand, the *number* of communes nonetheless participating in these exchanges is much larger for the hypothesis of a dynamic transport network. For this reason only assigning flows to arcs makes it possible to visualize it.

It remains to demonstrate the evolution of the transport system during the 100 iterations of the dynamic simulation, in order to compare side by side the two evolutions of the system of cities observed. We present here the transport network configurations for several milestone dates (Figure 4.7). Since political decisions regarding transport are taken according to the intensity of circulation flows, the situation after 20 iterations is logical; since considerable flows of alternating migrations converge towards the prefecture of the theoretical area, the capacity of the corresponding axes is systematically exceeded in the mornings, which otherwise constitutes the maximum rush hour of a weekday.

The 60 km/h road partial network prolongs the star diagram of 70 km/h roads. At this level, if the definitive migration flows converge towards the center of the area – not being able to easily carry out a daily commute towards the employment pool using the 50 km/h roads transport network – on the other hand, the towns located at the periphery of the prefecture, which were recently connected to a network of 70 km/h roads, can from now on accommodate definitive migrants searching for a more pleasant lifestyle, more affordable land or attracted by the possibility of home ownership.

This pattern makes the intercity towns more attractive and continues to spatially widen according to the movement possibilities provided by the quality of the existing infrastructure.

The connection between the bordering prefectures and the center of the area is more visible after 80 iterations, except for the southern part, in the form of successive sections of yet unequal quality due to the unfinished construction of the highway between these large cities. At the end of the simulation they are fully connected by the highway and other sections converge towards the prefecture of the studied area from the initially largest towns in a radius of approximately 30 to 40 kilometers. We have to point out that the improvement or creation of an infrastructure is final for the duration of the simulation; in other words, a hypothetical reduction in the population of the connected towns may lead in the long term to the network being under-used.

Thus, transport infrastructures are built for the long term; "those that will be decided upon and financed at the end of the 20th century, will have to be used for generations, even centuries" (see [SEA 94], p. 208). It is thus from this point of view that the evolution of the transport system during the 100 iterations had been "creative", in the sense that the infrastructure improvement or construction projects were implemented irreversibly.

In any case, the distribution of the town sizes at the end of this simulation is fully explained by the increase in the polarization of the area around the prefecture; but the transport networks contribute to spatially widen this perimeter, by means of the concentration process, i.e. suburbanization. The improvement of these networks thus facilitates maintaining certain population diffusion around the main pole.

By using only the mechanisms of population dynamization – generation of definitive migrations – and transport networks – delayed temporal adjustment shifted from supply to demand – we quantify the role of transport networks configuration for the spatial extension of the city (in its broader meaning of an agglomeration) and the greater population diffusion within the theoretical area under consideration.

We will not draw any conclusions here on the realism of the consequences highlighted by comparing the two areas, with and without modification of road networks. Indeed, the increasingly marked peripheral extension of cities on areas that had previously been agricultural or presented natural or historical value, clearly poses the problem of deterioration and induced effects, such as the costs of decontamination, in particular.

On the other hand, the participation of the towns affected by this sort of demographic trends towards the periphery can be a contributing element to ensure the conservation of certain day-to-day services and thus of employment, in an area larger than just the agglomeration, if the PLU (a French official urban planning document) or, failing that, the MARNU are coherent with a controlled urbanization and an active management of natural or historical areas.

4.4. Conclusion

Finally, we may also draw certain conclusions with respect to several other elements which are particularly visible in Figure 4.8. Apart from the already mentioned strong convergence of the alternating migration flows towards the prefecture of the studied area, we also observe a fairly clear pattern of directed flows following a circular configuration of the periphery/periphery type, which indicates the importance in terms of employment of zones external to the city center and the exchanges occurring between them[14]. In addition, the considerable size on the map of the upper semicircles (sum of incoming flows) for nodes located in the primary crown around the main town also proves their economic importance, which is also an observable pattern in real areas, such as the towns of Joue-les-Tours or Chambray-les-Tours located in the immediate periphery of Tours in Indre-et-Loire.

14 We recall that on the map considered, the direction of home-work commute flows follows the rule of right-hand circulation; in this context, bold lines represented on the "right lane" of certain arcs connected to the city center indicates a strong convergence of morning commutes towards it.

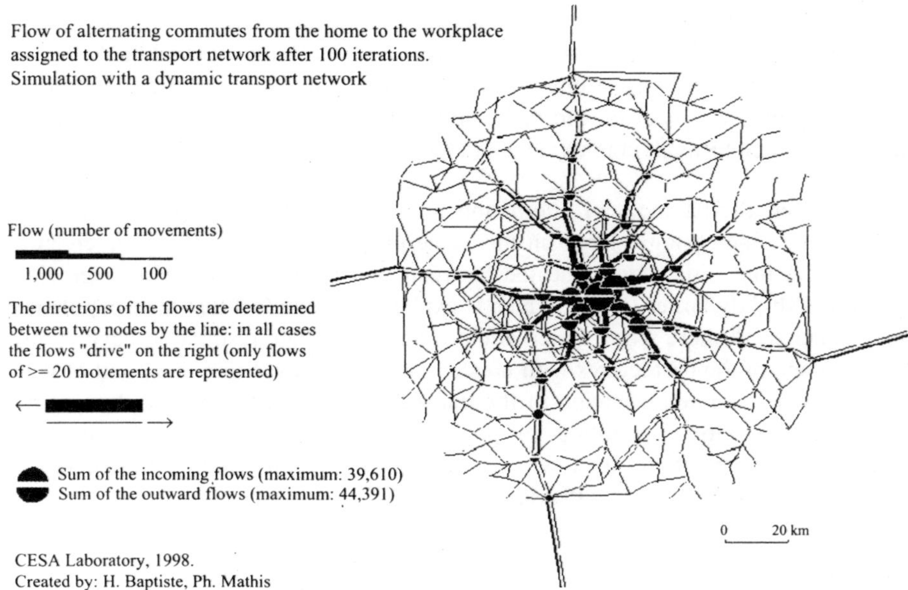

Figure 4.8. *Alternating migration flows assigned to the transport network*

A transport policy based exclusively on a short-term vision of adjustment to the transport demand really helps to explain the elements stated in the introduction with respect to the evolution of the transport system configuration: from this point of view and excluding any political will controlling the territory, the city "copes" well with transport, the city models the spatial configuration according to its requirements – strong demand – manifested through the continuity of the highway network of the prefecture under consideration towards the four bordering cities of the same size and radially towards its "hinterland". Besides, we note in Figure 4.8 the attraction exerted by these cities on the studied area thanks to the boldness of the lines converging towards them.

Although the number of daily incoming and outward commuters of the towns connected to a highway interchange is quite considerable during the morning journey, we note a real variety of situations in terms of balances[15]: regardless of the highway networks considered (connecting the prefecture of the area to each of the four cities in the bordering departments) the migratory balances are sometimes positive, zero or negative. No regularity can really be discerned thus verifying the assumption, according to which linear causality between infrastructure and

15 Let us specify that each commune located along the highway has an entry and exit interchange.

economic development must be rejected and no automatic reaction can be declared[16]. The highway can successively generate development (highlighted by the attraction of migrants, or a strong positive balance), provoke a significant number of alternating migrant departures to make up the deficit by reciprocal flows (reflecting a supply related to weak employment and therefore a low number of activities).

However, the balances of alternating migrations, whether they are positive or negative, are increasingly large on this scale in towns with a highway interchange than in the interstitial zones located between two of these axes. Moreover, these towns are slightly larger in size, which is explained by the fact that their populations may either work there when the creation of an interchange resulted in development or reach one of the prefectures with greater ease every day. From this point of view, there clearly is an amplification of the transport network impacts, which are either positive or negative, resulting in population gains or losses, more intensive alternating movements for the towns connected by a rapid transport infrastructure, as successfully demonstrated by Plassard [PLA 77].

4.5. Bibliography

[ARB 92] ARBELLOT G., *Autour des routes de poste: les premières cartes routières de la France, XVIIe-XIXe siècle*, Bibliothèque Nationale/Musée de la Poste, 1992.

[BAP 99] BAPTISTE H., Interactions entre le système de transport et les systèmes de villes: perspective historique pour une modélisation dynamique spatialisée, Thesis, Tours, 1999.

[BRA 90a] BRAUDEL F., *L'identité de la France: Les Hommes et les Choses*, volume 1, Flammarion, 1990.

[BRA 90b] BRAUDEL F., *L'identité de la France: Les Hommes et les Choses*, volume 2, Flammarion, 1990.

[BRE 98] BRETAGNOLLE A., PUMAIN D., ROZENBLAT C., "Space-time Contraction and the Dynamics of Urban Systems", *Cybergeo*, no. 61, 1998.

[CHA 97] CHAPELON L., Offre de transport et aménagement du territoire: évaluation spatio-temporelle des projets de modification de l'offre par modélisation multiéchelles des systèmes de transport, Thesis, Tours, 1997.

[COQ 72] COQUAND R., *Routes: circulation, tracé, construction*, Livre 1: circulation, tracé, Eyrolles, 1972.

16 This aspect of automation, of simple causality between infrastructure level and socio-economic development was the object of numerous publication, such as: [VOR 84, PLA 92], sometimes purely and simply calling into question the very concept of effect: [OFF 96] p. 52, where Offner speaks of congruence by considering that the infrastructure arrives at the moment when the socio-economic context of the newly connected city generates needs that cannot be satisfied that by this infrastructure.

[DOC 69] DOCKES P., *L'espace dans la pensée économique: du XVIe au XVIIIe siècle*, Flammarion, 1969.

[FIE 86] FIERRO-DOMENECH A., *Le Pré Carré: Géographie historique de la France*, Robert Laffont, 1986.

[FLA 89] FLAMANT M., *Dynamique économique de l'histoire: deux siècles de progrès*, éditions Montchrestien/E.J.A., 1989.

[JOU 95] JOURNAL OFFICIEL DE LA REPUBLIQUE FRANÇAISE, lay no. 95-115, 4 February 1995 d'orientation pour l'aménagement et le développement du territoire, 5 February 1995.

[LHO 97] L'HOSTIS A., Images de synthesis pour l'aménagement du territoire: la déformation de l'espace par les réseaux de transport rapide, Thesis, Tours, 1997.

[MAT 78] MATHIS P., Economie urbaine et Théorie des systèmes, Thesis, Tours, 1978.

[MER 91] MERLIN P., *Géographie, économie et planification des transports*, PUF, 1991.

[MER 94] MERLIN P., *Les transports en France*, La documentation Française, 1994.

[NAV 84] NAVARRE F., PRUD'HOMME R., "Le rôle des infrastructures dans le développement régional", RERU, no. 1, p. 5-22, 1984.

[OFF 96] OFFNER J.M., PUMAIN D. (ed.), *Réseaux et territoires: significations croisées*, éditions de l'aube, 1996.

[PLA 77] PLASSARD F., *Les Autoroutes et le Développement Régional*, PUL/Economica, 1977.

[PLA 92] PLASSARD F, "L'impact territorial des transports à grande vitesse", Derycke P.H. (ed.), *Espace et dynamiques territoriales*, Economica, p. 243-262, 1992.

[PUI 94] PUIG J.P., THISSE J.F., JAYET H., *Enjeux économiques de l'organisation de l'espace français: 1. Polarisation et concentration*, CESURE, 1994.

[REY 93] REYNAUD C., OLLIVIER-TRIGALO M., Tendances du transport européen et besoins en infrastructures, INRETS, 1993.

[STU 95] STUDENY C., *L'invention de la vitesse: France, XVIIIe-XXe siècle*, Gallimard, 1995.

PART 2

Graph Theory and Network Representation

Chapter 5

Dynamic Simulation of Urban Reorganization of the City of Tours

In Tours, at the city's "Maison des sciences", there is a group of researchers comprising archaeologists, geographers and urban developers, whose objective is to compare the points of view, the contributions of various disciplines on the same subject in a very free, exploratory and possibly iconoclastic way. This group chose the evolution of the city of Tours as a theme, on the one hand, because the known data is immediately available and, on the other hand, because the subject concerns all three disciplines. The selected area is illustrated by the plans of 1619, the map of 1770 (Figure 5.1) and the enlargement of the city of Tours on the same plan. These three plans demonstrate the evolution of the network, but also the stability of urbanized space.

The general hypothesis of this work is as follows: it is possible to model the changes of the urbanization axis of the city of Tours. The traditional East-West axis developing along the Loire is gradually being supplanted by the new North-South axis during the second half of 19th century, whereas the Cluzel opening exists since the 18th century.

That implies taking into account:

– construction of the road of le Pont de Pierre and of the Grammont dam as a precondition for the considered reorganization of space, as shown by the map of 1770, at which date the city still had its former structure;

Chapter written by Philippe MATHIS.

– evolution of road and river transits, as well as a population increase;

– a very detailed description of space on the basis of circulation networks and localization of populations.

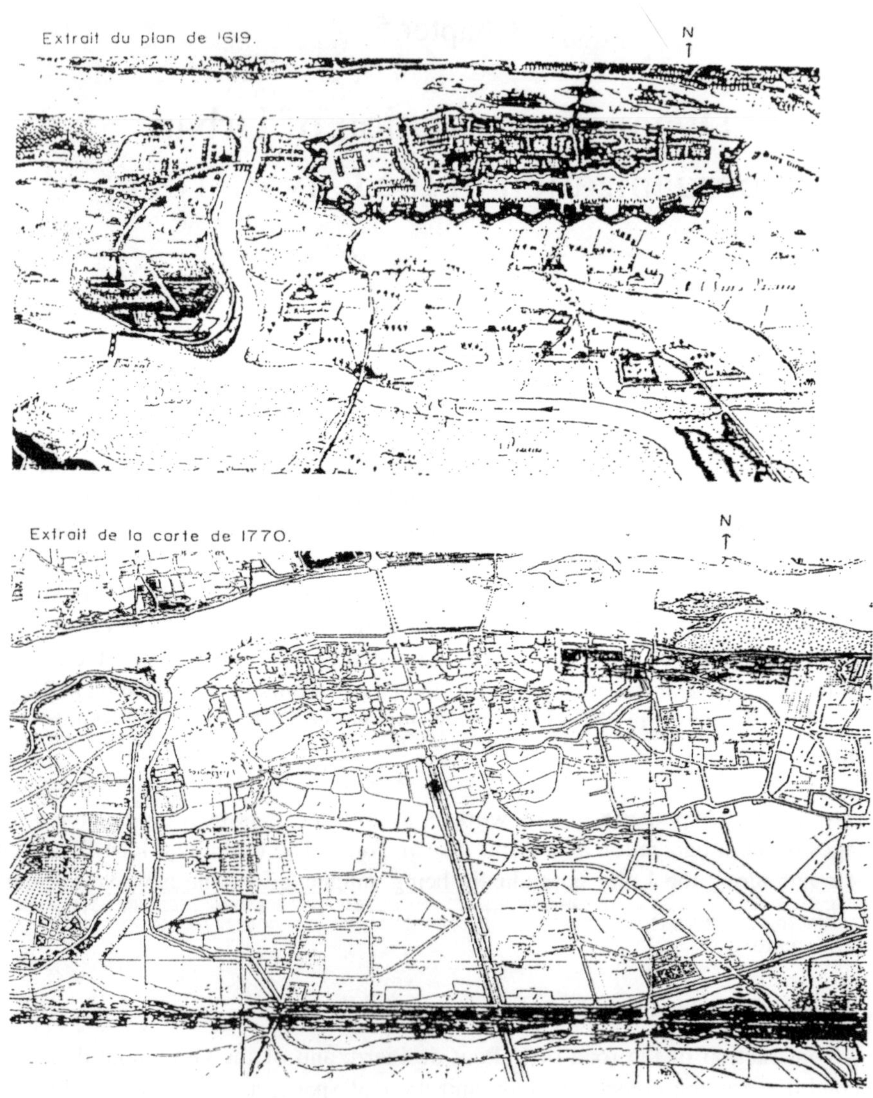

Figure 5.1. *1619 and 1770 plans of the city of Tours*

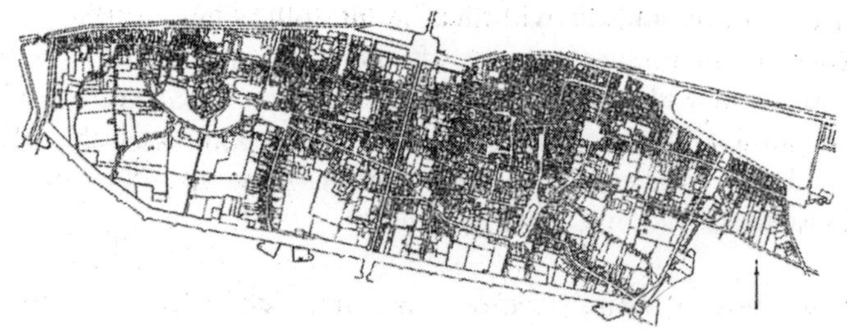

Figure 5.2. *Land parcels in the city of Tours, Napoleonic land register of 1836*

In short, it is necessary to answer the question: how do we go from what we call state 1 of the system, which is described above by the Napoleonic land register (Figure 5.2), to state 2 of the system reflected in the drawing from 1855 below, where we already see the buildings on the present day Jean Jaurès square, which is to become the center of the city of Tours, the unloading dock or first Tours train station built outside the city, at the edge of the mall that replaced the fortifications. However, this urban expansion remains very feeble and we may note that many intra-mural spaces are still free of constructions.

More precisely, the general hypothesis involves the following specific question: does merely taking into account flow modifications, populations and their locations make it possible to account for the phenomenon by means of a simple dynamic simulation model?

We then transform this specific question into a work hypothesis enabling a rationalization by means of land values, which enables one of the possible formalizations of the problem.

Figure 5.3. *1855 drawing: oblique wiew of Tours from a tethered balloon*

The precise work hypothesis will then be the following: a differential evolution of transit flows involves an evolution of the land price patterns, which in turn induces a modification of population localization. This localization modification in turn involves an evolution of local flows, which reinforces the effect of the differential evolution of transit flows and so on; the very type of circular causality of explosive nature described by Denise Pumain.

Technically these various modifications are relatively easy to include in D.LOCA.T[1], which is the simulation model used here. However, that raises several problems related to the data:

– the precise localization of the population, the distribution of census data;

– the problem of lands whose values we do not have immediately; it would be possible to collect them by using notary records, but that is an arduous job for a historian that does not fall within our abilities;

– the problem of traffic which again points to the lack of immediately usable data.

We will thus deal with the problem of data in the first part and in the second we will deal with the model and its transformations, so as to expose in the third part the most interesting results of our simulations.

5.1. Simulations data

As far as the population is concerned, we have used the results of censuses published by the INED[2]. We have the censuses from Year II to our days and prior to that data concerning fires.

This data is naturally global. To distribute it spatially we have the Napoleonic land register of 1832 and the works of the UMR Archeology determining the coefficients of land use on the basis of this diagram (Figure 5.4).

1 D.LOCA.T *Dynamic of Territorial Localizations*, Mathis, patented software, is the prototype of a certain number of models of the CESA laboratory. It originates in various GAST models [MAT 78].

2 INED population census.

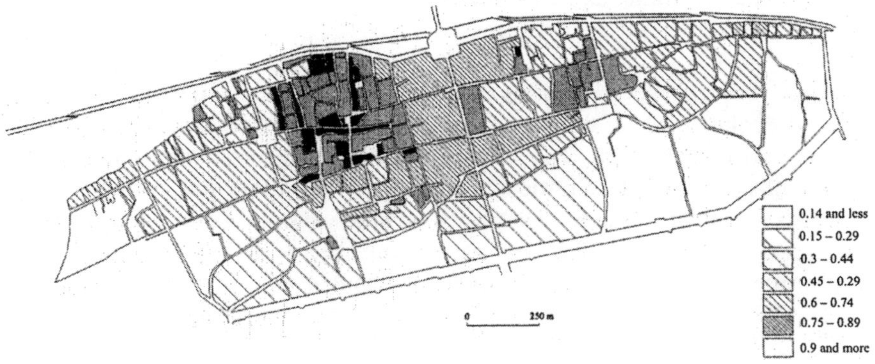

Figure 5.4. *Coefficients of land use established by the UMR Archeology under the direction of Galinié*

Thus, upon analysis with our archeologist colleagues, there was a possibility of decentralized population distribution in the urban area whose approximation would be relatively operational. The hypothesis on which this spatial distribution of the population of the city of Tours rests is as follows: the population is proportional to the coefficient of land use according to the Napoleonic land register. This, of course, remains a simplification that does not take into account a certain number of factors, such as the difference in density related to incomes and the status of the people or households, the existence of mixed surfaces, housing and activity etc. However, albeit simplified, this spatialization seems acceptable as a basis for a simulation, all the more since we had no other possibility in the short or medium term.

Regarding transport, we knew of the existence of several networks according to the old plans, in particular, the map of 1770 for the paths outside the city and the land parcel chart of 1836, which was corroborated by the drawing of 1855. The present network has already been modeled by the CESA Laboratory[3].

We were then faced with the difficulty of choosing a network: should we take into account only the network in existing 1850 or could we retain the current network? In the second case we tend to induce the result that we wished to simulate, whereas in the first case we complicate the simulation problem by adding another

3 EA 1376 then EA 211; see Chapter 13.

dimension: that of simulating network generation[4] with all the possible delay or anticipation problems that this implies.

Initially we adopted the second solution. Indeed, the current network is the product of evolution and if this introduces temporal auto-correlation, then we avoid distortion and complication. The primary object of the model is to simulate the evolution of population localizations according to the differential evolution of the two largest means of transport and exogenous interventions. If we add the network development simulation, which was carried out afterwards[5], the problem clearly becomes more complicated and we avoid doing that at first, taking into account possible substitutions to measure the importance of a particular phenomenon.

The populations were localized at network nodes. The population has thus been distributed through the current network according to the land usage coefficients determined on the basis of the 1836 Napoleonic land register. This does not introduce practically any bias for the old town, which has remained identical until the rebuilding after the Second World War. In fact, the nodes punctually localize a population that actually occupies a certain area, whose intake capacity can be known and which can be determined by taking the dual of the graph comprising the road network[6].

Duality makes it possible to pass from the network to the surface, when the holding capacity of a surface (here a set of pieces of land) is saturated. We slip from a transport problem to a percolation problem, with stock taking increasing importance in terms of significance, at the expense of the flows that become of secondary importance, which is normal when the observation period is extended.

Once the hypotheses have been specified, the origin of the simulations was fixed by using the 1855 drawing[7] reproduced above: the city of Tours seen from a balloon. In this document we can note that the evolution of the city has just started at this date, whereas the unloading dock, i.e. the old Tours train station, is already constructed, as well as the railway connection to Paris.

---

4 The problem of network extension by auto-similarity had not yet been solved at the time of this simulation.

5 See Hervé Baptiste above for the development in capacity and speed, and Philippe Mathis for the growth of networks by using fractal properties: see below in Part 3.

6 The other solution would have been to localize the populations at the nodes of the dual graph of the road network, which would have corresponded to localize them at the "centroid" of the surfaces of the road network. This solution would have been preferable, since the inhabitants are localized in the small islands and not in the networks, but the algorithmic solution has not been found for this case yet either. Actually, these are, among others, the difficulties involving the more theoretical aspects exposed in Part 3.

7 1855 drawing.

The stability of the urban structure until this time encouraged us to modify our specific work hypothesis and to simulate the evolution of the city only by starting from 1851, since from this date we have networks, localization of populations and, most importantly, three means of transport instead of two: the river, the road and the railway.

Moreover, transport has led us to partially reconsider our point of view. Indeed, we have a good indicator of population localization and thus of an estimate of land values but, on the other hand, we do not have at our immediate disposal precise data on traffic volumes and especially on their differential evolutions. To take this fact into account we reversed the problem: instead of simulating the effect of differential transport evolutions on urbanization, we sought the hypotheses to keep in order for the model to really simulate urban evolution, such as it was retraced by the evolution of the population. With the work hypothesis specified, in the following section we quickly expose the model used for this simulation and its adaptations.

5.2. The model and its adaptations

For this new type of simulations we have used the D.LOCA.T model, which was appropriately modified for this experiment.

5.2.1. *D.LOCA.T model*

The D.LOCA.T model developed by the author is the predecessor of the laboratory models, since its premises date back to the 1970s. Written in Fortran it was, on the one hand, modified and adapted to the changes in the capacity of the following computer generations and to the considerable modifications in the programming language from Fortran IV to 90 and, most of all, it was used systematically to demonstrate the possibility of simulating a particular type of problems. However, quite unlike its efficiency in calculation, its lack of ease of use caused the development of models in Visual Basic, which are better adapted to the current standard with a strong graphic dimension.

At the start of this research this model comprises six stages: a space description module using graph theory and taking into account the multimodal aspect of the transport system, a module of calculation of the "shortest" paths in terms of distance, journey duration, cost or a combination of these characteristics, a gravity module generating the origin/destination matrix, a modal assignment module, a space assignment module and a module of evaluation of certain consequences[8].

8 See Chapter 8.

Using dynamic allocation the model adapts to any type of space: from the city center to Europe. It is dynamic, simulating by successive iterations from period to period. Its present limits are, on the one hand, the capacity of the computer used and, on the other hand, the absence of graphic aspects. The latter point is mitigated by the use of Laboratory models and, in particular, of MAP[9]. D.LOCA.T was used for calculations and, in particular, successive iterations, where its speed of calculation could be used to the full. After each iteration it wrote on disk the files with the results in the standardized laboratory format; files that were read and transformed into maps by MAP. Thus, we did not mitigate the absence of graph capacity in Fortran 90, but rather a greater difficulty to use them as compared to the highly user-friendly capabilities of Visual Basic in this area. Both types of models were then fully and logically used.

The description method of the area is multimodal and multiscale. It is based on the zoom method with quasi-fractality[10]. In the present research the possibility of multiscale justifies the general hypothesis: we can effectively simulate the two levels simultaneously, which implies, in fact, studying a certain number of consequences, such as the auto-transformation of flow variables, as well as the "state variables"[11].

With respect to the module of calculation of the "shortest paths", which is based on Floyd's algorithm, the present version takes into account primarily road traffic at a uniform intra-urban speed, which corresponds to the reality of the time: horse-drawn or pedestrian circulation at practically identical speeds.

The gravitating model is a model of general access, an extension of the Reilly model used since 1973, whose efficiency has been recently confirmed by tests conducted by Chapelon [CHA 97] and Baptiste [BAP 99].

Spatial assignment is easy thanks to Floyd's algorithm that simultaneously provides the shortest path and the node preceding the destination node for all pairs of nodes. From there on the procedure is very simple: from one previous node to the other we reach the origin.

The modules of analysis of the consequences were not taken into account for this application, since the important consequences that we wish to simulate, the modification and extension of the residential zones, were not yet tackled then.

---

9 Alain L'Hostis.

10 Laurent Chapelon. See also in Part 3.

11 In the sense of J.W. Forrester, 1969, *Urban Dynamics*, MIT.

5.2.2. *Opening of the model and its modifications*

Firstly, the population develops considerably during the 1851-1936 simulation period: it goes from 30,000 to 150,000, which corresponds to an average annual rate of increase of 2% in the city of Tours. The use of an average annual rate is a simplification which can be avoided, because the model can run with an automatic annual increase or a rate that can be annually set by the user, just as the majority of other characteristics.

River, as well as road or rail traffic is taken into account, primarily, by creating a flow vector for transit as well as for movement originating at or bound for Tours. Firstly, we measure its hypothetical effect on the city, due to the inversion of the problem and, secondly, it can be integrated as data or calculated at the higher level of the system of the cities where it is endogenous, as interurban and city-countryside relations.

The hypothesis of constancy of flows does not hold in our case and especially for this period. Thus, we integrated a differential evolution coefficient of modal flows: river circulation decreases since 1850 until it disappears before the First World War, whereas road traffic grows in a relatively constant way. Moreover, the railroad appears at the beginning of this period. The Tours train station has already been built in 1855 and develops exponentially until the end of the century, but it generates intra-urban or extra-urban traffic from the freight station. These flows therefore cross the city only partially, but create or consolidate the development pole to the south of the old town.

In our model land values depend on accessibility and this simple hypothesis was corroborated in our former work. Accessibility itself depends on distances, modal flows, the evolution of the population and its location.

Lastly, the possibility to construct in the area, which depends on how easily a certain section in this zone can be flooded naturally or artificially, the political will to manage city planning, the existence of various road networks or land retention were taken into account by means of an urbanization vector whose value is equal to zero when, at a determined point, construction is impossible, equal to 1 when it is neutral and more than 1 when there is a will to manage the development and to construct in this area. This aspect has to be developed.

These modifications are neither truly important, nor heavy in the model, but they lead, as we shall see, to rather radical changes in the basic mathematical tool.

5.2.3. *Extension of the theoretical base of the model*

Indeed, the nature of data like that of the followed objective makes the simulation develop from a problem of flows solved by using graphs to a problem of percolation where the objective becomes the analysis of the modifications of space localizations: variables of state.

Firstly, it should be stressed that the description of space by graphs is merely partial: only the locations of poles and the relations that exist between them are described. The quasi-fractal method of space description developed at CESA very clearly develops it using the zoom system.

There is an internal homothety whose principles are constant and apply to all the levels of modeling of transport networks. When applied to the city this enables the transition of the taking into account of isolated poles to quasi-continuous spaces.

There is a semantic and theoretical glide: nodes represent a set split into zones. The dual graph defined on the basis of the primal that constitutes the road network becomes a percolation grid whose primal graph describes the passage from one cell to another and the problem of quadriconnectivity or octoconnectivity no longer arises in this case. Space then becomes quasi-continuous on a specific scale.

What becomes essential to follow are no longer flows in the primary sense, but rather the dynamic differential evolution of populations localizations and, in particular, localizations of population increases, knowing that for these periods densification is limited by the construction method and the available space.

To take into account the physical impossibility of too high a concentration at a point in space we find a population maximum per node. If this maximum is reached, the node in question is no longer taken into account during the process of relocation of the populations moving at each iteration.

The map of population locations reveals the networks: the primal graph and the delimitation of zones around each pole and the dual graph. The latter becomes a percolation grid whose graph describes the possible passage from one grid to another. This passage is not necessarily performed contiguously, with order 1 connectivity, but, on the contrary, may be carried out in jumps. The model is capable to precisely retrace these various movements for each cell where there are simultaneous population arrivals and departures.

Thus, we can, on the one hand, spatially analyze the total population movement and, on the other hand, follow it punctually which, once we segment the population in the next version, will make it possible to follow each category or population group specifically. This aspect presents particular interest in monitoring the effect of urban development measures and the influence of amenities.

5.3. Application to Tours

Applying the model to the case of Tours obliged us to confront certain difficulties during simulations before leading to convincing results.

5.3.1. *Specific difficulties during simulations*

In fact, these difficulties were, *a priori*, foreseeable. There are two aspects linked to the spread of a gravity model. Indeed, even if the elasticity assigned to the population is equal to one, the simple fact of spatially spreading the population has two consequences. Firstly is that the result, i.e. the total attraction of the spatially spread population, is not equal to the attraction of the same population concentrated in a same point because internal distance is introduced and this internal distance is necessarily equal to at least 1, otherwise an undesirable multiplicative effect is obtained. Thus, the distance between two connected points of the second order (with an intermediate point) will be equal to at least 2 on a grid, by simple respect of triangular inequality, otherwise it is no longer a distance. The simple fact of initially distributing a population therefore has a considerable effect which decreases its total influence with respect to the value obtained when it is concentrated in a single point. This is a utopian situation in the full sense of the expression, even if this latest hypothesis is very often used for simplification.

A second phenomenon comes into play at this level: the influence of inequality of graph faces. If, indeed, the localized populations are punctually different for each node of the primal or each face of the dual, the simple mechanics of the dynamic model means that after a certain number of iterations the imbalance between populations by sector increases very strongly. It is therefore necessary to ensure a relatively homogenous or relatively "normal" distribution over all of the contiguously inhabited zones. The possible maximum population per point makes it possible to solve this problem, since when a location becomes very attractive, its population grows and when it reaches the maximum, the attraction benefits the connected zones.

This difficulty is found on the scale of the town, i.e. between Tours and the peripheral towns. Indeed, if we spatially disaggregate the main city while considering the populations of the peripheral towns which are concentrated in a single point of their territories, even if these populations are small, they are then often clearly greater than a single peripheral sector of Tours. The same phenomenon of growing imbalance is reproduced to the benefit of external centers in the latter case. Thus, we are obliged to either spatialize the peripheral towns to a non-negligible distance or simulate only the city of Tours itself.

For the simulated period we chose to focus solely on the city of Tours given that the growth of the peripheral towns started later, primarily after the 1950s. For the period considered (1850-1950) the influence of the peripheral towns remains fairly negligible in terms of attraction and possible relocations, and thus our simplifying choice is acceptable.

5.3.2. *First results of simulation*

The results presented here are among the very first obtained, the model not having yet been subject to systematic digital exploration. However, two main simulations have been carried out: on the one hand an ad hoc evolution regarding the transport flows for each mode as constant and on the other hand, an evolution integrating a differential variation of modal flows (decline of river transport and growth of the road transport).

The first simulation describes the evolution dynamics for the stable case, all else also being equal, i.e. without changes at the infrastructure level or technical progress of any kind. This simulation serves as a point of comparison for the other series of hypotheses.

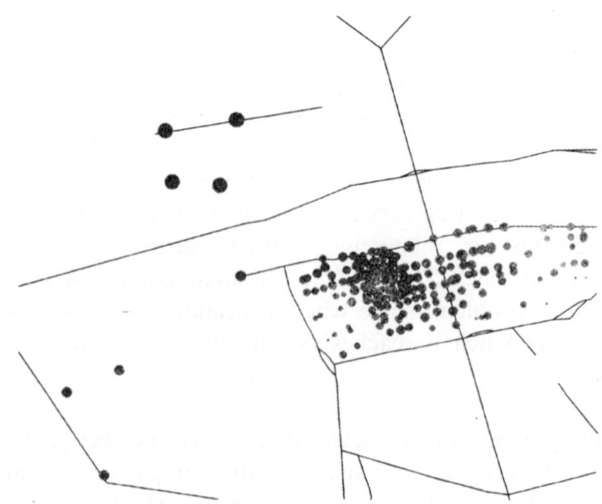

Figure 5.5. *Pattern of settlement in 1851 according to the coefficient of land usage (Napoleonic land register, 1836)*

The second simulation integrates structural modifications and very fast technical progress for road and, especially, rail transport, which will quickly supplant river transport and make it disappear within half a century.

The population is unequally distributed inside the city walls, it is concentrated mainly in the west and, more particularly, in the Saint Martin's district where the density is highest (Figure 5.5).

In both cases considered the annual population growth rate is 2% with respect to the population in 1851, which corresponds to the average growth rate between the 1851 and 1936 censuses, a period during which the city increases from over 30,000 to more than 150,000 inhabitants.

5.3.2.1. *Evolution simulated with constant importance of river and road traffic*

The simulation proceeds without opening of additional locations, considering that the intake capacity of *intra-mural* Tours is largely sufficient as shown by the drawing from 1855 where many unbuilt areas may be seen. The following map (Figure 5.6) shows the absolute differential evolution: populations localized at the beginning of the period (shown in gray) are superimposed on populations localized at the end of the period (shown in black).

We note very clearly, first of all, the very strong influence of the proximity of large transport arteries: the river and rue Royale[12]. The influence of the river remains largely dominant in this simulation and the traditional East West axis is extended and reinforced in a district which was traditionally oriented towards river-related activities: the old port of Saint-Pierre-de-Corps is in the immediate vicinity of the Loire-Cher canal which is located between the towns of Tours and Saint-Pierre-de-Corps. This canal replaces ruau Saint-Anne, which previously ensured communication between the two rivers, as shown in the maps of 1619 and 1770, and gradually became silted up.

The second observation is a tendency for rebalancing to the east and very slightly towards the south. Indeed, as mentioned above, we introduced a population maximum per node to avoid local concentration that would exceed the reception capacities of the frame. This maximum by node was fixed at the level of the population located between Saint-Martin and the Loire, considering that it corresponded to the maximum coefficient of land usage in the Napoleonic land register.

12 Rue Royale is the old name of the present-day rue Nationale.

Figure 5.6. *Development over 10 years without additional locations (influence of river and road traffic axes, population growth of 2%)*

However, this hypothesis does not correspond to the reality of location and relocation movement at the time.

5.3.2.2. *Evolution simulated with "differential growth" of means of transport*

Under this hypothesis we kept the above coefficients with two important changes: the influence of river and road means of transport is no longer constant, therefore river transport decreases by 5% per annum and road traffic, on the contrary, intensifies at 5% per annum. The trends of this differential development conform to reality, since the river traffic disappeared at the beginning of the 20th century and road traffic shows a strong growth on the Paris-Chartres-Tours-Poitiers-Bordeaux axis.

The map of population growth below (Figure 5.7) showing, as previously, the superimposition of populations at the end and at the beginning of the period enables us to immediately note a very weak relocation close to the river, except in the east. However, there again it is definitely less strong than under the previous hypothesis. On the other hand, we note a much stronger growth on the North-South axis of Tours, the street which became rue Nationale.

It is clear that taking a differential rate into account, i.e. growing for roads and decreasing for the river, modifies the relocations or new localizations.

The phenomenon appears much more clearly when we take into account the absolute variation of populations by site, which is illustrated in the map below (Figure 5.8). We can observe that all considerable absolute variations take place along two axes: decrease at the river, which loses its attractiveness, and at the same time growth along rue Royale. Over the period considered the population growth is not strong enough to spatially diffuse outside the limits of the old ramparts and the dominant phenomenon are still quite clearly relocations.

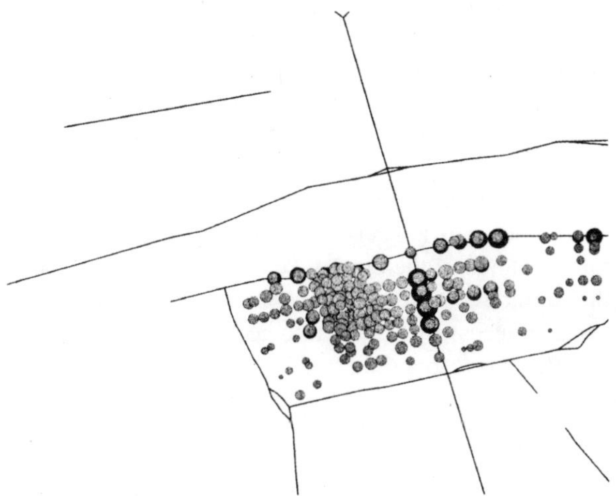

Figure 5.7. *Development over 10 years without additional locations (increasing influence of the road to Bordeaux and decreasing influence of the Loire, annual population growth of 2%)*

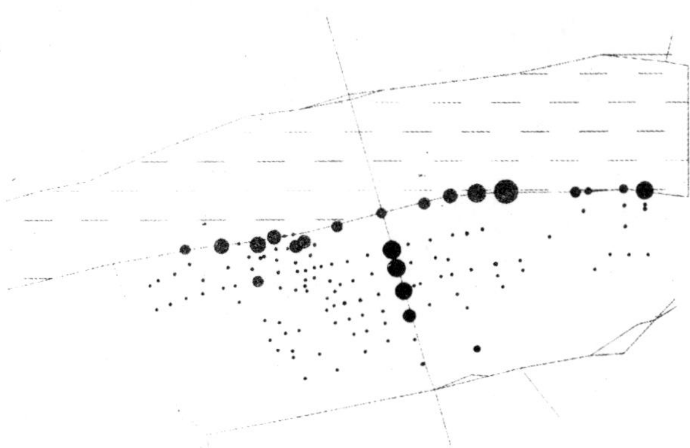

Figure 5.8. *Absolute differences between the hypotheses with differential rates of development of the means of transport: urban inversion emerges*

However, this makes it possible to note new tendencies which are masked by absolute values: relative growth in space, albeit still discrete, and, in particular, in the south and the west, as if the population started a process of diffusion or rather of rebalancing from the densest zones at the beginning of the period. It is a relative increase in density in the southern zones close to the ramparts or of reallocation of areas formerly used as parks or gardens by the religious congregations, which were established at the periphery of the old town, as may be seen from the 1855 drawing.

However, *intra-mural* space is still far from being saturated! The negative values appear as lighter circles and the positive values as darker circles. The figure of relative growth (Figure 5.9) enables us to specify the phenomenon: it confirms the preceding observation regarding the importance of movement along the two transportation axes.

These initial simulations thus tend to corroborate the modified work hypothesis and the general hypothesis: it appears that, indeed, it seems possible to model the phenomenon of axis change in the city of Tours, and it is not enough to merely take into account the attractiveness of transport routes to arrive at the observed development, it is also necessary to take into account the differential development of the two types of traffics corresponding to the historical reality.

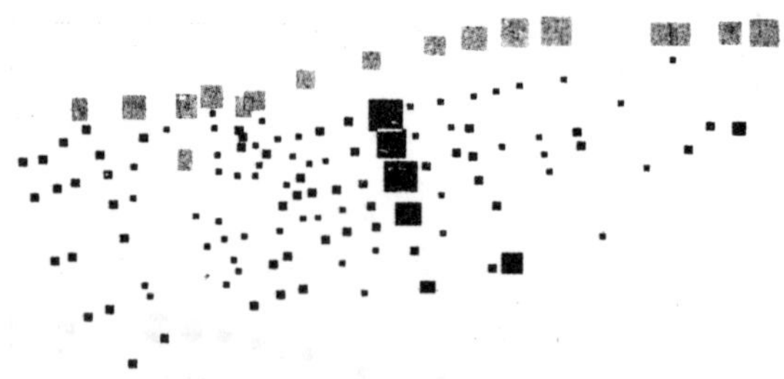

Figure 5.9. *Relative differences between the hypotheses with and without differential development of the means of transport (positive evolution darker and negative evolution lighter)*

5.4. Conclusion

We are aware that it is only a very first attempt at clarifying the problem and that the model is still neither completely explored, nor even complete for this specific type of simulations. However, we may note as of now that the opening of the model, possibly with respect to two levels, and the extension of its capacities through the introduction of minor modifications, two vectors and three constants tend to change its nature at the level of the theoretical base of the tool by gliding from one concept to another, from graph to percolation. In terms of methodology, it appears that the inversion of the problem that we were obliged to perform due to the unavailability of data proves to be effective.

With respect to the field of application we pass from a praxeologic model to a cognitive historical model. From a certain point of view, following the example experimental archeology, we introduce experimental history by modeling.

A result obtained indirectly is the existence and temporal localization of a change in the spatial evolution of the city and this is shown, if not demonstrated.

A general result of the theory of systems is corroborated with regard to reproduction: by using the model we show that the system can reproduce itself differently, at least with respect to its spatial form, although the reproduction mechanism remains identical.

Lastly, in terms of "explanation", we go from "everything happens like this" to "what is needed within the framework of the logic adopted for everything to happen like this". In this case the model produces second degree knowledge!

5.5. Bibliography

[BAP 99] BAPTISTE H., Interactions entre le système de transport et les systèmes de villes: perspective historique pour une modélisation dynamique spatialisée, Thesis, Tours, 1999.

[CHA 97] CHAPELON L., Offre de transport et aménagement du territoire: évaluation spatio-temporelle des projets de modification de l'offre par modélisation multi-échelles des systèmes de transport, Thesis, Tours, 1997.

[MAT 78] MATHIS P., Economie urbaine et Théorie des systèmes, Thesis, Tours, 1978.

Chapter 6

From Social Networks to the Sociograph for the Analysis of the Actors' Games

As has been illustrated up to now, it is rather common to use graphs to deal with transport and some other technical networks. We will now attempt to approach the social sphere by means of social networks. Although, usually it is a field of interest for certain sociologists, we will see how they can specify such concepts as "system of actors", "game of actors", "network of actors" or that of governance, which are often present in contemporary debate and publications on urban planning. However, unlike these paradigmatic concepts that are often used to mask the thorny issue of power and politics in our discipline, we will choose a more operational point of view, which, in a certain manner, conforms fairly well to the general spirit of those who use graphs.

"A social network is by construction a *set of relations* between a *set of actors*" [FOR 97]. Behind this use of "social network" there is a paradigmatic model, which poses that the behavior of actors is determined by their relations and their attributes. The term "actors" is generally used to designate individuals, but it can also refer to households, associations or any other group form. They are mutually engaged in varied relations of power, friendship, distrust. The set of these relations is properly known as the *social network*. Finally, actors are distinguished from each other by attributes, i.e. the values of variables such as age, resources, socio-professional category, religion, which are supposed to generally directly explain their behaviors.

Chapter written by Sébastien LARRIBE.

Moreover, a "system of actors" has to be understood as a set of explicit or implicit rules that organize an activity (in fact, involving actors in it). The "game of actors", in turn, is the manner, in which the actors take on these rules, interpret them, constrain themselves accordingly or circumvent them in order to achieve their respective goals and carry out their projects [MER 92]. From this point of view, the social network is therefore a means for actors to play inside a system.

Sticking with the game metaphor, we can say that when there are only two players involved (duel situation), the situation is perceived by each of them directly, since it is clear whether the other player has adopted a cooperative or an uncooperative attitude (in other words, if he is an adversary or an ally).

However, when there are three or more players everything changes. Cooperation attitudes and generally lasting alliances and coalitions may appear and render the tactics less direct, more ambiguous and fluctuating. Consequently, the entire game changes because everyone may be obliged to establish a "shifted" relationship with the others, which is a parallel relationship at the margins of the "real game". To some extent a "meta-game" with vague rules is created. For some the true interest of the game lies therein. For others, on the contrary, these attitudes are comparable to cheating, which may lead to situations of rupture, exclusion, or appointment of a referee.

If we wish to study them, it is convenient to represent these game and cooperation-conflict situations using the formality of graphs. We obtain figures comprised of nodes representing individuals and edges or arcs between two nodes which represent the relations, if they exist, between two individuals.

Here we propose to analyze precisely this exercise. Passing by sociometry and social networks analysis, we will dedicate greater attention to sociography, whose viewpoint largely exceeds the connotation or even "illustrative" usage which is sometimes too attached to graphs.

6.1. The legacy of graphs

When only two actors or players and the fact that there is or there is not a relationship between them are considered, the situation is common and the graph that can be plotted is no less so: the two individuals are either connected or not by an edge. However, still with just one type of relation possible and three non-interchangeable individuals, there are already eight possible situations. If we then consider the case where two relations are used to describe the operation of this tripod and that these two relations can exist simultaneously between all three pairs of players, 64 possible arrangements are obtained (Figure 6.1).

Figure 6.1. *64 organizations of a tripod with 2 possibly probable cumulative (but not directed) relations*

From this figure we can deduce the potential role of an actor relative to his position in the structure (it is exactly the sense, which is usually given to "hierarchical position", for example). It is clear that this role results at the same time from the relations that an actor does or does not maintain with the others and, simultaneously, from the relations that the others do or do not maintain between them. This observation does not, however, come without a cost.

On the one hand, it necessarily implies the construction of relatively broad and global figures and views to represent the analyzed situations. Moreover, it leads us to think that the reading process needed for the analysis of these graphs has to be multiscale, thus leading us to gradually decipher the operation of binomials, tripods and sub-groups of any other size. Later on we will return to consider these characteristics, but suffice it to say for now that they lie at the heart sociography.

Let us suppose that in Figure 6.1 the gray edges symbolize a hierarchical relation between A, B and C and the black edges symbolize a friendly relation. Figures only containing black edges are traditional organizational diagrams. We also talk of arcs and not of edges by considering that hierarchical relations are not symmetric.

When we transfer our attention to the figures comprising gray and black edges simultaneously, we understand quite quickly how the gray edges "disturb" this hierarchical organization. In other words, the gray subgraph superimposed on the black subgraph, obviously, considerably modifies the interpretation that can be made of the graph on the whole.

It is also necessary to pay attention to the absence of edges, which does not necessarily mean absence of relationship. We are talking here of the property of transitivity. For example, in all the configurations that contain two edges, gray and black, we may think that there is no relation between two of the three nodes. However, if A and C are friends (connected by a gray edge) and A is the superior of B then, consequently, and although the relation between B and C is not visible, we can suppose that, functionally, C has a "means" (A) of influencing B.

In all this, the question of power in general, in such dimensions as authority, potentiality and, especially, influence, is central. The general aim of the exercise is to know how the sociogram (informal relations) of an organization or a company disrupts the structure of the organizational diagram (formal relations), how it makes it possible for certain actors to perform "hierarchical jumps" of several levels, how actors are not, in reality, "in the place" which we could have believed to be theirs. It is, of course, in this sense that the term "network" recovers its somewhat negative connotation. It conveys the idea of avoidance, bias, maneuvering that would enable one individual to "reach" another by means of intermediaries or by playing on another relational register, whereas structurally and formally he has not been given this possibility.

We owe the first analyses based on data collected about interpersonal relations to Moréno [MOR 70][1]. It gave rise to sociometry defined as "the science of measuring

1 "Sociometry has the aim of studying the psychological properties of populations mathematically; to this end it implements an experimental technique based on quantitative

relations between humans". Among the tools introduced and used by Moréno for data processing we are only directly interested here in the *sociogram.* It is a graph constructed exclusively on the basis of emotional relations which are recorded between individuals and considered by the author as the real maps of "psychological geography" [MOR 70][2].

The structural method[3] introduced by Moréno has seen many developments, in particular, stemming from anthropology and field of organizational research. In a certain manner, the organizational chart is for Fordism what the sociogram is for the current of Elton Mayo's human relations, and it is perhaps to Michel Crozier that we owe one of the first attempts to bring these two movements closer. Although schematic, since the beginning of the 1980s this constant evolution constitutes and organizes a specific field dedicated to the analysis of social networks.

This latter type of analysis is no longer limited to the psychological level and, therefore no longer limits itself to the sole study of emotional relations between the members of a group. In principle, any relation between members can be processed, even though, in general, we limit ourselves to taking into account the relations of collaboration, support, advice, influence or control.

As for the processing, it may be a question of producing individual scores (scores of centrality, prestige, etc.), of determining the make-up of sub-groups in order to have a structure emerge in a vast network of individuals, of highlighting particular structural configurations (determining the equivalent positions in the structure), or revealing the indirect relations between the sub-units of a group.

From this point of view and in general, the method of behavior analysis following the use of the conceptual "network" model is organized in two stages. The first consists of determining a structure in the set of data about "actors" and "relations". To this end we primarily use sociometric indicators (density, centrality, prestige, etc.) and methods of determination of cohesive sub-groups (i.e. where the members maintain more relations between them than with members of another group).

---

methods and exposes the results obtained through the application of these methods. Thus, it performs a methodical investigation into the development and organization of groups and into the position of individuals within groups…" [MOR 70], p. 26 and XLI.

2 "By psychological geography we mean the graphic representation of the interrelationships that link the members and groups of a same community, a) with respect to their topographic location and b) to the psychological currents that circulate between them" [MOR 70], p. 26 and XLI.

3 Sometimes referred to as "neo-structural" to take into account the "structural" approach of Claude Levi-Strauss [LAZ 98], p. 3.

Actually, all the formalism of network make-up and of the tools of graph theory may only be applied here on the condition that the result can yield a social interpretation[4]. For example, the degree of a node becomes "centrality" in the sociometric context, whereas the internal half-degree may be compared to "prestige". Using the same idea, a complete maximum subgraph becomes a clique (or 1-clique), i.e. all the actors are directly connected thus indicating a very cohesive group. If a subgraph admits n as the maximum length of the paths connecting all the nodes, then we speak of an n-clique (n becomes the number of intermediaries). If a node in a subgraph does not have relations with at most k other nodes, it is indicated as k-plex. Conversely, in a k-core each node is attached to k different nodes. A clique (A, 1-2-3), a 2-clique (A, 1-2-3-4-5, the path between 4 and 5 passes by 6), a 1-clan (A, 2-3-4-5-6), a 3-plex (B, at least one node does not reach three others directly) or a 2-plex (C) (Figure 6.2).

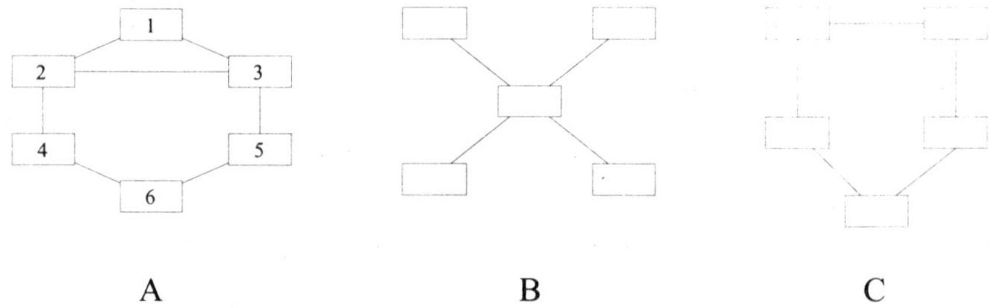

Figure 6.2. *Graphs or analysis of social networks?*

The second part of the analysis consists of trying to connect this structure with the observed behavioral phenomena, possibly by using information relating to the attributes. Strictly speaking, we qualify the first stage as *network analysis* and the second as *structural analysis*. As far as we are concerned they are not dissociable and that confers all its specificity to the procedure.

Most of all, we should not lose sight of the fact that in this approach the network analysis serves structural analysis. The first objective is clearly to understand, analyze and even *anticipate the behavior of an actor*. By behavior we understand here the exercise of power, i.e. again, the faculty or tendency to be able to make and impose a decision.

This type of graph has two limits: the organizational diagram and the sociogram. The organizational diagram reflects only the hierarchical relations, i.e. the power

4 Which is what has been carried out, at least partly, in [FLA 68].

distribution in the organization. As for the sociogram, it reflects only the emotional relationships of selection or rejection between the members of a group. The aim of strategic analysis is then found admirably transcribed in these two graphs: start knowing the relations of power (the organizational diagram), consider the informal relations (in a certain manner, the sociogram) and, thus, reveal a specific action context.

6.2. Analysis of social networks

Strictly speaking, in order to carry out a structural network study, we require, first of all, to collect three types of information. Firstly, we have to identify the individuals and their relations. That can be done by means of sociological scale surveys, by determining a type of relation which is supposed to characterize the analyzed system (we speak of a "relational variable", i.e. a criterion that makes it possible to connect the actors of the system), or by using "name generators"[5]. The second type of information consists of actors' "attributes". These are variables likely to suggest why an actor would be in a relation to a particular person (age, gender, social status, etc.). Finally, the last type of information can be collected regarding the behaviors of actors that are thought to be connected, through analysis, to the position of the individual in the network (mobilization, decisions, performances, etc.).

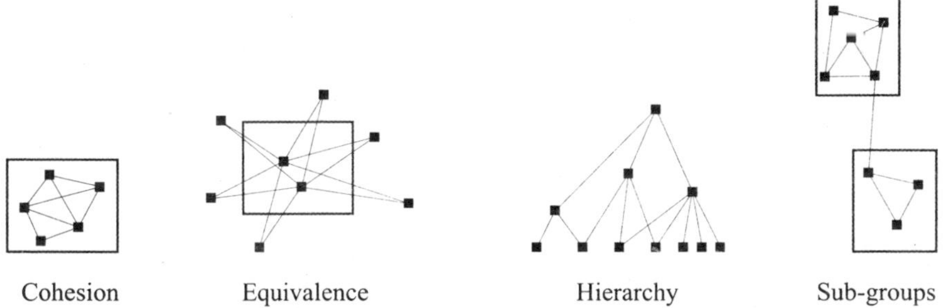

Figure 6.3. *Reference structures for the analysis of the social networks*

5 Name generators are questionnaires prepared with the goal of recreating a social network. It may be simply a question of the "with whom do you perform a particular activity?" type. In these proceedings we only avoid formulations that may cause various interpretations by the questioned individuals, such as the "who are your friends?" kind, since the definition of friendship may be rather variable. Questions about activities do not have this disadvantage, since it is the investigator who determines the associated relation type.

However, when we consider their implementation, these methodological elements have a direct feedback effect on the sense and even the use of the analysis. As an analysis tool, the graph is at their disposal to suggest a group functioning mode which comes in all the nuances ranging from a clique to the hierarchical tree structure (Figure 6.3).

However, we can also imagine an "in play" tool of network analysis, in other words, information about the network which is used and exploited by a particular actor. The point of view is strategic here because the actor in question may seek to work out his own game tactic by this means, for example, by wanting to know if there is a path between this actor and another, "how to reach so-and-so" and which intermediaries "to use" if this relation is not direct.

Under this new light, and if we wish to preserve all the sense of the game as such, it is not possible that a player sets up systematic protocols of data collection envisaged by strategic analysis. Because, of course, and yet again, to preserve all the interest of the game, some of this data is not intended to be revealed and may even stem from very personal appreciations or even indiscretions or other confidences collected under quite various circumstances. Would that be reason enough not to take an interest in it? Within the framework of a scientific work, perhaps. Within the framework of personal tactic and according to the teachings of Sun Tzu or Klausewitz, certainly not.

Thus, if we wish to preserve the possibility of network analysis in this particular context, it is quite necessary for us to admit:

– that it is up to each player to get information about the others and the relations that may exist between them without this being able to be done in an official and systematic manner;

– that information processed will be thus, by construction, biased if not partial and can only correspond to his own point of view at a given moment;

– that taking into account the unquestionable inaccuracy of this data and its approximate nature, it would seem inappropriate, at least, to use too fastidious a formalization (k-plex, k-core and others) to finally describe the game in which he participates.

Under these conditions, it is not truly social network analysis that each player engages in, but an analysis of his own representation of the network in which he is inserted[6]. And since we deliberately do not place ourselves in a deductive process,

6 However, it is not a question of analyzing his "personal network" because in the analysis of social networks the formula draws a radiating relationship diagram centered on a subject (the ends of the rays are not connected between them).

the construction of such a representation may only proceed by successive iterations. Indeed, it is quite necessary to imagine that, when confronted with the first impression, the player-analyst will quite quickly have additions to bring, individuals and relations to add. In this sense network analysis will be for him a means of identifying his lack of knowledge about the system and of mitigating it by means of many return trips between his periodic obligation to play and the preparations that he may make for the following step by taking into account what cold have occurred in this interval of time.

However, if we now extend the concept of a player to the more general one of an actor, we can undoubtedly consider this process of objectification of reality on other scales or in other contexts that hardly dissimulate the game metaphor. On the basis of elements that may sometimes be relatively uncertain or emotional, an actor could be brought, from result to result and from assumption to assumption, to change a certain number of his opinions, impressions and prejudices. By these means, which are altogether rather simple, there is a possibility of estimating or simulating the consequences of the existence, transformation, or discovery of new relations between two or several players. Being confronted with a representation of his knowledge, it finally becomes possible for him to reveal, formalize and specify his ignorance of the reality, in order to better fill in the gaps. And, at the end of the day, in this particular meaning of the "networks of actors", the determination of the *concrete context of action* of an actor is at stake.

By starting from the analysis of social networks we clearly see how this approach preserves certain aspects and how much it deviates in some others. On the one hand, the structural perspective seems to us particularly adapted to the description of a true game in progress and well beyond what the rules would enable us to understand. On the other hand, the conditions of data access and the ambitions of the analysis are specific to us. Everything that we propose to name as "sociography" is based on this distinction. Its methods and tools enable us, perhaps, to propose an original interpretation of graphs.

6.3. The sociograph and sociographies

To us "sociography" thus indicates this reflection method which is applied more specifically to relational data. We say "graphy" because we believe that graphs in general, as visual representations, at any rate at our level and with a little practice, can render a sufficiently good account of the operation of a network without it being necessarily useful to resort to overly elaborate mathematical singularities.

The project seems to us all the more accessible, since Bertin's graphic semiology already deal with networks [BER 68, p. 263-283] and since in graphs we can find all the principles and the advantages attached to representation and information communication systems [BON 88], to images and graphics [BON 75].

Is it not the case that a graph enable us to grasp very intuitively the specificities of a clique (Figure 6.2, A, 1-2-3), a 2-clique (Figure 6.2, A, 1-2-3-4-5, the path between 4 and 5 passes by 6), a 1-clan (Figure 6.2, A, 2-3-4-5-6), a 3-plex (Figure 6.2, B, at least one node is not directly connected to three others) or a 2-plex (Figure 6.3, C)? Following this design or scheme we clearly see the difference between organizational structures and we already "sense" that they cannot function in the same way. By a notable kind of shortcut: "you can see it".

We say "socio"-graphy to recall that the data which are handled simultaneously concerns formal and informal relations. "Network images" or the graphs that we obtain are thus a mixture of "shares" of organizational diagrams and sociograms, which is different from case to case.

If a camera makes it possible to take photographs, a sociograph is a "machine to take sociographs". It is thus more of a tool than a diagram. It must enable a representation of a database containing individuals and the relations that they can have between them. It must help the diagnostic of actors, but certainly not replace it. For that it must be relatively interactive and flexible in use because it is used in an iterative trial and error process, which seems to us to be the most effective means to approximate our reality. If graphs must appear – and they undoubtedly do – they are "exploratory graphs". However, we may also imagine that it can represent the attributes of actors in order to facilitate behavioral interpretation. Thus, we obtain a doubly "strategic" image. Strategic, because that is what the position of an individual in a network inevitably is; strategic, because this position really only becomes that when coupled with resources, at least potential ones (attributes).

We do not forget either that the network is still considered from the point of view of a given actor, in consideration of which the "sociographies" are by nature "self-centered". It is not, however, an "ego-network" as understood by sociometricians or analysts of social networks, i.e. a network of relations of an individual that could be represented in the shape of a star.

From the moment when edges can connect two points, one of which is not the center of the circle, the personal network becomes only a particular case of sociography, more precisely a subgraph. However, let us come now to a more concrete presentation of our project.

Thanks to computers, we can have a tool for automatic design and interactive exploration of graphs, in other words, our sociograph. Relational information can be stored in any kind of spreadsheet or database management system. The functional basic unit comprises binomials, for which two actors and a possibly valuated relation are indicated (Table 6.1). By convention, we can state that when the relation is directed, the direction is indicated by the order of appearance of the actors in the same line.

Actor	Actor	Relation	Valuation
A	B	R1	1
A	C	R1	1
C	D	R2	2
C	B	R3	1
...	...	...	...

Table 6.1. *Relational binomials*

As far as the attributes are concerned, the input of information is not more complicated. It is possible to specify a parameter list for each actor supposed to qualify him or to single him out. However, it is also possible and, perhaps, more useful to consider from the start the attributes which could possibly become exploitable later (Table 6.2). Starting from this stage, which is already partially carried out, the graph can be plotted. At the start, it is a global graph therefore representing all the existing relations. Since network analysis privileges structural data with respect to category data, the attributes are not represented (Figure 6.5a).

Actor	Studies	Age	Marital status	Chaser	...
A	F1	24	Single	Not	...
B	F2	37	Married	Yes	...
C	F1	21	Single	Not	...
D	F2	57	Divorced	Yes	...
...	...	...	...	...	...

Table 6.2. *The classification of individuals according to their attributes*

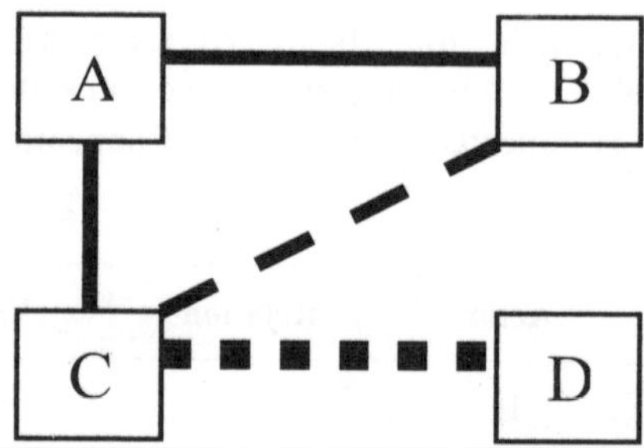

Figure 6.5a. *Generating a graph with relational networks*

This graph is the starting point for the exploration, which consists of moving the nodes to limit the tangles as much as possible and to reveal a readable structure. In other words, it is a question of approaching a planar graph as well as possible. Or even to locate sub-groups which are made up of nodes having more relations between them than with any others. There are automatic placement algorithms that may facilitate this operation but we have not yet felt the need to implement them.

This structural exploration may be made easier by successively representing each possible type of relations (Figure 6.5b). With the addition of a new relational type, the user observes new permutations and repositions the nodes accordingly. It also happens that we observe that this addition does not change anything in the representation. That is, two modes can be redundant to some extent, in which case either one of them is removed, or a "composition" of them both is performed.

Here we do not deal with partial graphs. Indeed, the fact of always representing all the nodes makes it possible to also locate those that "do not belong to the network" in a given relational mode.

Lastly, once a structure has been highlighted, displaying one or more attributes may make it possible to discuss the reasons that could have led to the observed forms. That can also lead to reposition certain nodes, if it is thought that the similarity of profile may favor a certain "bringing together".

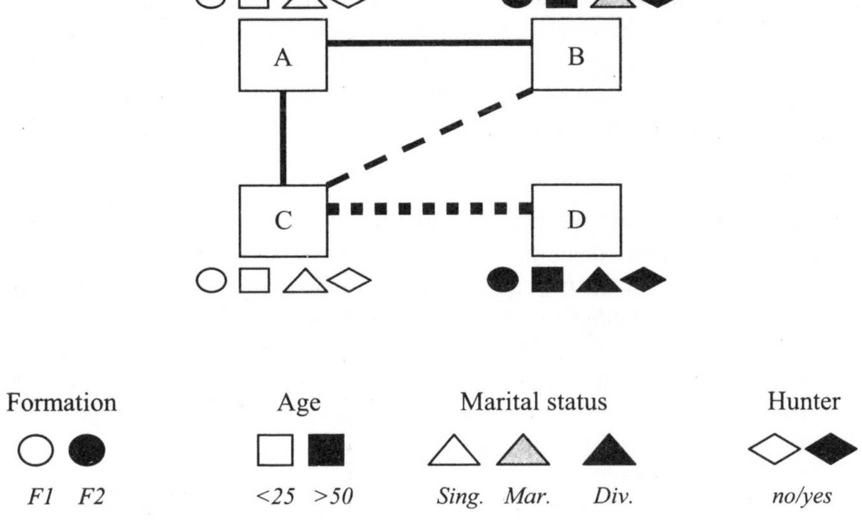

Figure 6.5b. *Plotting of the graph of relational networks according to attributes*

These two figures illustrate the guiding principle of sociography. The first is made up only on the basis of relations that are, nonetheless, distinguishable. It is complemented in the second figure by a possible representation of the attributes that highlights "resemblances" between nodes. Everyone is then free to interpret these similarities in terms of potential cooperation-opposition relations.

Krackplot 3.0 is a software which makes it possible to plot graphs by giving the possibility of distinguishing nodes by text color and geometrical forms, that is, two attributes[7] (Figure 6.4 in the color plates section).

This completes the first iteration of the process. The highlighted graph or sociography makes it possible to return to the database in order to supplement or correct the initially entered information in terms of binomials. From there a new graph can be generated on the basis or not of the preceding configuration. With time, as alliances and dissensions are formed, the process of real-time objectification of the game of actors is engaged.

7 Source: David Krackhardt. http://www.heinz.cmu.edu/~krack.

Thus, this is what we indicate as sociography. The exercise is altogether rather simple, which, in our opinion, does nothing to diminish its operational character. Even though as this explains that. Since, indeed, let us not forget that it is a work that must be able to be undertaken quickly, in real-time, since the game is in progress. Simplicity and speed in this context are thus, undoubtedly, assets that are at the very least non-negligible.

However, from here on we can, perhaps, consider a less game-related and more concrete implementation than the one that we have just described. Indeed, the readers would have all clearly understood that the term "game" is used as a metaphor to describe situations of actors in very diverse contexts. In particular, that has already for some time been the case in the field of environment and this application has spread more recently to the disciplines of urban planning and development. It is thus, of course, in a more specific situation of this type that we will try to develop our approach later on.

However, before proceeding further, we clearly need to specify that our aim and our analysis are predictive and not prescriptive. Unlike game theory and some other attempts, our ambition is neither to determine nor to even approximate an optimal organizational solution for a group of actors placed in front of a given problem. Several authors have already underlined the vacuity of such an ambition in the field of urban planning and development and we will not return on that subject there. Our own ambition is rather to try to specify the decisional context of a project by rendering more complex the organizational framework that it seems to have. In this sense, it is a question of limiting as much as possible the set of possible or imaginable decisions by restricting it in a certain manner within the mesh of a graph. As we can transmit onto a map the geographical limits of a project, so we try to transmit onto a sociography the contours of the decisional space that has to lead to its implementation.

Here is the type of figure (Figure 6.6 in the color plates section), from which sociographic exploration may begin. All the relations between individuals are represented without distinction. The placement in circle is arbitrary but rather convenient for a first attempt. Here, the star shape suggests that the operator was integrated in the relational base, but the observable relations in the lower right quarter show than it went beyond the ego-network. Since each edge represents a relational mode whose legend is found in the left-hand column, the graph is at most a 12-graph.

It is there that we finally see the great interest of the structural approach and social networks analysis for our problem, since the method "aims to systematically contextualize the behavior of actors" (see [LAZ 98], p. 7). We do not specifically seek to understand the sense of an individual's action, but we try specifically to position an actor in his network in order to *estimate* his potential of action, his opportunities, as well as his constraints. The method aims at the predicable not the predictable. In other words, the idea of this approach is to try to represent the "playing field" where we watch our actors move. In a certain manner it is, first of all, about strategy rather than about trying to clearly define the game environment.

If we have thus somewhat given up on the idea of really being able to understand the strategy of an actor, we think that the only alternative is to encourage reflection, even imagination, by trying to determine all of his resources or attributes. It can be a question of financial capacity, time, access to information, membership in various associations, i.e. a whole set of parameters constituting a potential for action. Obviously, interpersonal relations form part of this potential and, for this reason, the distinction from "attributes" is sometimes thin. It is primarily due to the potential character of the resource. Let us say that the relation is really observed in practice, whereas the attribute is a capacity that is considered to be available for mobilization.

We try to represent these two dimensions in the sociograph. In its usual form the graph represents the relational capacity of both. By means of various formalizations we try to represent the resources-attributes.

This figure (Figure 6.7 in the color plates section) was obtained by "purifying" the previous figure. All the relations in the modes that do not make it possible to distinguish between nodes have been eliminated. The nodes in Figure 6.7 have been moved by the user to obtain the present figure (Figure 6.8 in the color plates section) without calling upon a systematic process or an optimization algorithm. However, we observe that the nodes connected by gray edges form a clique, that the exchanges are also very intense between the individuals in the "red group", or that individual number 3 has a pivotal role between these two sub-sets.

To this end, it is necessary, first of all, to represent the resources upon which, at least partly, the action can be based. Then we have to render this representation dynamic, since, by nature, such dissemination of the resources is not immutable.

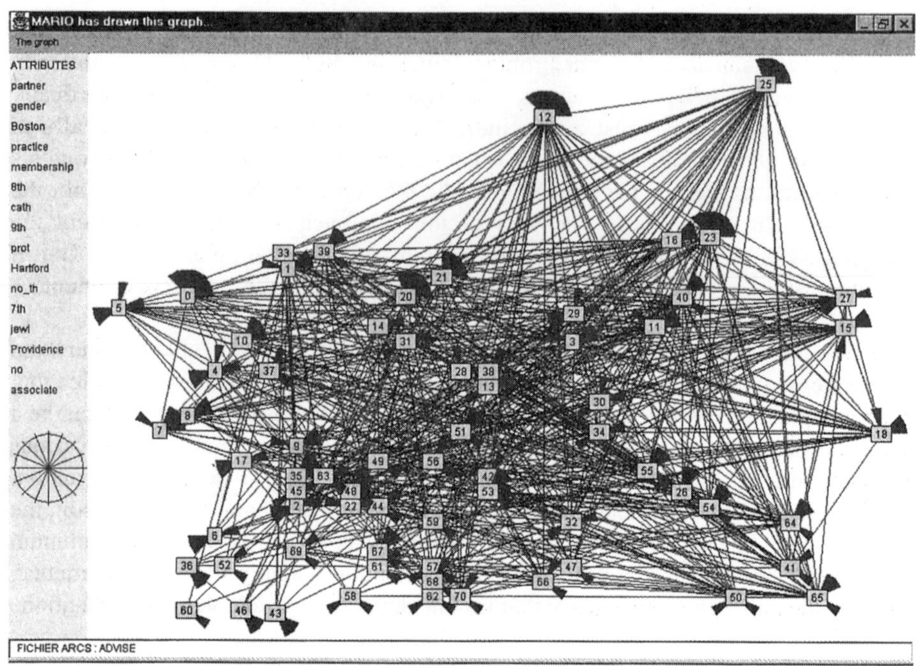

Figure 6.9. *Requests for advice and attributes of lawyers in a firm*

In Figure 6.9 we finally illustrate the difficulty that may exist in plotting a graph comprising a significant number of nodes and a significant number of attributes. The nodes are placed on a normalized reference scale whose origin is located in the top left. The x-axis represents the advice given: the more an individual is located towards the bottom, the more advice he gives. The y-axis represents the requests for opinion: the more an individual is located towards the bottom, the more he requests advise from others. The circle slices represent the possible attributes of the individuals. They highlight a certain similarity between the three individuals in the upper right quarter (who give out a lot of advice but hardly receive any). However, they also make it possible to observe individual singularities, such as node 16, which is in the same zone with an "attributes profile" that is extremely different from the three others or, on the contrary, node 0, which, despite its attributes, does not have the position that we would have believed.

6.4. System of information representation

Consequently, and in spite of criticisms that we could formulate in its respect, it is obvious that sociography is quite close to the theoretical framework of "limited" or "procedural" rationality. It is different, however, in the sense that while focusing on the dynamics of information representation-interpretation, it tries to avoid as much as possible any consideration of the actual motivation of individual action.

To solve such difficulties, that is, to contribute to decision making in this environment, three large logic types can be developed: that of multicriterion, a more critical approach of rationality in the organization and, finally, "information representation systems", among which we place our sociographs.

Another more recent attempt to sweep the field of the possible games of actors is to build a multiagent system (MAS). By "agent" we understand the active elements that are like our actors, but also passive agents that can be perceived, created, modified or destroyed by the former. In principle, it "is enough" to formalize the rules of behavior of each of these agents and then to let data processing "do the job" in order to observe what occurs when several agents are found in the same environment. To date MAS were used experimentally as an aid to public decision-making or for problems of territorial management [FER 98]. In our case we would have to create as many agents as actors with their own rules of behavior in order to observe the result of their interactions and we would have to restart the process as many times on the basis of assumptions about the individual strategies of the agents.

By adopting a more descriptive point of view, other authors on organizational sciences criticize the very principle of rationality of the decision. We group a little hastily in this set the work of the human relations school, the group dynamics of Kurt Lewin, those on organized collective action, i.e. studies that showed how the human element came "to disturb" the mechanical operation of industrial companies. The model of the "concrete system of action" (CSA) by Crozier and Friedberg is one of most famous on the subject [CRO 77, FRI 93]. It states that an organization creates a context of action by imposing formal relations of power between individuals, but that the latter change its composition through their informal relations, which finally yields a CSA. In this sense, however, it should be clearly understood that the CSA is not meant to generate situations of conflict and opposition between actors, but enables them, on the contrary, to solve a sum of concrete, daily and detailed problems, so that the organization may really engage in its functions. In this system individuals act to preserve an influence or minimal power and this influence in turn depends on the ambiguity of the role that they have in the organization, or that they believe to have. However, all these power games may on certain occasions be opposed to the principle of rationality, which is really why it is described overall as "limited" or sometimes "strategic".

The third possible way that we wished to explore is that of the "interactive data visualization" tools. Here, the focus is on the acquisition and exploitation of information, since these two stages should be used as a pivot and to stimulate reflection. In this case, it is a question only of shaping (often graphically) this data in order to enable the user to detect in this heterogenous information mass the singularities, which he would then be able to analyze more thoroughly.

Web Path is an interactive data visualization tool. It is a file manager, also called Visidrive, which shows a tree structure of data in 3D (Figure 6.10 in the color plates section). There are several types. Here, the file labels appear in a sort of a tube, then on a wall in perspective[8].

From this point of view, strictly speaking, it is not a model of analysis, but at most a set of tools or means. However, these are tools that do not yield direct and clear results. Contrary to many tools, data input into the sociograph is not homogenous. Consequently, it is not possible to reduce it to some kind of indicator in the hope of then determining the optimal solution to our problem.

On the other hand, by using the sociograph, i.e. by giving this problem the shape of a graph, it is, perhaps, indeed possible to initiate an analysis. Presented in the form of a relational database or various tables containing the attributes of the actors and their relations, the decision-making problem cannot be perceived in all its dimensions and therefore cannot be solved. We substitute this with just one image, a sociography, which is able to open perspectives for discussion and thus, possibly, (but that almost exceeds our ambition) of resolution.

For this reason a sociography is clearly a graph with all its operational logic. Indeed, the graph could not solve the Koenigsberg problem anymore than it could directly solve the problem of the shortest path. On the other hand, it is, indeed, from the moment when the problem is presented in this form of arcs and nodes that solutions to these questions can be conceived and then generalized. Sociography is no more ambitious when it proposes to apprehend a decisional problem.

Our sociograph is an attempt to take on directly these "additional" questions, these small "asides" that are still regarded by many as disturbing elements of a robust and reasonable practice, but which for us are a specific and irreducible dimension of the *praxis* of urban developers. We were largely inspired for its implementation by the data processing tools used by social networks analysts. However, we associated with it a certain number of specificities which are better suited to make it usable *in situ* by urban planners or developers. It is still not certain

8 http://davinci.informatik.uni-kl.de/vis98/archive/tp/papers/webpath.html.

whether the former and the latter have a real will to apprehend this dimension of the *praxis*.

To conclude, it is necessary to consider that a sociography is altogether quite a commonplace object: a non-planar p-graph enriched at the nodes by some figures symbolizing the attributes. As far as we are concerned we have not yet found more effective means of accounting for these relations and attributes. Having said that, it is perhaps this simplicity or this minimalism wherein lies the power of images in general and, thus, of graphs in particular.

6.5. Bibliography

[BER 68] BERTIN J., *Sémiologie graphique*, Gauthier-Villars, Paris, 1968.

[BON 75] BONIN S., *Initiation à la graphique*, EPI, Paris, 1975.

[BON 88] BONIN S., "La cartographie, processus de communication: méconnaissance et sous-utilisation des images graphiques dans l'enseignement et dans la presse", *Bulletin du comité français de cartographie*, no. 1, issue no. 115, 1988, p. 19-31; and the special edition of the Comité Français de Cartographie, "Théorie de l'expression et de la représentation cartographiques et exercices pratiques", *Bulletin du comité français de cartographie*, issue no. 117-118, p. 19-31, 1988.

[CRO 77] CROZIER M., FRIEDBERG E., *L'acteur et le système*, Seuil, Paris, 1977.

[FER 98] FERRAND N., DEFFUANT G., "Trois apports potentiels des approches 'multi-agents' pour l'aide à la décision publique", p. 373-385, CEMAGREF-ENGREF, *Colloque Gestion des territoires ruraux: connaissances et méthodes pour la décision publique*, Clermont-Ferrand, 27-28 April 1998.

[FLA 68] FLAMENT C., *Théorie des graphes et structures sociales*, Mouton/Gauthiers-Villars, 166 p., Paris, 1968.

[FOR 97] FORSÉ M., LANGLOIS S., "Réseaux, structures et rationalité", *L'année sociologique*, 1997, 47, no. 1, p. 27-35.

[FRI 93] FRIEDBERG E., *Le pouvoir et la règle: dynamique de l'action organisée*, Seuil, 1997, Paris.

[LAR 99] LARRIBE S., Représentations auto-centrées et interactives d'un réseau d'acteurs en aménagement, PhD Thesis, Tours, 1999.

[LAZ 98] LAZEGA E., *Réseaux sociaux et structures relationnelles*, PUF, Paris, 1998.

[MER 92] MERMET L., *Stratégies pour la gestion de l'environnement: la nature comme jeu de société*, L'Harmattan, Paris, 1992.

[MOR 70] MORENO J.-L., *Fondements de la sociométrie*, PUF, Paris 1970.

Chapter 7

RESCOM: Towards Multiagent Modeling of Urban Communication Spaces

7.1. Introduction

In the present chapter we aim at developing a data processing tool within a systemic individually-centered approach dedicated to the study of the spatial structure of urban communications. The utility, which we would like to attach to this tool, consists, first of all, of setting up a method of graphic representation of urban communication spaces, so as to deduce several explanatory indicators with the ultimate goal of assisting decision-making in urban development.

To arrive at our goal, we employ a theoretical framework enabling us to study the city as a socio-cultural communication system where individuals exchange information in specific spaces whose design can create favorable conditions for the initiation and maintenance of the communication flow between them. These spaces are described as phatic.

However, we have observed that to simulate a permanent social process integrating multiple behavior modes, the multiagent system (MAS) approach offers the methods necessary to achieve our objective. The MAS consists of representing in computerized form the behavior of entities (commonly referred to as agents) that act in the world with the objective of making a phenomenon emerge on the basis of their interactions. Each one of these interactions develops according to rules, conventions, etc. which are determined by a certain "interaction protocol". The

---

Chapter written by Ossama KHADDOUR.

definition and installation of this protocol bring into play a great number of mechanisms and properties which relate not only to the protocol itself, but also, more specifically, to the communication medium as well as the regulation mechanisms that may exist at the overall system level. In our work concerning an urban system, agents are individuals with the desire, the need, the obligation or prohibition (according to a socio-cultural code) to use phatic spaces as communication (interaction) media.

Called RESCOM, our data processing tool has been developed in a CAD environment by using a functional programming language: Lisp. The development of this tool enabled us to represent graphically on two- and/or three-dimensional maps a set of indices which look further into our knowledge of phatic spaces in the city. We will describe, in the following pages the main functions of RESCOM which show the utility that this tool could have both on the epistemological and operational level of the urban development.

It should be noted that all data that we will use to illustrate the iconographic RESCOM products are the result of *in situ* work near the inhabitants of the town center of Latakia in Syria.

7.2. Quantity of information contained in phatic spaces

To understand what a quantity of information may represent it is necessary, first of all, to recall that a piece of information indicates, by definition, one or more events among a finite set of possible events. Thus, there is an inversely proportional relationship between the frequency of appearance of an event and the information which it brings.

In a practical sense we may use the following example: if when looking for a type of building in a city containing a set of buildings (N), we specify that this building is a "mosque", we provide information that reduces the search time proportionately to the number of mosques (n) in the city. This information is all the more interesting, since it reduces the number of later possibilities, in other words, it *reduces uncertainty*. We are thus led to define the quantity of information as a growing function of N/n: the smaller the number of mosques, the higher the ratio N/n and, thus, the higher the quantity of information carried by the statement "it is a mosque". According to Shannon, we can express this definition mathematically:

$$QI = -\log p_i \tag{1}$$

where "QI" is the quantity of information carried by the statement and p_i is the probability of event i.

On the basis of the mathematical Shannon's communication theory we propose to apply this formulation of quantity of information to a set of phatic spaces (e_1, e_2, ...e_n) by basing ourselves on descriptive criteria (category, type, time, etc.) of each space. Let us take the example of a city holding six phatic spaces whose descriptive criteria are established in the following table.

Space	**Category**	**Type**	**Time**
e1	religious	mosque	1940
e2	religious	mosque	1980
e3	religious	church	1940
e4	religious	church	1920
e5	religious	synagogue	1920
e6	public	market	1900

Thus, space *e1* has a probability of appearance of 5/6 in its category (five religious spaces among six spaces studied), 2/6 for the type (mosque) and 2/6 for the time (1940), thus a quantity of information equal to $-log(5/6*2/6*2/6) = 3.4$. According to the same calculation, space *e6* represents a quantity of information equal to 7.7.

On the basis of relation (1) and by using the table above we may calculate the quantity of urban information contained in each of the considered phatic spaces.

Space	Category	Type	Time	Probability			Quantity of information
				Cat	Type	Time	
e1	religious	mosque	1940	5/6	2/6	2/6	–*log* (5/6*2/6*2/6) = 3.4
e2	religious	mosque	1980	5/6	2/6	1/6	–*log* (5/6*2/6*1/6) = 4.4
e3	religious	church	1940	5/6	2/6	2/6	–*log* (5/6*2/6*2/6) = 3.4
e4	religious	church	1920	5/6	2/6	2/6	-*log* (5/6*2/6*2/6) = 3.4
e5	religious	synagogue	1920	5/6	1/6	2/6	–*log* (5/6*1/6*2/6) = 4.4
e6	public	market	1900	1/6	1/6	1/6	–*log* (1/6*1/6*1/6) = 7.7

In terms of data processing the algorithm associated with this function can generate the quantity of information following one, several or all the descriptive criteria adopted for the spaces. In terms of programming, this algorithm is inbuilt into an interactive instruction that makes it possible to choose the criterion or criteria, according to which the user wishes to calculate and then to represent the quantity of information contained in each studied phatic space.

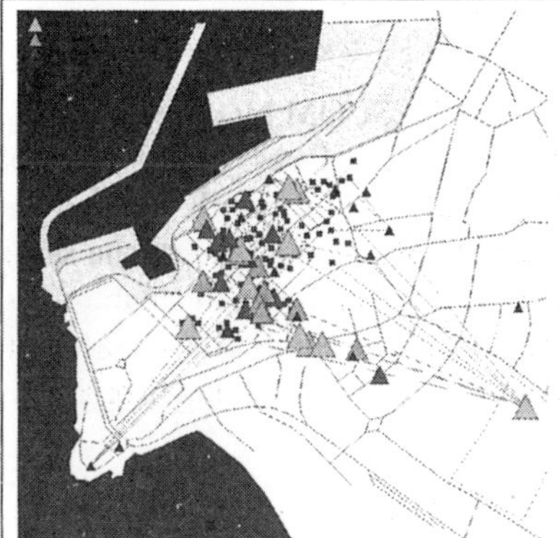

On the right: map of the quantity of information contained in the phatic spaces mentioned by the inhabitants of the town center of Latakia. On the right: various views in perspective projection with a rendering showing the town center and the quantity of information in the phatic spaces in Latakia. The values are represented by cylinders. Everything is represented against the cartographic background of topographic services.

Design and plotting: O. Khaddour, Ph. Mathis. CESA, 2003.

The quantity of information contained by each phatic space constitutes a component of its semantic contents and will thus form part of the graphic entity (the point, for example), which represents it. This enabled us to use it in other explanatory indices of the socio-cultural communication network in the city. Thus, in RESCOM we have developed functions that calculate and represent phatic centrality according to Levaitt and the degree of cognitive dissonance according to Festinger [KHA 04], which we will not be described here in order to leave enough space to demonstrate the usefulness of RESCOM on the operational level of the urban development.

7.3. Prospective modeling in RESCOM

We tried to concretize the prospective dimension of RESCOM by developing the modules concerning the representation of phatic attraction surfaces, on the basis of which we proposed a "game" simulating the behavior of the individuals looking for sources of information and communication in an urban environment. From the same point of view, we have assigned to RESCOM a functionality which enables the development of requests in pseudo-SQL relating to data used in RESCOM and, finally, an inference module which helps observing the changes in spatial practices after the establishment of a new phatic space.

7.3.1. *Phatic attraction surfaces*

Other possibilities to study phatic spaces on a cellular basis are also possible in RESCOM, in particular with respect to graphic analysis of spatial data. Within this framework we have tried to explore the potential of the gravity model to limit the attraction surfaces of each phatic space. The interest of the gravitational model for the estimation of the attraction surface had been pointed out in the 1920s by the economist Reilly.

Reilly's objective was to determine how the unsatisfied demand of an intermediate place (m) located between the two larger cities (a and b) would be distributed. Thus, he starts with two poles considered without dimension and studies their influence on space [MAT 71a].

$$Va / Vb = (Pa / Pb) * (Dam^2 / Dbm^2)$$

where Va and Vb are the sales of a and b in m, Pa and Pb the "Weight" or population of a and b, and Dam and Dbm the distances from a and b to m. By generalizing Reilly's statement we can calculate the attractions of all the points a, b, … n at the point m.

$$A_{ij} = P_i / (\text{Distance}_{ij})^2 \tag{2}$$

We then probabilize linearly by dividing each attraction by the sum of the line, in order to pass from a ratio of attraction between two points to a generalization between n points, which enables us to find Huff's formulation

$$P\{Ar_{ij}\} = P_i \, D_{ij}^{-2} / \Sigma_k (P_i / D_{ij}^{-2})$$

In the case of phatic spaces where the weight is expressed in terms of quantity of information we can, in all spatial points j, look for the phatic space that exerts the strongest attraction on this point and thus trace the spatial extension of the maximum influence areas of each space. As a result, the attraction surface of a geographical site (phatic space in our model) draws a hyperbolic cone, which when projected onto a map is represented by concentric circles of decreasing intensity around the site concerned [PUM 01].

The use of cellular space and thus of the attributes of each of its cells led us towards a specific formulation based on the observations of Philippe Mathis in connection with "the fields' equation" [MAT 71b], whose simulations were limited to two-dimensional projections (the computer equipment in 1971 did not make it possible to plot three-dimensional representations). Along the same lines, by using a CAD system we set up a technique, which made it possible to generate three-dimensional representations of attractiveness in a cellular space.

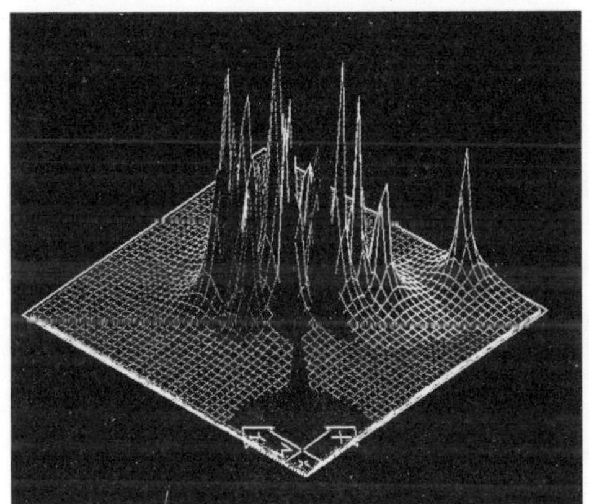

Figure 7.1. *Various representations of the attractiveness of phatic spaces according to Reilly's model*

The first results obtained from this simulation do not coincide with the networks obtained following the investigations. Thus, the most characteristic spaces in the Latakia area, i.e. the University (in the right corner) and the Corniche Sud (in the bottom corner), seem to become less attractive (since they are at the periphery of the city) with respect to other spaces that are geographically more accessible in terms of distance.

In fact, the use of quantity of information as "mass" revealed to us the need for a *specific development of the distance* covered in search of this rather particular "good", since it is clear that for certain people neither the distance nor its cost play the decisive part in their choice of phatic spaces.

We have thus set up a prospective module in the form of a game aiming at simulating the behavior of an individual who is subjected to the constraints of belonging to a socio-cultural category (gender, age, profession, etc.) in his choice of reference phatic space. Below we will explain the functioning of this module, as well as its ramifications.

7.3.2. *Game of choice*

In parallel to the thermodynamic laws of Boltzmann and Shannon's formulation, we will suppose that a blank cell contains informational potential if only by its "unique" localization in the grid constituting the geographical space of the city. That is, in a grid comprising 12 cells where any cell contains a quantity of information equivalent to: $-log\ 1/12 = 0.2$.

Consequently, in the case of movement within a grid of "n" isomorphic cells, each cell that we pass through brings a quantity of information equal to (log 1/n). The usefulness of this formulation is proven when the cell corresponds to a distance so that the distance covered in the grid is expressed in cellular unit (cellular distance). The equivalent of this distance in terms of information will be:

$$QI_d = -\log (d_c / N)$$

where (d_c) is the cellular distance and N is the total number of cells. By applying this concept of cellular distance to a space containing several sources of information (p_1, p_2, ..., p_n), which correspond to various quantities of information (Q_{I1}, QI_2, ..., QI_n), we can simulate the behavior of an individual traversing this space by consciously or unconsciously looking for a maximum of information with a minimum effort and cost.

For our game we suppose that the individual placed in this space must evaluate all the available sources of information in order to choose one of them. However, the fact of moving to get to the node corresponding to the individual's search adds a quantity of information generated by the distance covered.

$$QIt = QIn + QId$$

where "QI_t" is the quantity of information that the agent could acquire during movement, "QI_n" is the quantity of nodal information that a source of information contains. Finally, "QI_d" is the quantity of information generated by distance. On the basis of this formulation we generated the second simulation which shows the difference between the effective choice of actors as revealed by surveys and the choice of actors according to the opportunity to choose a space ensuring an optimum quantity of information. The following figure shows, in particular, the great variation of distances covered by individuals to get to various phatic spaces of their environment.

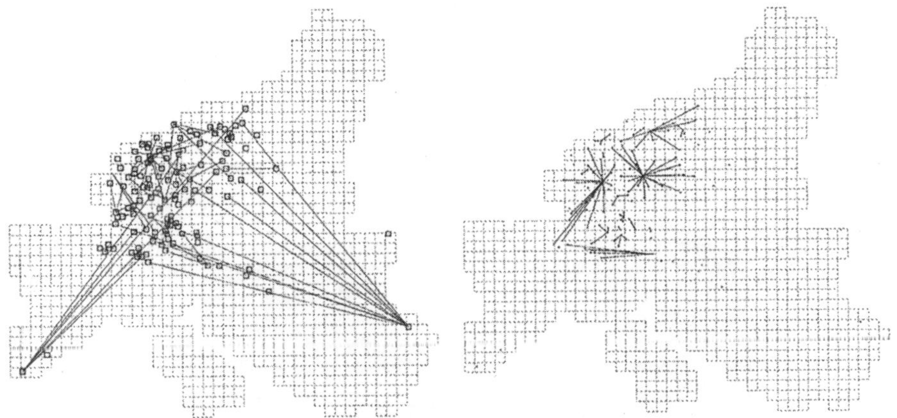

Figure 7.2. *Comparison between the network according to surveys (left) and the network of the simulation according to a game of choice (right)*

If we take a closer look at the statistical indicators of these two maps, we will clearly see that, on the one hand, the distance covered is greater in reality than according to the simulation and on the other hand, the scattering of values is also more considerable in reality than in the model.

The summary table provided below therefore only represents the "temporary" results showing the need for additional adjustment of the algorithms integrated into RESCOM.

Indicator	According to surveys	According to RESCOM
Maximum distance	3391.72 m	875.85 m
Minimum distance	29.09 m	29.09 m
Average distance	628.1 m	240.53 m
Standard deviation	906.3	197.92

By analyzing the link between distance and the quantity of information, which we calculated on the basis of data received from the inhabitants of the town center of Latakia, we noticed that the quantity of information increases with distance. The logarithmic function in the algorithm specifying the calculation of the quantity of information generated by distance can partially explain this phenomenon. Otherwise, we see in this link an echo of Webber's observation on the relationship between the quantity of information received per hour and per person, the level of specialization and the distance separating the urban domains [WEB 94].

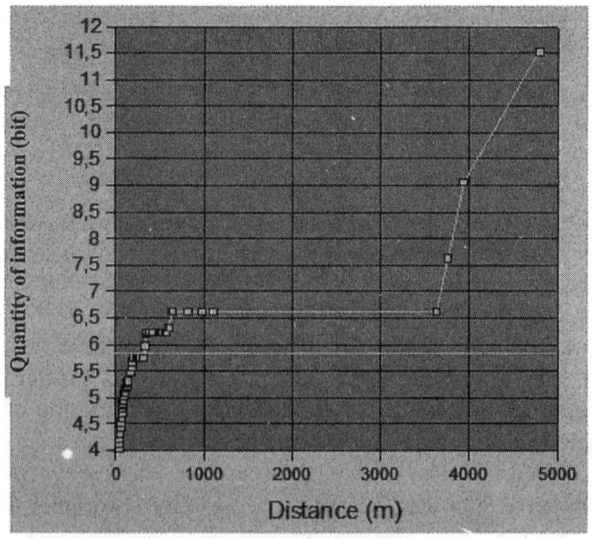

Figure 7.3. *The link between distance and the quantity of information for the inhabitants of the town center of Latakia*

However, the following graph poses several questions about the nature of thresholds where the quantity of information seems invariable despite the increase in the distance covered (between 700 m and 3,500 m). However, this threshold groups together the majority of the choices expressed by individuals questioned during the surveys. Within the framework of this chapter we can only propose partial explanations by leaving the door open to future projects tackling this question.

Below, we will test several hypotheses that may lead to simulations that are closer to the "limited rationality" of individuals in their choice of phatic spaces.

Let us recall that an individual with a perfect knowledge of his environment really does not exist. For this reason we propose to incorporate in our simulation a set of constraints that will be imposed on the mobile agents: proximity of a central point, agent-specific resources, maximum distances to cover. We can also introduce the relation to individuals "residing" in the neighboring cells, so that, if "the majority of my neighbors go to a certain communication space, I go there too".

$$QIt = QIn + QId + Qic$$

where

– "QI_t" is the total quantity of information;

– "QI_n" is the quantity of nodal information contained in a phatic space;

– "QI_d" is the quantity of information generated by distance where the distance covered must be equal or lower than a maximum distance;

– "QI_c" is the quantity of information representing the number of individuals who choose a communication space. ($Qi_c = log\ n/N$, where n is the number of individuals who choose a specific space and N is the total number of individuals simulated.)

Lastly, we can add a multiplicative parameter of the binary order α – called dummy variable by economists. This parameter makes the frequenting of a given space by a certain category of individual either possible (α=1) or impossible (α=0).

$$QIt = (QIn + QId + QIc)*\alpha \quad (4)$$

The binary parameter works in situations where a certain socio-cultural code almost completely reduces to zero the possibility that a Muslim man goes to a church or that a Christian woman goes to a mosque. Consequently, we can simulate situations where girls from districts known as "traditional" cannot go to nearby phatic spaces and prefer to go further away in order to escape certain codes of conduct (the prohibition of the mixing of genders).

Once the quantity of information has been calculated according to the criteria mentioned above, and in order to put into action a few hypotheses, we can simulate the possible behavior of agents in search of sources of information and communication in the urban environment. To this end we use a variation of Reilly's law, which has been widely used in geomarketing.

7.4. Huff's approach

Huff's model proposes taking into account the uncertainty regarding the exact localization of the borders of areas of influence by defining a relational probability, which is the relationship between the influence of a site and the influence of all the sites. Huff's problem can be formulated as follows:

– In a set of sites k, each site with a mass M and a location j represents, for the individual located at a point i, an opportunity O_{ij} that can be evaluated using Reilly's formula: $O_{ij} = M_j/(D_{ij})^2$.

– The probability of choosing a destination is equal to the opportunity of this destination divided by the total sum of destination opportunities: $P_{ij} = O_{ij}/\Sigma_k O_{ik}$.

By substituting mass with quantity of information we propose to simulate the behaviors of individuals involved in a search for information and communication in the city, by using the following formula:

$$P_{ij} = [QI_j/(D_{ij})^2]/[\Sigma_k QI_k/(D_{ik})^2] \quad (5)$$

where (QI) is the quantity of information as previously defined.

By limiting ourselves to the geometrical criteria of individuals (location in the geographical space) we implemented the algorithm which is necessary to test the formulation suggested above. The results obtained using this algorithm reveal a considerable attenuation of the decisive character of Reilly's law and reveal behaviors that hardly fit a *homo economicus*.

Apart from the representation of the potential attraction of phatic spaces, this simulation revealed to us other aspects of the socio-cultural network of communication in the city. From the phenomenological point of view, by comparing networks of individuals and their phatic spaces on several scales we discover auto-similarity [MAT 03]. This leaves the door open to future research aimed at thoroughly investigating the modeling of socio-cultural communication networks in the city by following the principles of fractal geometry.

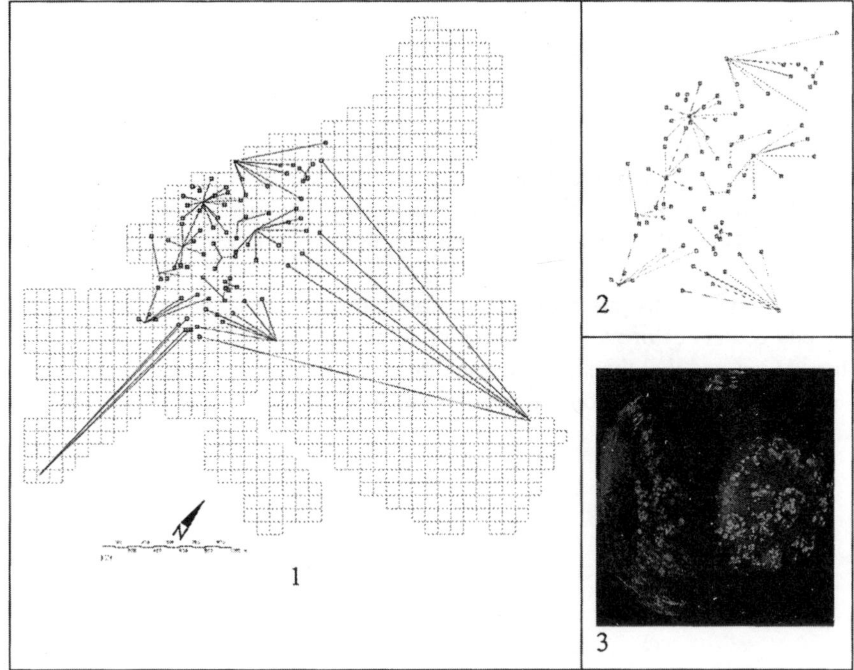

Figure 7.4. *1. The attraction of phatic spaces according to a variation of Huff's model, 2. zoom on the network showing the planar "aspect" of the graph, 3. cartography of the global communication network*

Up until now we showed the possibilities that RESCOM possesses to develop and help to interpret the spatial structure of communication. However, we marked out the limits of this chapter by aiming primarily at representing socio-cultural communication networks in the city, and secondly at the possibility of intervening as a designer and/or a decision maker in the operation and the organization of these networks. This leads us to the "inference" module that we have integrated in RESCOM.

7.5. Inference

We prefer to use the term inference, although it is not modern, to indicate (according to the French "Petit Robert" dictionary) a "logical operation, by which one admits a proposal in virtue of its connection to other proposals already held as true".

By doing this it is possible to simulate the behavior of individuals in the event of the opening of one or more new phatic spaces. The characteristics of this space, geometrical (location) as well as semantic (descriptive criteria), will play an important part in its potential to attract individuals searching for an information and communication space.

By keeping the example of the town of Latakia, we simulated a scenario of the opening of several shopping malls trying to respect a spatial distribution covering four sectors of the city: the city center (C2), the northern periphery (C3), the eastern periphery (C4) and the unused port grounds in the western part of the city (C1).

The following illustration clearly shows that, in fact, the shopping malls located in the city center sector would be more attractive than those located in the periphery. It should be noted that the shopping mall that we proposed on the unused grounds of the port (C1) would be more attractive compared to that located to the east of the town center (C2).

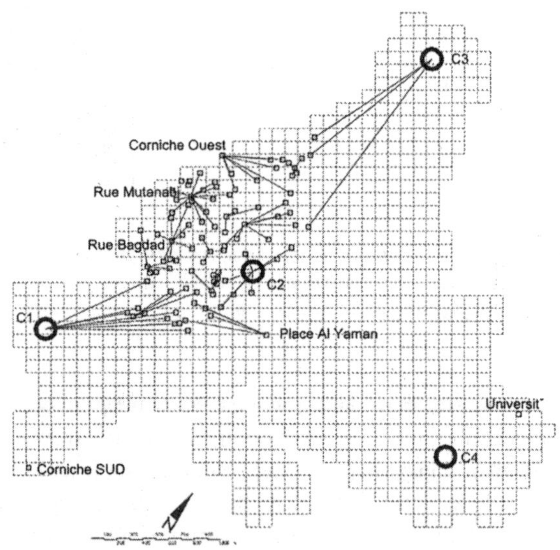

Figure 7.5. *Illustration: Simulation of the attractiveness of the new phatic spaces suggested for a project. Design and drawing: O. Khaddour, Ph. MATHIS. CESA, 2003. Urban planning*

On the other hand, the introduction of these new phatic spaces did not reduce the performance of the already existing spaces, in particular, Rue Baghdad and rue *Mutanabî,* which coincides with the results obtained by using Huff's approach. This leads us to believe that these spaces deserve careful consideration, first of all, to reveal their architectural and urban characteristics, but also in order to propose the necessary interventions in terms of urban development operations which are aimed to improve their communication nature.

7.6. Conclusion

We believe that the quantity of information, such as we have formulated it, is still far from having revealed its entire potential, due to the fact that we could not make a comparative study between several cities, because the mathematical formulation of this function leads to a specific notion of entropy which makes it possible to define the specificity of each type of city. Consequently, the quantity of information would differ for each city depending on its entropy.

We also believe that it is possible to use the model of indifference of Nicos Devletoglou to reveal more characteristics of the areas of influence of phatic spaces in the city. Set in a duopoly, this model allows the consumer to make choices according to the distance to cover. If the difference in distance between two poles decreases (while remaining equal or larger than a certain value), there will arrive a moment when the "customer" regards them as equidistant from his residence: the difference in distance then becomes imperceptible. That leads Devletgolou [DEV 65] to propose the concept of perceived minimal difference which corresponds to a zone covering the areas of influence of two poles, inside which individuals have no spatial preferences, but are rather submitted to a kind of mode effect pushing them to follow mass behavior in a larger zone [BER 71]. We generalize by calculating a zone of indifference between two dominant attractions which enable the cartography.

In fact, if we generalize Reilly, we obtain Huff's formula, but each point of space is still subjected to a dominant attraction except in the case of triangular or square distribution of strictly identical poles or in extremely improbable cases where the difference in weight would be strictly compensated by a difference of the square of the distances. In reality a point is always located in a single dominant zone or at the border or the intersection of two borders, although then it is necessary to take into account "the thickness" of the border. The generalization of Develtoglou introduced by Mathis [MAT 71a] (generalization to any exponent of the distance n and to poles of different weights) made it possible to reveal indifference zones which have very different or even dissymmetric form.

Thus, we finish these few illustrations on the manner in which we chose to use RESCOM. Let us note that a developer, a town planner or an architect continuously propose virtual, hypothetical places that might well be accepted by reality as well as refused or neglected. Consequently, all information drawn from simulations can be validated or rejected only through experience which generates a new perspective. RESCOM is restricted to the role of a "machine of questions" striving unceasingly to acquire new data and generate knowledge by means of new descriptive criteria of new phatic spaces and/or their users.

7.8. Bibliography

[BER 71] BERRY B., 1971: *Géographie des marchés et du commerce du detail*, Paris: Armand Colin, pp. 153-154.

[DEV 65] DEVLETGOLOU N., 1965: "A Dissenting View of Duopoly and Spatial Competition". *Economica*. Vol. 32. pp. 140-160.

[KHA 04] KHADDOUR O., 2004: *Modélisation individu-centré des espaces de communication dans la ville. Vers une modernisation de la représentation de l'environnement urbain au Levant*, PhD Thesis, University of Tours.

[MAT 71a] MATHIS P., 1971: *Introduction à une théorie unitaire des implantations commerciales*. PhD Thesis, Paris I University, p. 234.

[MAT 71b] MATHIS P., 1971: *op. cit.* pp. 290-291.

[MAT 03] MATHIS P., 2003: "Une méthode de simulation et de visualisation multiniveaux" in MATHIS P. (ed.): *Graphes et réseaux, modélisation multiniveaux*, Paris: Hermes, p. 286.

[PUM 01] PUMAIN D., SAINT-JULIEN T., 2001: *Les interactions spatiales*. Paris: Armand Colin, p. 37.

[WEB 94] WEBBER M. M., 1994: *L'urbain sans lieu ni bornes*. Paris: Aube, p. 82.

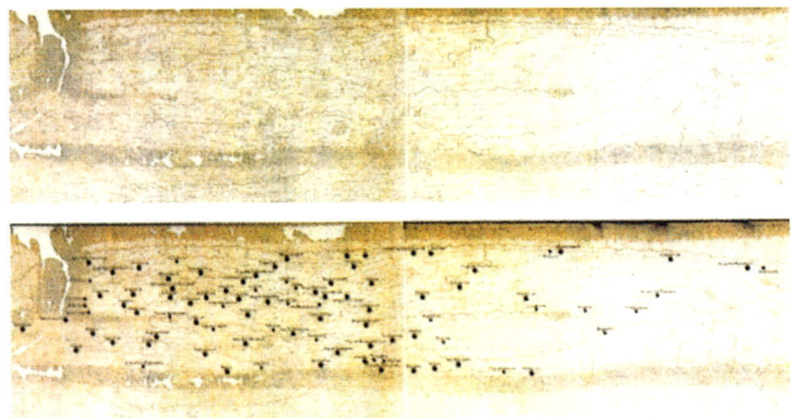

Figure 10. *Peutinger table*

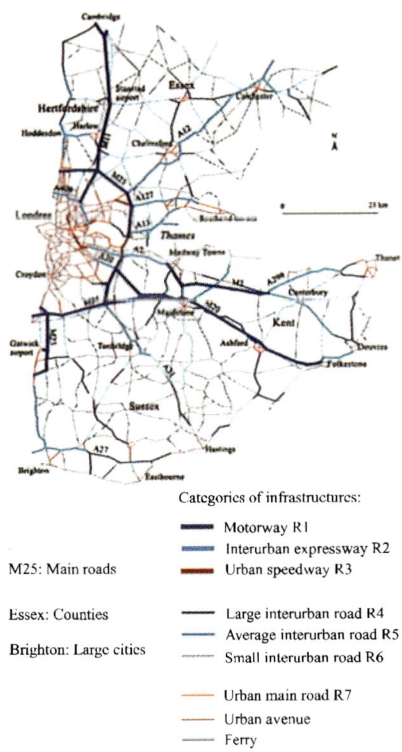

Hertfordshire
Londres
Croydon
Essex
Thames
Kent
Sussex
Categories of infrastructures:
Motorway R1
Interurban expressway R2
Urban speedway R3
Large interurban road R4
Average interurban road R5
Small interurban road R6
Urban main road R7
Urban avenue
Ferry
M25: Main roads
Essex: Counties
Brighton: Large cities

Created by M. Appert, L. Chapelon, 2001, UMR 6012-ESPACE, Montpellier
NOD/MAP software: L. Chapelon, L'Hostis, Ph. Mathis, CESA, 1993/2002

Figure 1.3. *Graph of the road network in the East Thames corridor*

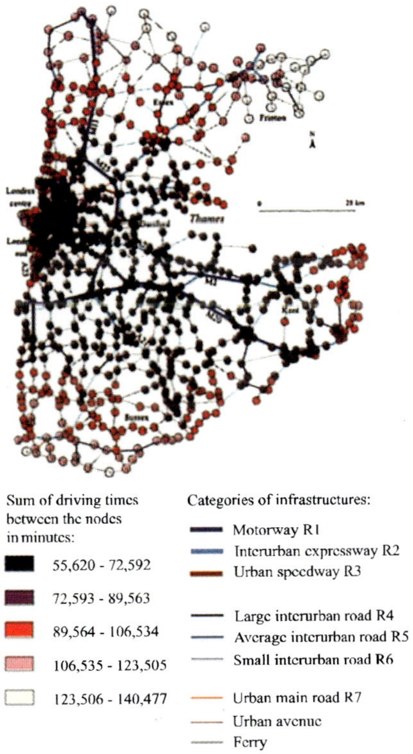

Thames
Sum of driving times between the nodes in minutes:
55,620 - 72,592
72,593 - 89,563
89,564 - 106,534
106,535 - 123,505
123,506 - 140,477
Categories of infrastructures:
Motorway R1
Interurban expressway R2
Urban speedway R3
Large interurban road R4
Average interurban road R5
Small interurban road R6
Urban main road R7
Urban avenue
Ferry

Created by M. Appert, L. Chapelon, 2001, UMR 6012-ESPACE, Montpellier
NOD/MAP software: L. Chapelon, L'Hostis, Ph. Mathis, CESA, 1993/2002

Figure 1.4. *Sum of minimal driving times between 5 and 6 pm*

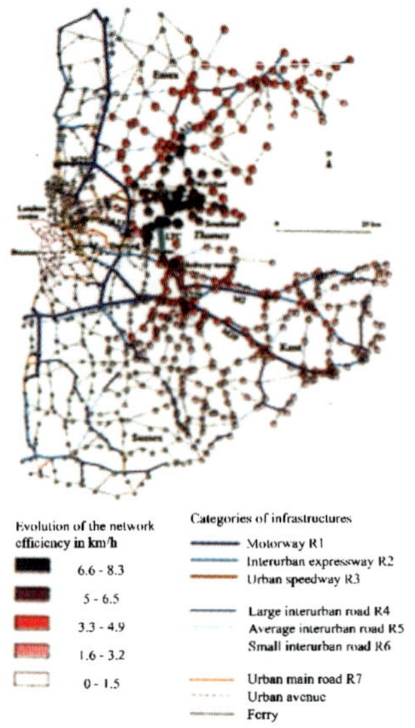

Created by M. Appert, L. Chapelon, 2001, UMR 6012-ESPACE, Montpellier
NOD/MAP software: L. Chapelon, A. L'Hostis, Ph. Mathis, CESA, 1993/2002

Figure 1.5. *Performance of the road network between 1 and 2 pm (off-peak)*

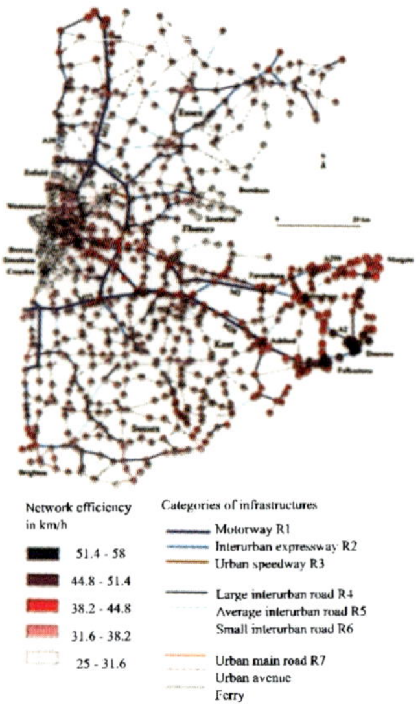

Created by M. Appert, L. Chapelon, 2001, UMR 6012-ESPACE, Montpellier
NOD/MAP software: L. Chapelon, A. L'Hostis, Ph. Mathis, CESA, 1993/2002

Figure 1.6. *Performance of the road network between 5 and 6 pm (rush hour)*

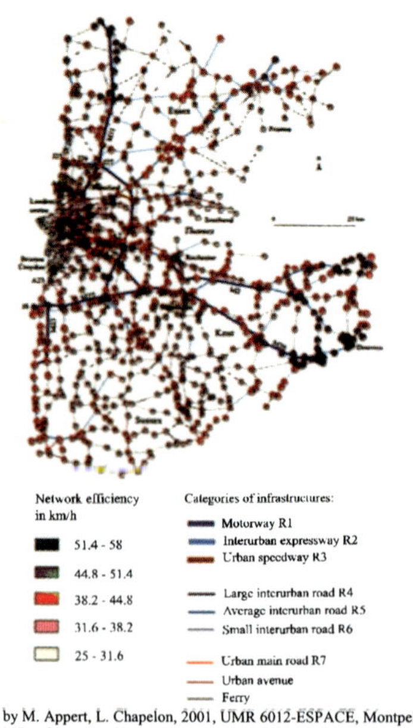

Created by M. Appert, L. Chapelon, 2001, UMR 6012-ESPACE, Montpellier
NOD/MAP software: L. Chapelon, L'Hostis, Ph. Mathis, CESA, 1993/2002

Figure 1.7. *Simulation of the performance results obtained with the construction of the Lower Thames Crossing in 2015*

Chapter 8

Traffic Lanes and Emissions of Pollutants

8.1. Graphs and pollutants emission by trucks

The formulae used to calculate the emissions of vehicles (regardless of the category) take into account the type of vehicle (catalytic or not, diesel, petrol, etc.), automobile park, weight (cars, utility vehicles or trucks), highway type associated with characteristic driving cycles (urban, rural and road). It appears that a vehicle's load has little influence on the emissions of cars as opposed to trucks, more so since it varies between load when empty and fully loaded vehicle. There is, therefore, a strong link between the load of a vehicle, the consumption of energy and emissions. The larger the load (increase in resistance to the movement of the truck) the more work is required to drive that mass at a constant speed (70 km/h, for example). This resistance depends on the load of the truck, but also on the gradient of the road, since the work required from the engine is not the same during ascent as during descent. Consumption as well as the fuel breakdown products are then greater or smaller depending on the needs.

We easily calculate (i.e., obtain a satisfactory evaluation of the emissions) the annual emissions of light vehicles according to vehicle, type of driving (urban rural, etc.) and annual mileage. However, this method is not easily applicable to the trucks (effect of mass), if the gradient of the road is considered. It is, thus, of primary importance to consider the routes (origin, destination, gradient, etc.) in addition to the method used for passenger vehicles.

Chapter written by Christophe DECOUPIGNY.

We will show in several stages how the use of the graph theory may or may not be justified: what does it bring in the evaluation of heavy vehicles emissions? We will also tackle the problem of the integration of the road gradient into a hitherto two-dimensional graph.

For that we will first present the method used, in order to offer a better understanding of our process and results. As the aim is not to comment on the method of calculating the emissions, we will limit ourselves to evaluating carbon monoxide (CO) and nitrogen oxides (NOx) for trucks lighter than 7.5 tons. Emissions will be calculated by using a programming protocol in Visual BASIC, which is added to the NOD model by Laurent Chapelon[1].

We will work in two stages. The first consists of observing the behavior of emissions depending on speed, slope and load on a section of the A28 highway. The second will enable us to extend the emissions calculation formulae to a road transport system graph.

Figure 8.1 shows the section of A28 developed on the basis of a digital terrain model [SER 00], whereas Figure 8.2 represents the graph of the French road network. This graph is based on a file with 300 points representing the largest French cities, either by population or by importance in the road network, and a file with 912 undirected arcs (or 1,824 directed arcs). This file is composed by using a national transportation system logic. There are five hierarchical levels of roads, to which a mean velocity for each type of vehicle is associated (that is, five functional binomials[2] for trucks lighter than 7.5 tons):

– highway: 110 km/h for cars and 70 km/h for trucks;

– expressways: 90 km/h and 60 km/h for the latter;

– roads: 70 - 70 km/h and 55 for trucks (in the majority they correspond to trunk roads);

– roads: 60 and 50 (40 and 30 km/h for trucks).

Several types of arcs may connect two cities. The graph of the transport system is a p-graph where width or colors correspond to the various partial infrastructure networks. The partial graphs defining the infrastructures are directed, since the nodes have an altitude, in order to calculate an average slope for the arcs.

1 "Software for space-time evaluation of transport supply modification projects", design and realization: Laurent Chapelon, Philippe Mathis, CESA laboratory version 4.2, August 1998.
2 See Chapelon, *ibid.*

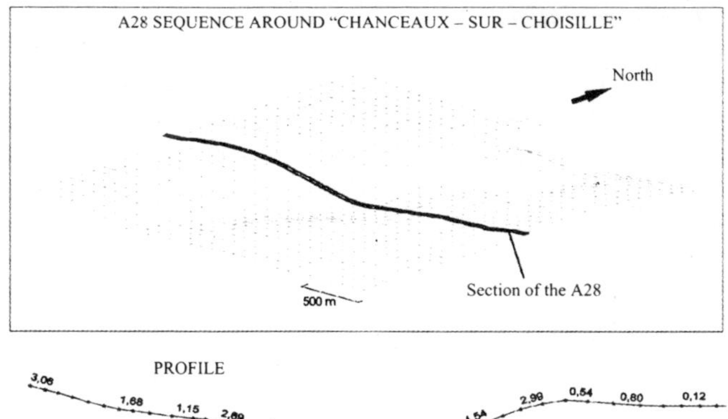

Profile and absolute value of the slopes

Ch. DECOUPIGNY, K. SERRHINI, CESA April 2000

Figure 8.1. *Profile of a section of A28*

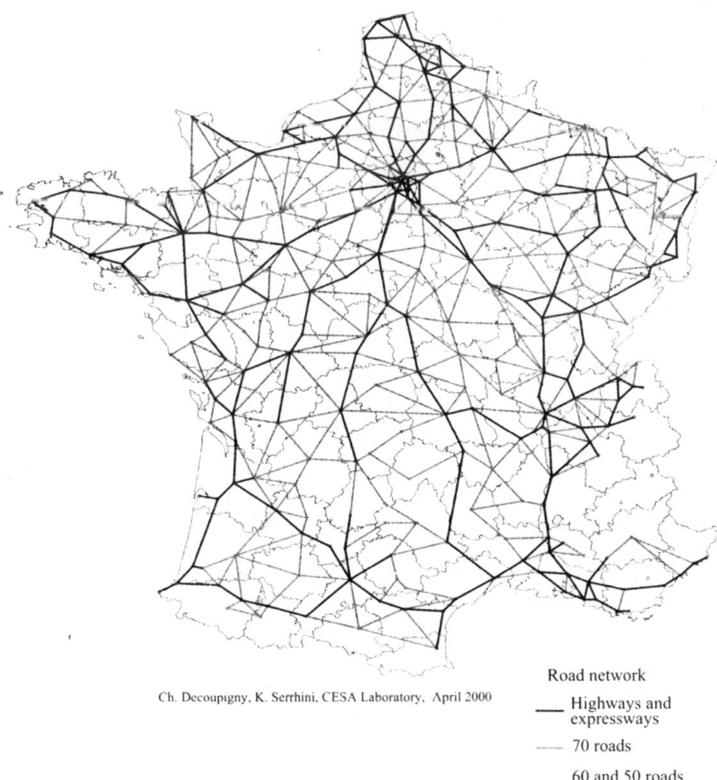

Figure 8.2. *Graph of the road network*

8.1.1. *Calculation of emissions*

We use formulae of the hot emissions for heavy-duty vehicles that take into account the weight of the empty vehicle, the road gradient and the load state of the vehicles [HIC 99]. They are derived using data from engine test-bed measurements according to the standardized driving cycles performed on vehicles that are representative of the vehicle fleet. They are expressed in grams per kilometer according to speed, the road gradient, the load and type of the vehicle.

8.1.1.1. *According to speed*

There are several cases depending on the pollutant considered and the speed of the trucks.

For carbon monoxide:

– speed ranging between 5 and 100 km/h

$$HotCO,j,k = 37.28\ Vk^{-0.6945} \qquad [8.1]$$

where:

– hotCO,j,k is the hot emission in g/km of vehicle j over an arc k;

– Vk is the average speed over the arc k.

For nitrogen oxides:

– speed ranging between 5 and 50 km/h

$$HotNOx,j,k = 50.305\ Vk^{-0.7708} \qquad [8.2]$$

– speed ranging between 50 and 100 km/h

$$HotNOx,j,k = 0.0014\ Vk^2 - 0.1737\ Vk + 7.5506 \qquad [8.3]$$

We then have the data for five pollutants i, carbon mono and dioxide (CO and CO_2), nitrogen oxides (NOx), particles (PM), and volatile organic compounds (VOC). There are four classes of vehicles j and five classes of arcs k, to which the mean velocities Vk are associated.

We then obtain the unitary emission in grams per kilometer for a pollutant and a type of empty vehicle for each arc of the graph. After that the slope factor can be introduced into the calculations.

8.1.1.2. *According to the slope*

Emissions according to the slope are determined by multiplying the emissions of an empty vehicle without slope by a correction factor [HAS 97]:

$$ASi,j,k = A6,i,j,k\ Vk^6 + A5,I,j,k\ Vk^5 + A4,I,j,k\ Vk^4 + A3,I,j,k\ Vk^3 + A2,I,j,k\ Vk^2 + A1,I,j,k\ Vk + A0 \quad [8.4]$$

where:

– ASi,j,k is the correction factor for a pollutant i of a vehicle j over an arc k with a slope ranging between – 6% and 6%;

– Vk is the average velocity over an arc k;

– A6, ... A0 are constants for each pollutant i of a vehicle j.

The emissions according to the slope are then given by the relation:

$$EChoti,j,k = ASi,j,k * Ehoti,j,k \quad [8.5]$$

8.1.1.3. *According to the load*

In the case that interests us the load of the truck has great importance. Indeed, fuel consumption and, thus, emissions vary with the load of the vehicle. As well as taking slope into account, there exists a correction factor function for emissions according to the load [HIC 97]:

$$\Phi(\gamma,v) = K + n*\gamma + p*\gamma^2 + q*\gamma^3 + r*v + s*v^2 + t*v^3 + u/v \quad [8.6]$$

where:

– K is a constant;

– n - u are coefficients;

– γ is the gradient in %;

– v is the average speed.

The emissions are then given by the relation:

$$\varepsilon = \varepsilon u * \Phi(\gamma, v) \qquad [8.7]$$

where:

– ε is the emissions rate corrected for the load and the gradient in g/km;

– εu is the base emissions factor;

– $\Phi(\gamma, v)$ is the load correction function;

– γ is the slope in %;

– v is the average speed.

We have the emissions in grams per kilometer for our two pollutants, according to the load of the vehicle and the slope for each arc of our graph. From here on each arc of the graph is defined by average slope, distance, speed of circulation and unitary emission in g/km for each pollutant.

8.1.2. *Calculation of the minimum paths*

For the calculation we use Floyd's algorithm. This latter is carried out in three stages: the first is the creation of the matrix of minimum arcs, followed by Floyd's preparation and, finally, the algorithm itself.

8.1.2.1. *The matrix of minimum arcs*

For any arc connecting two nodes i and j we determine the one with the smallest value of the selected character. It may be the shortest distance or driving time, etc.

8.1.2.2. *Floyd's preparation*

This is a simple but important stage because it enables the transition from the concept of an arc to that of a path. For any pair of items i and j, if there is a minimum arc connecting these two nodes, then the path between i and j is implemented with the value of this arc, otherwise we assign the value 32,000 to the path, given that regardless of the pair of points, the value of the path is less than 32,000.

Thus, we have the matrix of paths which are not yet minimum, where all the values of the paths between two items i and j are defined, but not determined. This simply means that when the path between two nodes is composed of only one arc, it is at the same time defined and minimum. On the contrary, if it is composed of several arcs, its value is automatically fixed at that of the largest integer, waiting for a lower value that would replace it, if this exists after Floyd calculates them.

8.1.2.3. *Floyd's algorithm*

This is carried out in three stages:

– calculation of a path between i and j;

– comparison with the already defined value;

– replacement if the newly calculated value is less than that already existing.

The matrix obtained is called matrix of precedents, since we have not only the values of the minimum paths between all pairs of nodes i and j, but also of the index of the previous node, which is the destination node of the path, thus making it possible to find the routes.

We note by PathMin (…) the shortest time value, for example, to connect city i to city j.

We then have, according to the parameters of minimum arcs selection, four matrices of precedents:

– according to time: PathMin (i, j, time);

– according to the distance in kilometers: PathMin (i,j, distance);

– according to the minimum emission of the pollutant considered: PathMin (i, j, emission);

– according to the percentage of emission (emission indicator): PathMin (i, j, % emission).

Once we know for each arc the distance in kilometers, the time of driving through the arc according to the speed and the unitary emissions, it is easy to determine for all the minimum paths, regardless of the selection parameter, the driving time, the distance covered and the emissions of all the paths.

We can thus compare the elements of the same matrix, but also the various matrices two by two. We have various methods of comparison:

– monopolar: we study, for example, the accessibility from all the cities to city i. We can also isolate two cities to study their relations;

– multipolar: we study, for example, the accessibility from any point to all the others.

8.1.3. *Analysis of subsets*

We have developed this method to study Programs one European Spatial Planning (SPESP)[3]. It made it possible to fill in the absence of the traditional nodal approach (monopolar or multipolar) by offering a correct definition of the corridors between large geographical areas (such as the Ruhr, the plain of the river Po or Ile-de-France), on the one hand, and the variation indicators of these corridors, on the other hand. It makes it possible to compare territorial subsets between them or with the rest of the area studied by proposing their specificities, their resemblances or dissimilarities.

8.1.3.1. *Implementation of the method*

Each node is defined by a name, a code, coordinates and a number of row i in the file:

number of row

Node(77).x = 537400
Node(77).y = 2383299 } coordinates
Node(77).z = 142

Node(77).name = Chartres

Node(77).code = 2800 code of the node corresponding to the department, here 28.

The fact that a node belongs to a subset is fixed by an index. By default, the absence of an index (or 0 index) means that the nodes are elements of our space of origin, but do not belong to any subset. Hereunder we define the nodes belonging to subsets that have to be studied by indices other than zero. In the example treated later on in this chapter we chose to select two sets: the cities of the Nord-Pas-de-Calais (index 1) for the first, and index 2 for the cities in departments on the Mediterranean coast.

---

3 *Study Programs One, European Spatial Planning, Working Group 1.1 Geographical Position*, December 1999.

EXAMPLE.–

Node(5,2).x = 846400.
Node(5,2).y = 1815199.
Node(5,2).z = 48.
Node(5,2).name = Marseille.
Node(5,2).code = 1300.
Node(50,1).x = 650099.
Node(50,1).y = 2626199.
Node(50,1).z = 51.
Node(50,1).nom = Lille.
Node(50,1).code = 5900.

→ Index of the subset

The number of subsets is limited to n-2 (n being the number of nodes). Indeed, if this number is equal to n or n-1, all the nodes are element sets, bringing us back to monopolar or multipolar nodal analysis. Similarly, the number of nodes included in a set is limited. Two cases are possible:

– if there is only one subset indexed 1, then the number of elements of this subset must be strictly larger than 1 and less than or equal to n-2;

– if there are several subsets (at least two indexed 1, 2), then each set must comprise at least one element and no more than n-nx elements (where nx is the number of elements already belonging to a subset).

Let us note that we exclude the possibility of an element belonging to two subsets.

– formal statement:

- let there be a graph G = [X,U],

- X set of vertices or nodes numbered i = 1, 2, 3, etc., N,

- U set u(i, j) of ordered pairs of nodes or arcs with $i \in X$ and $j \in X$.

We define a subset as a subgraph GA = [Xa, Ua] such that $GA \in G$, with the following.

	1,1	2,1	3	4,2	5,2	N	Σ
1,1	0	Val (1,1;2,1) P (1,1;2,1)					Σ [1,1;N]
2,1	Val (2,1;1,1) P (2,1;1,1)	0					Σ [2,1;N]
3	Val (3;1,1) P (3;1,1)	Val (3;2,1) P (3;2,1)	0				Σ [3;N]
4,2	Val (4,2;1,1) P (4,2;1,1)	Val (4,2;2,1) P (4,2;2,1)		0			Σ [3,1;N]
5,2					0		Σ [5,1;N]
6							Σ [6;N]
N R						0	Σ [N;N]
Σ	Σ [N;1,1]	Σ [N;2,1]	Σ [N;3]	Σ [N;4,2]	Σ [N;5,2]	Σ [N;N]	Σ . Σ [N;N]

Σ [N, n1] (under columns 1,1 and 2,1) Σ [N, N2] (under columns 4,2 and 5,2)

where Val (etc.) is the value of the minimum path and P (etc.) is the preceding node

Table 8.1. *Matrix of precedents and subset*

where the index of the subset numbered A = 1, 2, 3, etc., is M with $1 \leq M \leq N - 2$ if $G1 \cap G2 \cap G3 \cap, ... \cap GM = \varnothing$[4].

And $Xa \in X$ the set of nodes of the subgraph numbered a = 1, 2, 3, etc., n, with:

If A = 1, then $1 < a \leq N - 2$.

If A > 1, then $1 \leq a \leq (N) - \sum X a$.

where $\sum Xa$ is the sum of the elements already belonging to a subset.

Each subgraph defined in this way is included in the graph of the original transport network with the same typology of infrastructure. The analysis of several subgraphs highlights relations included in the original graph.

We thus have our various territorial subsets of our original space, upon which we employ Floyd's algorithm to obtain the following matrix of minimum paths (see Table 8.1). In this matrix we observe the minimum paths from a point to any other point, which are organized in three groups: from light gray to dark gray:

– the first: relations of set 1 that are either at the destination or at the origin of all the other nodes;

– the second: relations of set 2 that are either at the destination or at the origin of all the other nodes;

the third: relations of set 1 that are either at the destination or at the origin of all the nodes of set 2.

In the case of pollutants emissions, $\Sigma[n,n1]$ and $\Sigma[n,n2]$ are the sums of the emissions of each path bound for the nodes of the sets 1 and 2, $\Sigma[n1, n]$ and $\Sigma[n2, n]$ are the sums of the emissions of the paths originating at the nodes of sets 1 and 2 and bound for all the nodes of the graph. In our example we defined two subsets but more can be created. We may choose to analyze the relations between a subset and the rest of the space (analogy with the method of traditional monopolar analysis) or of several subsets, either between them or with respect to the whole of the graph.

The relations between the various elements of our graph can be defined according to four strategies of movement between two nodes.

---

4 In the example developed below we decided to prohibit the intersection of subsets. However, it is possible for an element to belong to two subsets. This possibility has the advantage of enabling us to widen the type of analysis suggested by comparing the relevance of an element – or a group of elements – belonging to a particular subset, for example, from the point of view of accessibility.

8.1.3.2. *Indicators used or choice of the selection parameter*

The objective being to quantify the emissions for the arcs, on the one hand, and for the paths, on the other hand, we chose to select:

– minimization of driving time;

– minimization of the network distance in kilometers;

– minimization of the emission for a given pollutant and type of vehicle (unitary emission in g/km multiplied by the distance covered);

– minimization of the emission of all the pollutant together for a given type of vehicle = synthetic indicator.

8.1.3.3. *Why a synthetic indicator?*

As we will see later on, pollutants do not exhibit the same behavior with respect to slope, load or speed. The least polluting path for carbon monoxide is not necessarily the least polluting path for nitrogen oxide. Therefore, how do we determine, if a path is more polluting overall than another? It is not enough to add up the quantities, since CO or NOx emissions are to the order of grams per kilometer, whereas the emissions of CO_2, for example, are in the order of hundreds of g/km. For this reason the following synthetic indicator is proposed:

For each pollutant we have the cumulated emission in grams per arc (unitary emission in g/km multiplied by the network distance of the arc):

– let Ei, j, k be the emission of pollutant i of the vehicle j over the arc k in grams;

– regardless of k $\rightarrow$ number of arc, $\exists$ at least an arc with the strongest emission of the pollutant noted by Emax i, j;

– let PEi, j, k be the proportion of the emission of pollutant i of the vehicle j over an arc k with respect to Emax i, j such that:

- regardless of k $\rightarrow$ number of arc, PEi, j, k = Ei,j,k / Emax i,j,

- we note by %Ej, k for an arc k and a vehicle j the percentage of emission of all the pollutants together such that:

- %Ej, k = (PEco,j,k + PEnox,j,k) * 100 / Nbi, where Nbi is the number of pollutants considered.

We then obtain for all the arcs of our graph the percentage of emission of all the pollutants together with respect to the maximum emissions of each pollutant without taking into account the unit of mass. Consequently, we can determine the paths with the minimum emission of all the pollutants together.

8.2. Results

We will start with simulations on the section of the A28 to highlight the relations between emissions, speed, slope of the arcs and load of the trucks. It will then be easier to identify the links between graph and emissions, first of all, for movement between two cities and then for relations between two subsets.

8.2.1. *Section of the A28*

The calculation of emissions requires certain modifications of the profile of the A28. The values of slopes were limited so as to lie between – 4% and + 4%, in order for the network speed defined beforehand to agree with the speed limits adopted in the formulae. From the digital terrain model (Figure 8.2) we work out the profile of the section, which is composed of 40 arcs defined by two nodes, a slope and a length. We then calculated the emissions for each arc (for various initial conditions).

The results presented below represent the emissions for the section in its totality according to the direction of movement – direction 1: movement from left to right, direction 2: movement from right to left– of the slope of each arc, speed (constant for all the arcs) and of the load.

Tables 8.2 and 8.3 recapitulate the cumulative emissions on the section of 2.9 km for three speeds.

Emission of co of an empty vehicle and without slope at 70 km/h	1.95 g/km 5.67 g
Emission of co of an empty vehicle and without slope at 55 km/h	2.305 g/km 6.7 g
Emission of co of an empty vehicle and without slope at 40 km/h	2.876 g/km· 8.36 g
Emission of NOx of an empty vehicle and without slope at 70 km/h	2.251 g/km 6.55 g
Emission of NOx of an empty vehicle and without slope at 55 km/h	2.291 g/km 6.66 g
Emission of NOx of an empty vehicle and without slope at 40 km/h	2.929 g/km 8.52 g

Table 8.2. *Emissions of empty vehicles and without slope*

We observe in these tables that the effects of the variables measured on CO or NOx emissions are the same:

– regardless of slope or load, the more the speed increases, the more the emission decreases;

– there is over-emission or under-emission with respect to the emission of empty vehicles and without slope when we take slope and load into account;

– the emissions are different according to the direction of circulation.

On the other hand, the impacts of speed, slope or load do not have the same intensity. Table 8.4 shows the variations of emissions according to speed.

	Function of slope: direction 1	**Function of slope: direction 2**	**Function of slope and load: direction 1**	**Function of slope and load: direction 2**
Emission of CO at 70 km/h	5.124	6.398	5.337	6.871
Emission of CO at 55 km/h	6.397	7.76	6.813	8.526
Emission of CO at 40 km/h	8.376	9.906	9.028	11.02
Emission of NOx at 70 km/h	5.715	8.352	6.77	10.26
Emission of NOx at 55 km/h	6.199	8.909	7.5	11.19
Emission of NOx at 40 km/h	7.981	10.844	9.685	13.725

Table 8.3. *Emissions (in grams) according to slope and load*

CO	Zero slope	Negative slope	Positive slope
Difference between 55 and 70 km/h in %	+ 18	+ 30	+ 19
Difference between 40 and 55 km/h in %	+ 19	+ 35	+ 26
NOx			
Difference between 55 and 70 km/h in %	+ 1.7	+ 13	+ 6
Difference between 40 and 55 km/h in %	+ 27.5	+ 41	+ 15

Table 8.4. *Variation of emission depending on the slope, in %*

This table indicates that for the same slope the effect of the reduction of the speed on the emissions is more marked during movement down a negative slope. However, the emission in grams per kilometer is greater for positive slopes than for negative ones. In addition, we can see that CO seems more sensitive to speed, whereas NOx seem more sensitive to slope (Table 8.5).

For CO, the emissions for a negative slope are lower than those for a zero slope, even more so when speed increases. Conversely, the emissions increase for positive slopes compared to zero slopes – as speed decreases. This behavior is easily understood through the study of forces applied to a mass moving on a tilted plane. The CO emissions are proportional to fuel consumption. On the other hand, the behavior of NOx is more ambiguous. However, we can say that the slope of the road affect the emissions of NOx more than those of CO.

CO	70 km/h	55 km/h	40 km/h
Difference between negative slope and zero slope	– 28	– 20.8	– 13.9
Difference between positive slope and zero slope	+ 31	+ 33	+ 35
NOx			
Difference between negative slope and zero slope	– 45	– 39	– 32
Difference between positive slope and zero slope	+ 70	+ 68	+ 56

Table 8.5. *Variation of emissions according to slope, in %*

Hereafter the slope – and thus the profile – developed here diverges during calculation of the emissions on the graph. Indeed, the slope associated with an arc will be calculated on the basis of altitudes of the nodes of the arc, and we will associate the path between an origin and a destination to a profile. For this reason Table 8.6 provides, on the one hand, the emissions according to an average slope calculated for the first and the last node of the profile and, on the other hand, the average emission – value in brackets – calculated for the emissions of each arc.

CO	At 70 km/h	At 55 km/h	At 40 km/h
Direction 1: slope of – 0.2 %	1.4 (**1.76**)	1.82 (**2.2**)	2.47 (**2.88**)
Direction 2: slope of 0.2%	2.56 (**2.2**)	3.06 (**2.67**)	3.86 (**3.41**)

Table 8.6. *Variation of emissions with load*

When we calculate the emissions for an average slope we observe in our profile that emissions are underestimated in the descending direction of the profile and overestimated in the ascending direction. Indeed, during calculation with the average slope we consider a constant slope for the entire profile without taking into account the rises and descents that may follow one another.

Thus, the behaviors of CO and NOx are different according to the pair (speed, slope). Moreover, it appears that CO is less sensitive to the slope than NOx. This explains the difference in emissions between direction 1 and direction 2. As a result, for our profile a movement from left to right (direction 1) can be described as downward movement, whereas from right to left (direction 2) we have an ascending movement.

at 70 km/h	CO	NOx
p = 0	+ 3.5%	+ 14%
p < 0	+ 3.5%	+ 13%
p > 0	+ 18%	+ 41%
at 55 km/h	**CO**	**NOx**
p = 0	+ 6.5%	+ 17%
p < 0	+ 4.3%	+ 13.6%
p > 0	+ 21%	+ 44%
at 40 km/h	**CO**	**NOx**
p = 0	+ 8%	+ 19%
p < 0	+ 5.6%	+ 15.3%
p > 0	+ 22.7%	+ 45.8%

where p is the slope

Table 8.7. *Variation of emissions according to the slope in %*

CO	70 km/h	55 km/h	40 km/h
Difference between negative slope and zero slope	– 28	– 22	– 15.8
Difference between positive slope and zero slope	+ 50	+ 51	+ 52
NOx			
Difference between negative slope and zero slope	– 45	– 41	– 35
Difference between positive slope and zero slope	+ 99	+ 107	+ 91

Table 8.8. *Variation of emissions according to the slope in %*

For both pollutants there is over-emission at full load (it is stronger for NOx), which is amplified for a positive slope (Table 8.7). The difference between vehicles with or without load on the same slopes increases slightly with the reduction of the speed. We also notice that the percentage of the slope accentuates the effect of load on the emissions (Table 8.8), especially in the case of NOx. The latter are more sensitive to the load of the vehicle than CO.

Taking into account the load of the vehicle does not modify the behavior of the pollutants, however, it intensifies the behaviors of the pollutants.

With this first approach we show that emissions are specific to the profile, according to the origin and destination of the movement and depending on the pollutants considered. We have also shown the sensitivities of the pollutants with respect to the initial conditions (speed, slope and load). We will now consider the case of graphs. However, before proceeding it is necessary to reflect on the relations of the trinomial composed of relief, infrastructure and the graph used.

We saw previously that the method used below underestimates emissions for negative slopes and overestimates emissions for positive slopes. Consequently, our approach can appear satisfactory in areas that are not too uneven (Rhone valley, Landes, etc.) and much less so in more hilly areas.

In addition, maximum slopes vary depending on the road infrastructure used. For highway structures, they are minimized with respect to the relief. Moreover, the kinematics of driving are very different for different types of infrastructure; on highways speeds are very close to the average speed, as opposed to trunk roads, for example, where speeds fluctuate more (entering an agglomeration, passing through a village, etc.). We deduce from it that emissions are better estimated for highway structures, especially if the profile is not very uneven. It is important to dissociate relief and profile in the case considered (the estimate of the emissions is finer for alpine highways which follow the bottoms of valleys than for hilly areas where descents and ascents follow one another). Consequently, the graph used is dominating. A simple but paramount question needs to be posed: how has the graph been elaborated? This question does not refer to techniques of digitalization or modeling, but rather to its later use: will we calculate accessibilities, plot Chronomaps [LHO 97] or estimate emissions? It should be specified that the graph used here was not developed to calculate emissions of pollutants from trucks, but for calculations of driving time. It was slightly modified for this new use.

8.2.2. *French graph*

Having studied the behaviors of pollutants for a profile we will generalize on a graph in two stages. The first consists of studying the Paris-Marseille and Paris-Toulouse routes, whereas the second is between two geographical subsets.

We do not have the profile along each arc and, nevertheless, we associate a path (or a route) to a profile.

Table 8.9 below shows the emissions of CO and NOx for the return journey according to three criteria of minimum paths selection.

	Paris-Toulouse		**Paris-Marseille**		**Toulouse-Paris**		**Marseille-Paris**	
	CO	NOx	CO	NOx	CO	NOx	CO	NOx
time	1.39	1.63	1.51	1.81	1.47	1.81	1.76	2
mileage	1.88	1.82	1.54	1.66	1.51	1.83	1.76	2.11
emissions	1.3	1.55	1.46	1.55	1.45	1.5	1.55	1.65

Table 8.9. *Cumulated emissions of CO and NOx in kilograms*

Figures 8.3 and 8.4 present the minimum paths bound for Toulouse and Marseille and bound for Paris, as well as the carbon monoxide emitted over each arc by a vehicle. Several comments can be made concerning carbon monoxide:

– the traffic corridors used are naturally the same (with different means) except for the journey from Marseille to Paris according to the emission;

– the south-north direction is more polluting;

– generally speaking, finding the optimal routes according to the minimum emission amounts to optimizing the driving time except for the journey from Marseille to Paris;

– the emission according to mileage is the most polluting.

These observations are easily understood. According to time we optimize the route according to the pair speed-distance, whereas according to mileage we optimize only the distance and for the emission we try to optimize the trio speed-distance-slope.

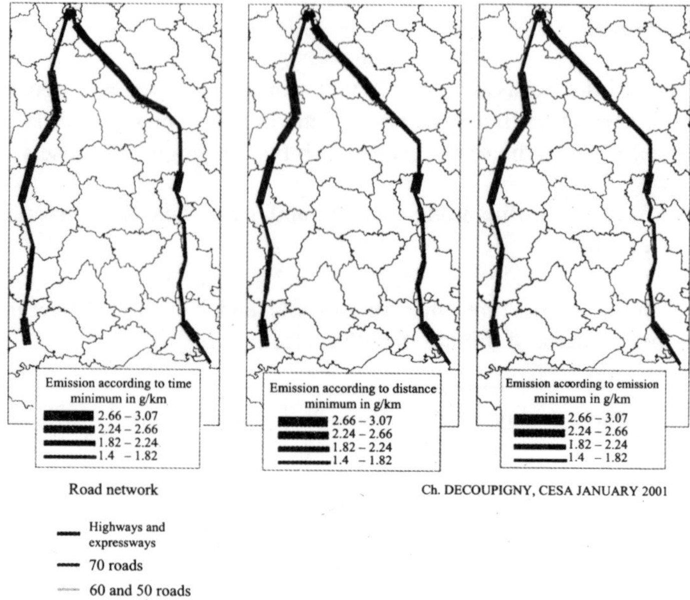

Figure 8.3. *Emission of CO per arc, of trucks lighter than 7.5 t and minimum paths, north-south direction*

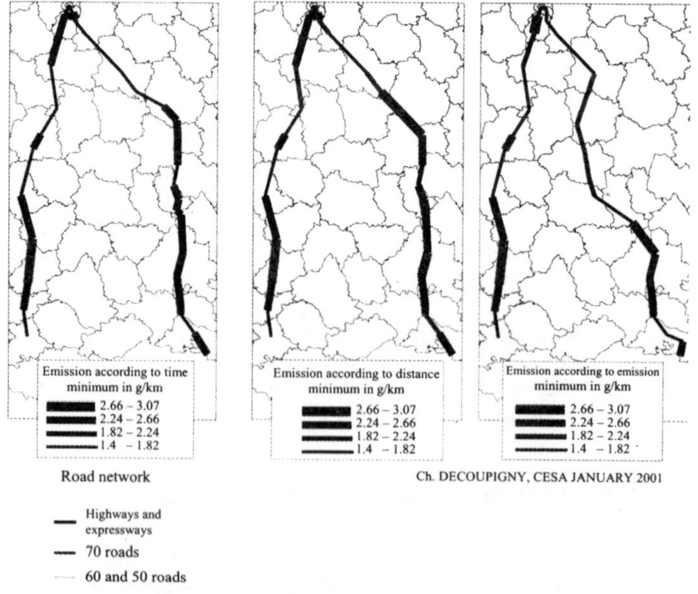

Figure 8.4. *Emission of CO per arc, of trucks lighter than 7.5 t and minimum paths, south-north direction*

Figures 8.5 and 8.6 show the minimum paths for the return journey, as well as the emissions of nitrogen oxides per arc. We observe that:

– the traffic corridors used are the same, as expected, for different means in both directions except in the North to South direction according to emission;

– contrary to CO, the difference between paths according to minimum emission and to time is a lot larger;

– the North-South direction is more polluting;

– the emission according to mileage is the most polluting.

There is greater variability of routing for NOx than for CO. We saw above that NOx is more sensitive to the slope than CO. We realize in this example that NOx is more sensitive to the trio speed-distance-slope than CO. NOx is thus more sensitive to the structure of the network and the initial conditions than CO.

By observing the unitary and cumulative emissions values for each path it appears that there exists little difference between an ecological approach seeking to minimize emissions and a transport approach seeking to minimize travel time (in particular for CO). We find here in a general manner the results presented for the profile of the A28.

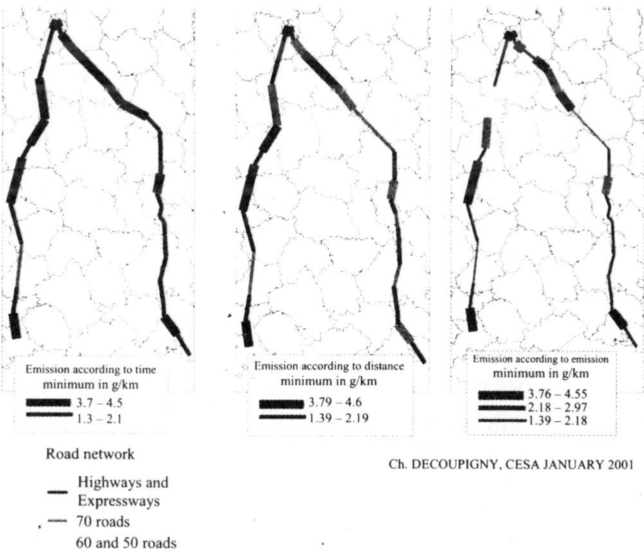

Figure 8.5. *Emission of NOx per arc of trucks lighter than 7.5 t and minimum paths, north-south direction*

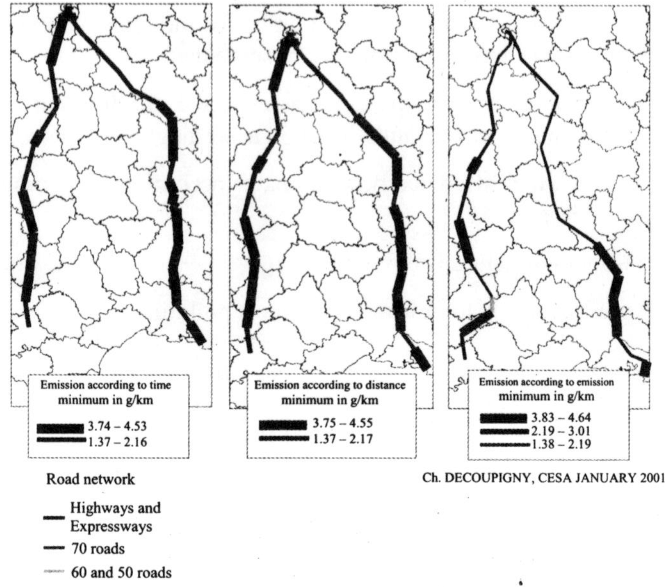

Figure 8.6. *Emission of NOx per arc, of trucks lighter than 7.5 t and minimum paths, south-north direction*

The contribution of graphs to this case makes it possible to locate the most polluting arcs and to highlight the heterogenity of emissions with respect to the network. According to the latter, the origin and the destination of displacement we obtain different emissions. It is on a local scale that the contribution of graphs will be the most relevant. This local scale may be integrated and thus analyzed, in particular, due to nodal zoom, which makes it possible to include a much more precise level n-1 graph into a level n graph [CHA 97].

8.2.3. *Subset*

Figures 8.7 and 8.8 show the relationships between the cities of the Nord-Pas-de-Calais and those on the Mediterranean coast. Four pieces of information are illustrated:

– the number of minimum paths per arc used;

– the quantity of pollutants (in kg) emitted per arc by all the vehicles using this arc;

– the unitary emission (in g/km) is the average of the pollutant emitted by a vehicle along the route starting at the elements of a subset and bound for an element of the second subset;

– the cumulated emission (in kg) is the sum of the emissions (in g) of CO or NOx of a vehicle along the route starting at the elements of a subset and bound for an element of the second subset.

We realize that paths with destination or origin in the south via Paris are identical to the paths found in the previous case; this shows that the study of subsets fits in well with the whole of the graph.

If we compare, for example, the CO emitted over the territory with the destination of Alès and Perpignan according to time, the cumulated emissions with the destination of Perpignan (28 to 30 kilograms) are higher than those for Alès (26 to 28 kilograms), since the kilometric accessibility of Ales is better. However, it is more polluting to go to Alès than to go to Perpignan. Indeed, taking into consideration unitary emissions, a vehicle will discharge on average two grams of monoxide per kilometer during its journey between a northern city and Alès, whereas this emission is only 1.8 grams per kilometer for Perpignan. This observation is valid regardless of the direction of movement, on the one hand, and the criterion of minimum paths selection, on the other hand. Therefore we demonstrate the smoothness of the connection between network quality and emission. The interest here is not to realize that going to a point A is more polluting than going to a point B, but rather to visualize using routes where the emissions are likely to be the highest. We, thus, clearly see the importance of the network for emissions.

In the case of a temporal approach to the minimum paths, there exists a main traffic corridor circumventing the Paris area via the A26 and the A31 and going through the Rhone valley. According to minimum CO emission, in the north-south direction there is a main corridor passing by Paris and following the Rhone valley and two secondary corridors, the first skirting around Paris, as previously, and joining the main corridor in Beaune and the other passing by Limoges. In the south-north direction there are also one main and two secondary corridors. The latter are very different from those found in the north-south direction. Moreover, we observe that according to driving time, the connections originating at or bound for Toulouse are achieved exclusively via Limoges, whereas other routes pass through the Rhone valley. This exclusiveness is not found in routes laid according to minimum emission. In the case of a bilateral connection, by considering emissions and the frequency of use of the arcs, it appears that with a temporal approach the emissions concentrate on only one main traffic axis, whereas there is a distribution of emissions over three large circulation axes according to the minimum emission.

Finally, the calculation of minimum paths according to time or minimum emission gives very similar emissions. On the other hand, the relationships between the two subsets resulting from the two approaches are very different, with very marked local consequences. We show therefore that these emissions have a strong sensitivity, and in particular those of NOx, to the initial conditions (speed, slope, distance), i.e. to the graph.

This traditional approach of graph theory by calculating emissions according to the slope between two nodes does not take profiles into account. The main problem is linked to the lack of reference to the profile of the roads. However, how can we take into account the profiles for a graph of national or international transport network without weighing down the representation or calculations of accessibility, for example?

In a graph each node is defined by coordinates (x, y, z) and a code. We propose to introduce the degree of the node (the degree of a node is the number of graph edges incidental to the node). From the profile of a road (or a digital terrain model) it is possible to create a node of the second degree at each point of inflection of the profile; these nodes are neither crossroads nor cities (which need to have a degree equal to or higher than 3). The aim is to transform the arcs into 3D polylines with ends at nodes of a degree equal to or higher than 2 and with nodes of the second degree as intermediate or inflection points. These polylines make it possible to take into account the profile of a road by calculating the absolute values of the slopes for each segment of a polyline.

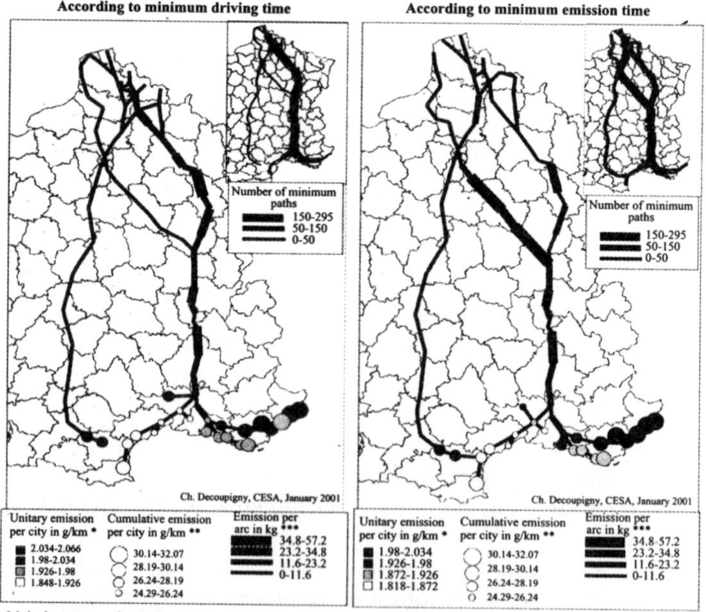

* it is the average of emissions in g/km emitted over a territory with an origin at every northern city and bound for the city considered

** it is the sum of emissions in g (emission in g/km * distance to cover) emitted over a territory with an origin at every northern city and bound for the city considered

*** it is the product for each arc of emission in g/km * the length of the arc in km * number of minimum paths per arc

Figure 8.7. *Emission of CO and minimum paths starting at Nord-Pas-de-Calais and bound for cities in the South*

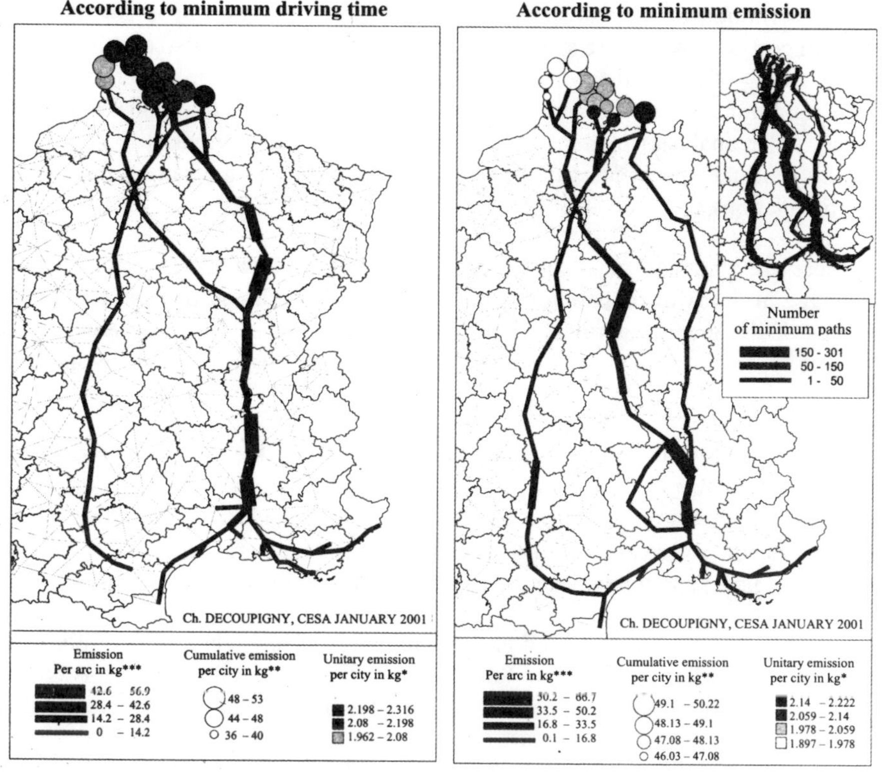

Figure 8.8. *Emission of CO and minimum paths starting at the southern cities and bound for Nord-Pas-de-Calais*

One of the advantages of introducing the degree of nodes is to be able to simplify the graph. During modeling it is easy to select only the nodes of a degree equal to or higher than 3 so as not to weigh down certain calculations.

Thus, we justify the advantage of using graph theory in the study of the emissions of the pollutants discharged from transport, in particular, from the prospective point of view. It is all the more justified when we introduce the slope factor into the calculation of emissions, which then prohibits the traditional evaluation of emissions (emission in grams per kilometer multiplied by the annual mileage of a vehicle). Indeed, the raw data on emissions between several cities, for example, does not make it possible to localize emissions and does not provide us with information concerning the spatial heterogenity of emissions.

The definition of subsets makes it possible to determine traffic corridors and emission corridors which are specific to the analyzed subsets. It also enables a better understanding of the modifications of the relations resulting from the change of the initial conditions, which are impossible to highlight by using a traditional monopolar or multipolar analysis (interspecific relations). Moreover, it makes it possible to shed light on the specificities of the elements belonging to the same subset (intra-specific relation).

Therefore, the use of graph theory in the calculation of emissions has an advantage in terms of modeling, on the one hand, (calculation of emissions for all the arcs of the graph regardless of the initial conditions or the study area) and in terms of representation, on the other hand since it makes it possible to spatialize emissions.

The cold emissions were not covered in this chapter; the data on this subject is scarce, although just as for cars there is over-emission for cold trucks. If we consider the cycles of driving imposed on the driver, it becomes clear that after nine hours rest there is cold over-emission. It is possible to locate the arcs where there is a stop (for the rest hour) and to evaluate the cold over-emission resulting from it.

Nevertheless, the method used has its limitations. They relate, first of all, to the graph and, secondly, to emissions formulae. For the graph a directed arc is defined by the nodes origin and destination, which is a means that is associated with a constant average speed over the arc according to a vehicle type, a distance and an average slope defined by the altitude of the origin and destination nodes. A route along the arcs does not correspond to the real profile (however, we saw that this problem could be solved by integrating polylines between two nodes of a degree which is equal to or higher than 3). There can also be variations in speed at the access to large urban conglomerations or cities (which can be approximated using nodal zoom). Indeed, we work with arcs by considering nodes only as origin and destination of the journey or as intermediate cities. However, there is a great diversity of relations between a node (a city) and the transport network, such as, for example, the existence or absence of a ring road, which introduces large differences at the local level. The sinuosity of roads has also not been taken into account, in particular, for minor roads.

The second problem relates to emission formulae and the lack of data on trucks. Most of the data is obtained for a test bed with standardized cycles and not *in situ*. However, this method has the merit of offering a first approximation by yielding minimum emissions according to the quality of the network (Figure 8.9 in the color plates section).

8.3. Bibliography

[CHA 97] CHAPELON L., Offre de transport et aménagement du territoire: évaluation spatio-temporelle des projets de modification de l'offre par modélisation multiéchelles des systèmes de transport, Thesis, Tours, 1997.

[HAS 97] HASSEL D., WEBER F.J., "Gradient influence on emission and consumption behaviour of light and heavy duty vehicles", delivery 9 of the MEET project, 3 TUV Rheinland, Cologne, Germany, 1997.

[HIC 97] HICKMAN A.J., "Emission functions for heavy duty vehicles", delivery 10 of the MEET *project, Project report SE/289/97*, Transport Research Laboratory, Crowthorne, United Kingdom, 1997.

[HIC 99] HICKMAN J., HASSEL D., JOUMARD R., SAMARAS Z., SORENSON S., "Methodology for calculating transport emissions and energy consumption", *Report SE/491/98*, Transport Research Laboratory, Crowthorne, UK, 1999.

[LHO 97] L'HOSTIS A., Image de synthèse pour l'aménagement du territoire: la déformation de l'espace par les réseaux de transport rapide, PhD Thesis, Tours, 1997.

[SER 00] SERRHINI K., Evaluation spatiale de la covisibilité d'un aménagement, sémiologie graphique expérimentale et modélisation quantitative, PhD Thesis, Tours, 2000.

PART 3

Towards Multilevel Graph Theory

Chapter 9

Graph Theory and Representation of Distances: Chronomaps and Other Representations

9.1. Introduction

Born from the resolution of practical questions which cannot be easily solved by using graphs (such as the problem of crossing the bridges of Königsberg (Figure 9.1), which was solved by Euler, or the problem of the four colors, which was introduced for the first time by the cartographer Guthrie in 1852), graph theory constitutes a mathematical framework that makes it possible to tackle problems in a very vast field.

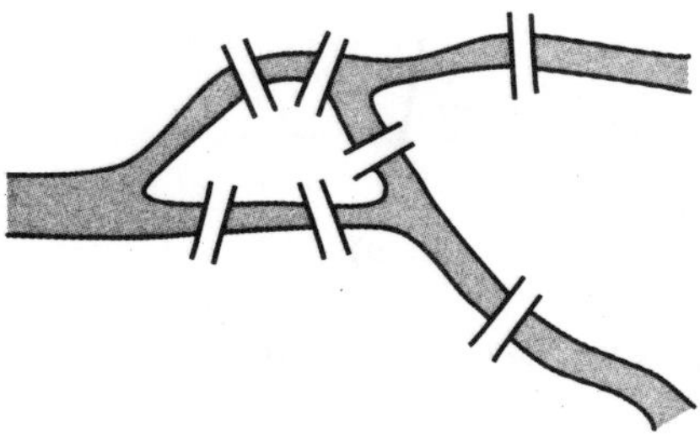

Figure 9.1. *Bridges of Königsberg*

Chapter written by Alain L'HOSTIS.

In the domains of geography, urban development and spatial planning, graph theory is used to tackle questions arising in the field of networks (whether they are transport and communication or even social networks). The general procedure is based on the translation of these problems into the mathematical language of graph theory so as to be able to treat them using, for example, the properties of connectivity or by exploring minimum paths. For this reason, in its applications to the social sciences, graph theory deals mainly with mathematical properties and much less with the graphic representation of nodes and arcs[1]. By refocusing on the graphical dimension, from which it was born, here we wish to develop the aspect of graph theory that relates to representation (plotting of graphs) in its applications to problems of spatial analysis.

Since graph theory is particularly adapted to the modeling of transport networks, the representation of the network graph may constitute an investigative direction that can answer the difficult question of the representation of distances.

A classical illustration of the difficult problem of the representation of distances is provided by Müller [MUL 79, p. 216]. In Figure 9.2 we laid out three cities noted by p, q and r. There is a highway connecting the cities q and r, and another between q and p, whereas r and p are connected by a small road. The shortest path by duration between the two latter cities passes by the city q and follows the highway. The question raised by Müller is as follows: "How is it possible to determine, on a map with a time scale, the position of points p, q and r when the shortest path from one point to another is no longer a straight line?" [MUL 79, p. 215].

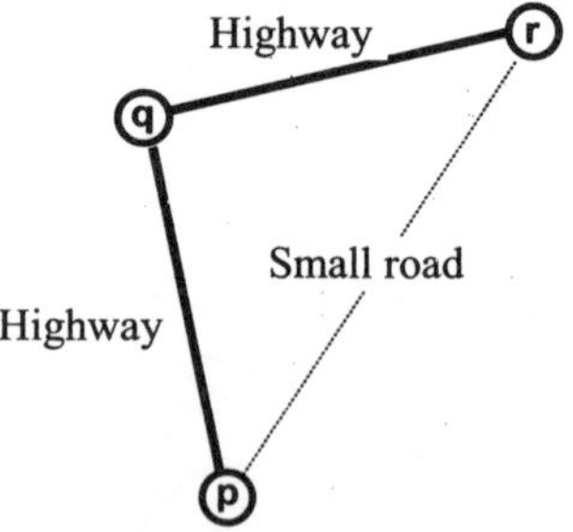

Figure 9.2. *Representation of the shortest paths*

1 When we study graph theory, one of the few properties that relates to their representation – their plotting – is the concept of a planar graph. A graph is known as planar if it can be drawn on a plane so that the nodes are distinct points, the edges are simple curves and no two edges meet apart from at their ends.

The straight line almost never corresponds to the general form of movement, i.e., to the general form of geographical distance. Consequently, as Angel and Hyman put it: "if we wish to study transport within a geometrical framework, we must develop non-Euclidean geographical theories" [ANG 72, p. 366]. In this spirit we will start by establishing a general form of distance on a graph on the basis of the basic definitions of graph theory. The goal is to give a mathematical form to a geographical measurement: the distance built in this fashion will make it possible to find support in mathematics to better envisage the spatial dimension associated with transport and movement and, finally, to question the cartographic representations used in geography.

To deal with the domain of representations we will create a method for reading distances on a map. The comparison of these measurements which are read with those provided by the geographical data will make it possible to characterize the various representations according to their capacity to represent geographical distances.

9.2. A distance on the graph

Mathematical distances are measurements between objects that respect the properties of positivity, uniqueness, separation and triangular inequality. From the mathematical point of view as well as from the point of view of geographical interpretation, one of the most fundamental properties of distances is that of triangular inequality because it introduces the idea of the minimum [LHO 97, p. 102 and 113]. The distance between two objects is associated with the measurement of the smallest gap that can be found. To illustrate this principle, on Müller's diagram the distance between points p and r can have two forms: if we measure it in kilometers, the distance is given by the small road, whereas, if we calculate it in duration of transport, the distance follows the highway route passing through point q.

In graph theory we have a particular application which associates a value to the shortest path: if a and b are two nodes of a graph G, the measure e (a, b) is given by the length of the shortest path between a and b.

If we consider the case of a valuated graph[2], the length of a path corresponds to the sum of the values associated with the arcs, of which it consists. With the measure e, we construct an application that associates a value in R to a pair of elements of the set of nodes S. It should be noted that in this definition there is no

2 In a valuation graph a numerical value is associated with each arc.

specification of the mathematical properties of the application. In particular, it is not specified, if the measure e is metric, i.e., if it respects the properties of distances.

In the general case of a non-valuated graph Flament indicates the construction of a metric on the basis of the measure e [FLA 68, p. 39]. The only condition for the graph is that the value associated with a link between two nodes is identical in each direction. The graph must therefore be symmetric. The measure between one point and another is considered zero. Defined on a symmetric non-valuated graph, with conventions for infinite and zero paths, the margin is metric.

On a valuated graph in order for the measure e to be a distance in the mathematical sense, we must verify one by one the metric properties (symmetry, positivity, separation and triangular inequality) for any pair of nodes. That leads to two conditions, which make it possible to establish a graph distance [LHO 97, p. 137]: the graph must be symmetric and for a valuated graph the values associated with the arcs must be strictly positive.

The distance criterion of minimum paths calculation can integrate the duration of transport, the cost, the length of the journeys, but also the various modes of transport (with a quality of connection that may or may not be homogenous) or even the nature of what is transported. If the graph considered is not symmetric, we can construct only a non-symmetric metric, a structure that belongs to the more general field of impoverished metrics[3]. Although non-symmetry is the general case for geographical distances (in time or cost) we will limit ourselves here to the analysis of symmetric distances by considering a simplification of the collected data.

The distance defined in this fashion is purely mathematical: it is only the connection between pairs of places with a measurement that takes the form of a measure e. It is a mathematical object that characterizes geographical data, but *a priori* does not have immediate and univocal translation in the sphere of cartographic representations.

9.3. A distance on the map

Although a geographical map primarily provides information about the location of places, it also informs us about the relations between these locations. In particular, reading a roadmap makes it possible to work out routes and determine the distances associated with them.

3 For more on these structures and their use in spatial analysis see [HUR 90, p. 225].

To define the function of distance that we have in a map Müller [MUL 79, p. 224] introduced the concept of "graphic distance". This distance is defined in a Euclidean form: we consider that the length of a path between two points on the map is directly proportional to the length of the rectilinear segment which connects them. However, movements carried out in the geographical space use transport infrastructures that have very little chance to follow a straight line between two places chosen at random. Thus, in order to measure a distance in kilometers on a roadmap the reader must transform the graphic distance mentally to obtain a useful measurement: the sinuosity of the road and the extent of detours along the way make it possible to correct, by scaling up, the Euclidean length of a route. In this context the Euclidean distance offers a first approximation of measurement. It is then adjusted by information about the infrastructure used. In fact, the concept of distance "as the crow flies", on which Müller's graphic distance is based, cannot be used to read a cartographic representation if we wish to understand the logic of movements.

For this reason it is necessary to introduce a different concept, that of "visual length" [LHO 97, p. 137], which is derived from Müller's graphic distance. The visual length of a connection between two points is equal to the length of the corresponding path. If, for example, the path takes the form of several segments placed end-to-end, the visual length corresponds to the sum of the lengths of these segments. Visual length is, thus, not a Euclidean concept. It should be noted that, according to this definition, visual length does not exist between two points unless there is a path. Visual length makes it possible, in particular, to provide a measurement of the routes and paths on the basis of a graph plot, i.e., on the basis of the map of a transport network.

By abandoning graphic distance in favor of visual length we lose the reference to the metric. The motivation of this choice becomes apparent when we refer to Müller's diagram (Figure 9.2). The visual length of the connection between p and q corresponds to the distance between p and q, if it is measured in kilometers. However, that is not true if the distance is calculated in journey duration. In this case the visual length read on the map along the small road is not a distance since it does not correspond to the minimum possible measurement in space-time.

By uncoupling the concepts of distance (geographical concept) and visual length (cartographic concept) we pave the way for a better understanding of the representations of distance.

Thus defined, the concept of visual length calls into question the usual use of map scale. Indeed, on a traditional geographical map the scale has the function to provide measurements of distance in kilometers as the crow flies: it is clearly a Euclidean notion of distance. This use of scale implies the postulate of verifying the isotropy and the uniformity of the areas represented: two properties that cannot be regarded as given when geographical spaces are dealt with.

Visual length establishes a measurement which, converted by using the scale of representation, produces a geographical distance. It is a non-Euclidean use of the map scale. A cartographic representation of distances is considered to be satisfactory, if the distances read are coherent with the initial data, that is, with geographical distances.

For Bunge there are two manners of representing distances: we may show on a traditional map the paths that can be sinuous, or we may represent distances in a simplified form but on a deformed map [BUN 66]. The second approach refers to the anamorphoses introduced by Tobler [TOB 63, p. 59-78] and consists of moving the locations of places according to distances[4]. This second approach is the one that interests us because it directly involves the representation of the transport network.

9.4. Spring maps

To illustrate the impact of the modes of transport on spatial relations Tobler considers the representation of distances in a mountainous zone [TOB 97]. The initial data set consists of measurements of real distances in kilometers that often follow sinuous routes. The degree of circuity of the associated graph (which is defined as the measurement of the difference between the distances in kilometers and the distances as the crow flies [KAN 63, p. 93-121]) is thus very high[5]. The spring map (Figure 9.3) is constructed on the basis of a set of places which are positioned at their real location [CAU 84, p. 40], i.e., their position on the plane of the map is not modified by anamorphoses.

The principle of spring representation is to depict the link existing between two points in space in the shape of a spring whose length is proportional to the length in kilometers. By reference to the Euclidean distance, connections which are longer than those "as the crow flies" are represented in the shape of a generally compressed spring. This graphic principle makes it possible to represent paths which are longer than the straight line without modifying the location of the ends.

To read such a map, it is necessary to get rid of our Euclidean reflexes because the distance between places is indicated by a path which is not a straight line. This distance can be understood only by using visual length because the graphic distance defined by Müller results in Euclidean measurements. The map is read by evaluating

4 On this subject see [CAU 96] (www.cybergeo.presse.fr).
5 For developments and analyses of the degree of circuity see [CHA 97, p. 330].

the intensity of the compression of the springs: the more compressed the spring is, the more difficult the connection (in other words it contains more turns and slopes due to the mountainous terrain, which lowers the average speed) and the more it moves away from the straight line. The representation is built around a direct reference to the Euclidean distance, but a reference that is implicit. Here we find the status of the Euclidean space as proposed by Cauvin, that is, as a reference space [CAU 84, p. 72].

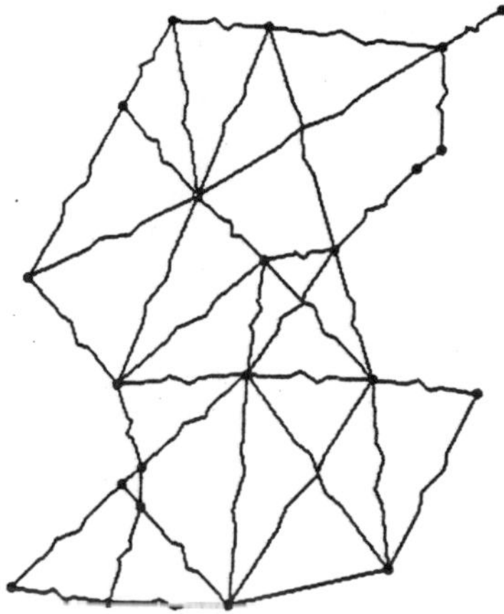

Figure 9.3. *Spring graph of the road network of Western Colorado*

The spring graph may be seen as the plotting of a graph. The graph is that of the transport network considered, which is valuated by the measurements taken for the network, and its plotting undergoes three constraints:

– the nodes of the graph are placed at the geographical position of the nodes, which they model;

– the length of the arcs of the graph is proportional to the associated value;

– the arcs of the graph are drawn in the shape of springs which ensure the proportionality of the lengths.

The distances in kilometers are considered here to be symmetric and they define a graph distance. The construction principle of the spring map makes it possible to establish a representation which is coherent with the associated mathematical structure. Therein lies the main merit of this type of map, which is to offer a representation of distances that remains coherent with real measurements, although they elude the Euclidean domain.

On the basis of the measurements represented on the spring map, Tobler then proposes three-dimensional anamorphoses. The adopted principle is to slide the location of the places on a vertical axis to stretch the distances "as the crow flies" in three dimensions. A regression using least squares makes it possible to find a configuration which minimizes the cumulated error for all the links (Figure 9.4).

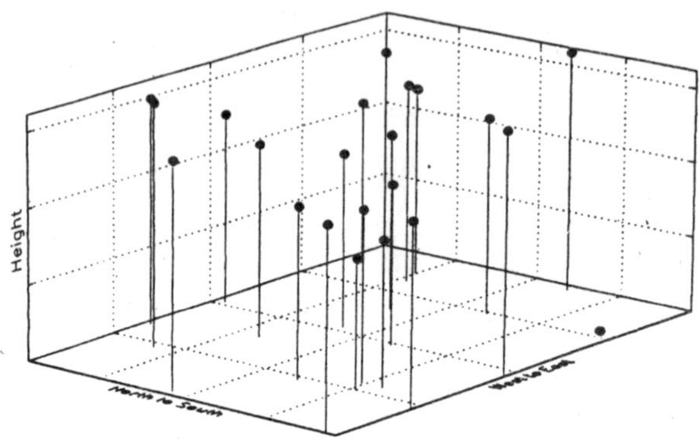

Figure 9.4. *Road distances approximated by displacing the nodes in the third dimension*

Then, a transport surface can be calculated by interpolation, by building a continuum based on the previously positioned sites (Figure 9.5). It is possible to find the distances between places by using a measurement along a two-dimensional plane.

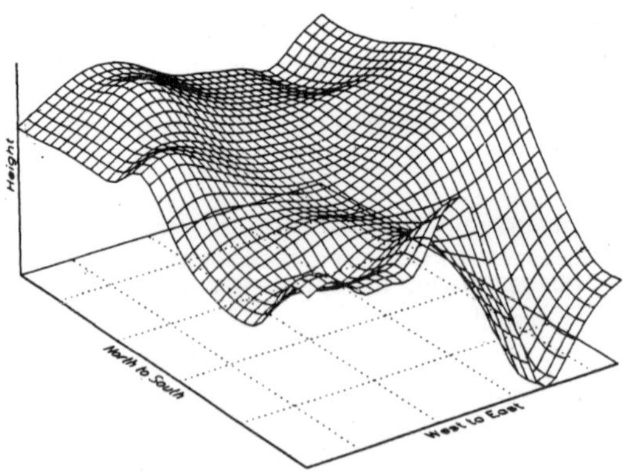

Figure 9.5. *Transport surface constructed by interpolation*

Let us imagine tracing the original transport network on this surface. This plotting of the graph would make it possible to find a coherence between the real distances and the represented distances, similarly to what is proposed by the spring map. If we consider that the initial data makes it possible to construct a graph distance (where positivity and symmetry are verified), the suggested map is coherent with this data, as well as with the mathematical structure that they imply and therefore also with the geographical reality.

The construction principle of the interpolated transport surface comprises, however, two important differences with respect to the spring map representation mode. Firstly, regression and interpolation introduce an error, which, even if minimized, makes the representation enter the domain of approximation, since the measurements taken have to integrate an associated error. Secondly, the use of the third dimension implies a projection system onto a two-dimensional surface of a paper sheet that carries the image. And there too we introduce an approximation, though of a different nature, due to the projection angles. This approximation can be largely reduced, if we help the reader to reconstruct the third dimension mentally. In Figure 9.7 the deformation of the grid, the hidden surfaces (in the south of the surface) and the three-dimensional coordinates system invite the reader to enter into the third dimension of the map and help him to reconstruct the surface topology.

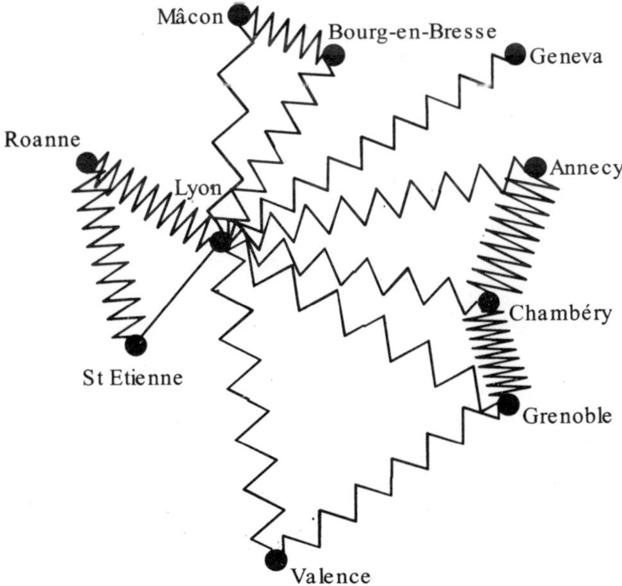

Figure 9.6. *"Spring" graph of the Lyon region*

The second application that we propose [PLA 87, p. 12] (Figure 9.6) shows the quality of service of the rail network in the Lyon region. The value of the arcs in the graph indicates the quality of service, which in turn is constructed on the basis of the quality of service of every line.

Visually, on spring maps, it is difficult to directly perceive the exact length of the arcs. We first see the generally "compressed" aspect of the spring which provides an indication of its length. The notion of visual length, which is given by the overall length of the spring, does not result here from a direct visual perception. In the case of the spring graph we are closer to the idea of a mental transformation carried out by the map reader than of a direct and proportional visual perception. However, does this mental transformation not proceed from the learning inherent to any non-traditional system of representation? As Tobler proposes, these representations may seem strange at first, but this is largely due to our more favorable attitude towards more traditional and also more familiar maps, which means that we tend to consider only conventional maps as realistic and correct [TOB 61, p. 164].

To answer the problem of the representation of distances, the plotting of a three-dimensional graph offers other latitudes and allows other modes of representation, which are to some extent connected with Tobler's explorations. We will now present the space-time relief maps, from the viewpoint of coherence with the initial data, but also by way of comparison with the systems presented up until now.

9.5. Chronomaps: space-time relief maps

Space-time relief maps, or Chronomaps, aim at representing a space of transport deformed by the coexistence of modes of transport with different performances [LHO 96, p. 37-43]. In the domain of passenger transport speed and cost per kilometer vary very strongly according to the transport modes, but also within a mode depending on the quality and type of the infrastructure used. An inhabitant of Aurillac who wishes to go by car to the prefecture of the area, Clermont-Ferrand, starts by driving on a trunk road with an average speed of 60 km/h until he reaches a highway entry, from where he drives at 110 km/h on average. In this example, the highway portion of the journey is covered almost twice as quickly as the portion on the trunk road. The possibility of swapping between the sub-networks of the transport system are very strongly conditioned by the nature of the networks and, therefore, the cartography of such a transport space must take into account the form of the networks and the heterogenous nature of the system. This difficult problem, which in anamorphoses is dealt with by dilating the space in a non-homogenous manner according to the quality of the connections, is solved on the relief map by employing the third dimension. To represent these distances, we do not change the location of the places, but the way of drawing the connections.

Chronomaps are constructed on the basis of a graph (Figure 9.7 in the color plates section, stage b). The graph representing the transport network must be planar and saturated in order to enable the construction of a three-dimensional surface (Figure 9.7, stage g). To respect the heterogenity of the network the graph adopts the shape of a p-graph[6], in which each sub-network appears as a partial graph[7]. Thus, between Nantes and Angers, the highway and the road are superimposed in the p-graph and each infrastructural network is considered as a partial graph of the p-graph. The nodes of the graph are again assigned to the main cities of the area considered, but also to the singular places, whose consideration is necessary for a thorough understanding of the transport network, such as the Essarts highway interchange located between La Roche-sur-Yon and Cholet. The choice of the density of nodes in the space considered is one of the determining elements of the form of the relief and, thus, of the produced image and the message that it conveys. Low density can make it possible to show the effect of networks with a broad grid (TGV and highway) over the entire area, whereas greater density makes it possible to express the complexity of local distortions.

Representing distances in a homogenous network with connections that are not far from the straight line is easy: in the plot in the plan of the highway graph (stage c) the length of the arcs plotted is directly proportional to the time-distance and the associated errors are small. On the other hand, in a heterogenous network, where the differential journey speeds (or costs) are considerable, the proportionality of the lengths of the arcs cannot be respected if they are drawn in the form of segments in the plan. In space-time relief maps the arcs with poorer performances are traced under the plane of the nodes in the shape of two segments, in such a way that their length is proportional to the journey durations. The worse a connection performs, the more an arc becomes concave and moves away from the straight line on the plane (stages d, e and f). During the last stage the coloring of the graph surfaces in different shades of gray makes it possible to reconstruct the transport surface and facilitates reading the relief.

The Chronomap is constructed as the plotting of a graph representing a multimodal transport network. The principle of plotting follows three rules that can be stated as follows:

– the position of the nodes in the graph corresponds to their real position;

– the length of the arcs in the graph is proportional to their effective length (be it in kilometers, hours, etc.);

– to enable the proportionality of the lengths, the arcs are drawn in the third dimension.

---

6 In a p-graph there can be no more than p distinct arcs connecting the same pair of nodes.

7 A partial graph G' of a graph G contains all of the latter's nodes, but only a part of its arcs.

From the point of view of graph plotting, the relationship of this representation principle with that of Tobler's spring maps is very strong. The location of the nodes is the same and the principle of proportionality for the length of the arcs is common. The only difference lies in the manner of tracing the arcs.

Another interesting point of comparison of the two models is their way of representing the continuum. According to Tobler, the passage from graph plotting to the transport surface is achieved by using a regression along a vertical axis followed by interpolation. In the space-time relief map the passage from graph plotting is established by tracing the facets of the graph on the basis of the arcs. Each triangular facet (the graph must be planar and saturated) is in turn divided into four triangles which are traced from the middle of the facets. The surface constructed in this manner then undergoes the deformation of graph arcs. The surface is associated with the road graph and not to the highway graph because access to normal space is only possible by using the road. This idea of universality of the road network [DUP 94, p. 146], defined by Dupuy, is expressed by Tobler by defining the border of the network: the border of the road network is constituted by the edge of the carriageway, whereas the border of the highway network is constituted by the access points [TOB 61, p. 89]. On the relief map the slope of the arcs directly translates the associated relative speed: the more marked the slope, the lower the connection speed is with respect to the possible maximum speed appearing in the plan of the nodes occupied by the highway network. By geometrical construction the slope of the facets in the graph is sharper than that of the arcs upon which they are based. This means that the performance of the interstitial sub-network that services the area of the facet is poorer than that of the connections bordering it. This principle is coherent with that of network hierarchy: in transport networks the performance drops with the size of the grid [PLA 91, p. 22]. The representation of the continuum, which is defined as the dual of the network, does not contradict the principle of relief representation. However, it is necessary to draw attention to the fact that the distances shown by the relief map are, first of all, those produced by the transport network. Here, the first function of the continuum is graphic: the shades of gray of the facets help the reader to imagine the third dimension, just as the deformed grid used by Tobler did.

Chronomaps do not show space in the form of plane layers, but of superimposed complex surfaces. The image in Figure 9.8 in the color plates section shows two sub-networks of the road transport network which are arranged in superimposed layers in the third dimension and coming into contact only at the sites of highway interchanges. This is a direct illustration of the tunnel effect of fast infrastructures.

Each space is registered according to its own relief: the highway is in the plane of the cities where the straight line draws the fastest path but where space is reduced to a set of points (network space), whereas the other networks follow a relief which becomes more marked as accessibility diminishes (banal space). The space of the road is a continuous but irregular surface, less accessible, but where the concept of proximity still operates. The map shows a space dualized by speed.

The large highway axes which serve the territory appear clearly inscribed in the plane of the cities. However, in the interstices of this network the space-time relief reflects the lack of accessibility due to the inferior quality of the infrastructure. The network space of the Riviera, from Avignon to Menton, which is well connected thanks to an extensive highway grid, strongly contrasts with the poorly accessible space-time of Alpes-de-Haute-Provence and Hautes-Alpes.

The Chronomap of the road and the highway renders readable the minimum duration transport routes by showing the degradation of the relative conditions of circulation outside of the highway type network. Thus, the Grenoble-Nice connection is ensured via the Rhone valley along a highway route that traces a vast loop whose visual length remains lower than that of the route closer to the straight line through Gap and Digne which passes through an irregular space-time relief.

In formalizing the construction principle of relief maps, the first constraint relates to the length of the arcs without taking their visual length into consideration. In this respect we may note a difference with the construction principle of "spring maps". In spring maps visual length is directly proportional to the effective length of the arcs, even though the visual length of springs is more inferred than directly read.

Relief maps are particular views of a general three-dimensional model. The length of the arcs of the construction principle is a constraint on this general model. The images, whether projected onto a screen or onto a sheet of paper, are two-dimensional. The concept of visual length may be used only with respect to the images. An image is created by the transformation of a three-dimensional structure – the model – into a two-dimensional form – the map. Just as on the interpolated transport surface of Tobler, the complex play of three-dimensional structure projections onto the plane of the map causes disturbances in the lengths of the arcs. These disturbances justify the fact that we only have an "approximate scale" of space-time in relief maps. The visual length which is available for images is not exactly proportional to the effective length present in the data. Mathematically, this results in the fact that a three-dimensional surface and the final image are not isometric: there is a distortion due to the projection on the plane. For arcs belonging to the third dimension, the distortion between visual lengths and effective lengths is not homogenous. It is therefore difficult to evaluate visually: it is necessary to find the position of the segments in three dimensions. It is the role of gray shaded facets, of viewing angle indication, to help to mentally reconstruct the relief in order to make the distortion understandable. Let us note that the general principle of relief maps construction places no constraints on images, meaning that the adjustment of the image generation parameters arises from the choices made by the cartographer. An infinite number of possible images correspond to the single model, according to the choice of the viewing angle, or also of the point of view[8].

Although the approximation due to the projection of the three-dimensional structure is common to that of the interpolated transport surfaces of Tobler, quite unlike the latter, from the point of view of the construction principle, space-time relief maps address the exact domain. This means that the length of arcs of the three-dimensional model is strictly equal to that coming from initial data.

9.6. Conclusion

Distance belongs to concepts applied in disciplines that are very far from each other. In Tobler (for whom any map results from a mathematical transformation whose parameters it suffices to modify), in Müller [MUL 79, p. 215-227], or in Huriot and Perreur [HUR 90, p. 225] the issue of the representation of distances, in geography as well as in other disciplines that concern space, insistently refers to mathematics.

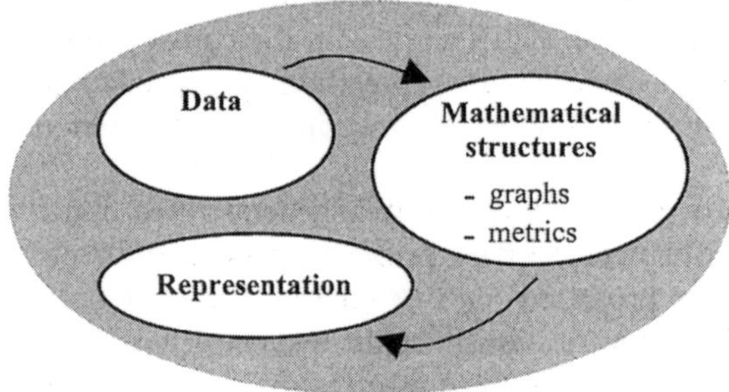

Figure 9.9. *The role of mathematical structures in the representation of distances*

In this spirit, the representation procedures exposed here are all based on graphs. The very object of this contribution (the issue of the representation of distances) is a call for a positioning in the field of metrics. Graph distance has been defined in this manner. The representations all are constructed on the basis of a graph plotting. Reading the map requires measuring instruments (which are the scale of representation and the graphic length), which make it possible to reconstruct metric space and understand its distance.

8 For relief maps of the Atlantic coast according to an unusual point of view see [MAT 96, p. 97-111].

Whether implicit or explicit, the reference to mathematical structures is necessary for the procedure of representing distances (Figure 9.9). The use of mathematical forms enables the clarification of principles and ensures the freedom of contradiction [BUN 66, p. 2].

The principles of representation clarified here show the essential contributions of graphs use in the cartography of distances and the understanding of spatial relations. All the avenues are far from having been explored and the models still hold important development prospects. Indeed, if the relief maps presented show the connections in the form of two broken segments, nothing prevents us from imagining other configurations using curves, for example.

The cartographies of distance can hardly be understood without instructions (a scale of correspondence and the way to use it), which sometimes disturb the established order of cartographic conventions.

9.7. Bibliography

[ANG 72] ANGEL S., HYMAN G.M., "Urban spatial interaction", *Environment and planning*, GB, vol. 4, p. 350-367, 1975.

[BUN 66] BUNGE W., *Theoretical geography* (revised 2nd ed.), Gleerup (Lund studies in Geography), Lund, Sweden, 1966 (1962).

[CAU 84] CAUVIN C., Espaces cognitifs and transformations cartographiques, Strasbourg, Thesis, 1984.

[CAU 96] CAUVIN C., "Au sujet des transformations cartographiques de position", *Revue électronique Cybergéo* (www.cybergeo.presse.fr), 1996.

[CHA 97] CHAPELON L., Offre de transport et aménagement du territoire: évaluation spatio-temporelle des projets de modification de l'offre par modélisation multiéchelles des systèmes de transport, Thesis, Tours, 1997.

[DUP 94] DUPUY G., "Réseaux", *Encyclopédie d'économie spatiale: concepts - comportements – organisations*, J.-P. Auray, Antoine Bailly, P.-H. Derycke, J.-M. Huriotb (eds.), *Economica*, p. 145-151, Paris, 1994.

[FLA 68] FLAMENT C., *Théorie des graphes et structures sociales*, Paris: Gauthier-Villars/Mouton (Mathématiques et sciences de l'homme), 1968.

[HUR 90] HURIOT J.M., PERREUR J., "Distances, espaces et représentations", *RERU*, France, no. 2, p. 197-237, 1990.

[KAN 63] KANSKY K., "Structure of transport networks", Chicago: Department of Geography, *Research Paper no. 84*, 1963.

[LHO 96] L'HOSTIS A., "Transports et Aménagement du territoire: cartographie par images de synthesis d'une métrique réseau", *Mappemonde*, no. 3, p. 37-43, 1996.

Chapter 10

Evaluation of Covisibility of Planning and Housing Projects

10.1. Introduction

The aim of this chapter is to represent the visible landscape. The study of the visibility of a linear transport infrastructure calls upon graph theory, which is a modeling tool of the studied terrain. The latter is considered on a *local* or even *microlocal* scale.

To achieve our goal, a quantitative decision-making aid tool specific to the object of managed landscape has been developed. In this context we propose to summarize the main functionalities of the computerized model: 3D-IMA. The latter reverts to the traditional GIS architecture and uses the principles of digital image synthesis.

On the basis of graph theory and within the framework of *landscape law* of January 8, 1993, 3D-IMA proposes two main complementary functions:

– multiresolution topography;

– localized visibility or *covisibility* (determination of the area visible to an observer from a single point; see below) and *generalized visibility*.

The first part of this chapter is dedicated to the method of multiresolution topography and its usefulness with respect to the surface and/or voluminal

---

Chapter written by Kamal SERRHINI.

representation of a given space. In the first place we will briefly present the model of *visible landscape* for *planning* (VLP), which is the true backbone of this chapter.

In the second part, we will explain in detail the principles and characteristics of the method of measuring the visibility surface of an object (highway, building, etc.). Some examples used for visibility results will illustrate the subject. We will then study the contribution of graph theory to the determination of various variants of covisibility.

After the thorough examination of the covisibility approach the last part is dedicated to a comparative analysis between the existing method and our method of measuring the visibility of an object in its spatial environment. Finally, we will conclude by describing the usefulness to space management of creating specific functions, such as covisibility and perspectives.

10.2. The representation of space and of the network: multiresolution topography

The 3D-IMA[1] tool is specialized in dealing with the thorny issue of quantitative evaluation of the visual impacts of large landscape management projects. Indeed, the implementation of an extensive project, such as a highway, a new TGV line, high or very high voltage lines, etc. should be preceded by numerous feasibility studies, comparisons with past situations, digital and/or modeled simulations, etc. as well as the evaluation of the visual impacts of the project on the landscape (*landscape law* 1993). Our work forms part of this framework.

10.2.1. *The VLP system*

The VLP system (see Figure 10.1) is a systemic formalization of the problems of this research: determining the landscapes which are most frequently visible from the construction site and vise versa, estimating management sites that are generally visible to an observer from the landscape.

We consider this diagram as the discussion thread, which will guide us throughout this chapter. By using colored arrows with different forms or aspects we

1 The main characteristic of this tool is the implementation of thematic and spatial analysis functions. This model bridges the gap between digital databases (vector data and/or rasters), on the one hand, and CAD software (AutoCAD®) for the restitution of results, on the other hand. Indeed, our aim is not to fully recreate a SIG, but rather to develop a specific model by taking into account previous achievements in geographical data processing.

will distinguish amongst four families of interrelations, which are at the origin of the regulation mechanisms of the VLP system:

Relations 1 and 3

These two relations symbolize the *transition* between two stages of different but successive levels or ranks.

Relations 2 and 4

They symbolize interrelations or, to be more exact, *loops* which are either inside the same subsystem or between two different subsystems. It is a process, which in certain or favorable situations will take place a certain number of times: it is the case, for example, of the determination of the generalized visibility of an installation.

Relations 5 and 6

They symbolize return actions or *feedback* [BER 73, p. 42]. Both relations 5 and 6 are the feedback emitted by the results subsystem (covisibility and cartography) to regulate either the *simulation-modeling* subsystem, or the *natural and/or anthropic* subsystems.

Feedback 5 could induce modifications, adaptations or adjustments of the *modeling-simulation* subsystem. It may act in order to modify certain constraints, among which is the observer's height, the simulation of various heights of forests (5, 10, 15 meters), etc.

Feedback 6, on the contrary, may act directly upon the *natural* and *anthropic* systems to integrate new geographical information (a new installation, a future ZPPAUP or ZNIEFF, etc.). This feedback may also act by removing existing information: comparisons between situations with or without a given infrastructure.

Relation 7

It symbolizes the contribution of decision-making aid elements in landscape planning by using the covisibility method. The decision of whether to set up or not to set up the future project depends on many systems or criteria of various natures: economic, sociological, legal, political, etc. and some remain still largely in the subjective domain (that of relations, of the game of each actor or groups of actors linked to the project in question). This latter idea, symbolized in relation 7 by a green arrow with a multipolar base, reverts to the multicriteria methods of assistance to decision-making.

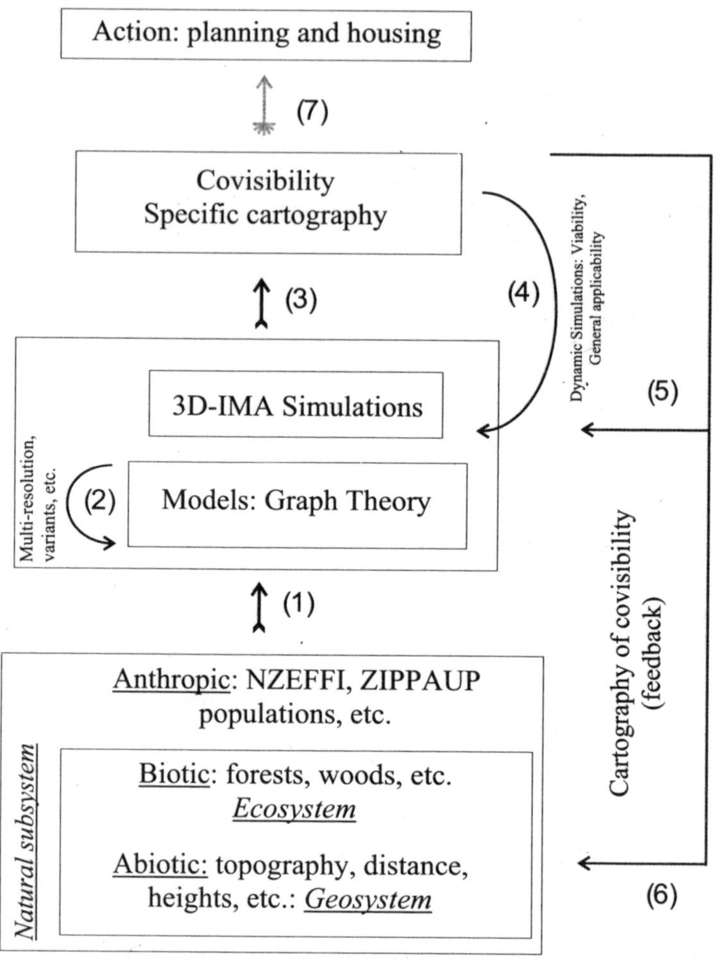

Figure 10.1. *VLP system [SER 00b]*

Figure 10.1 summarizes all of the subsystems (abiotic, topography, water, biotic, forest, anthropic, ZNIEFF[2], ZPPAUP, populations, flow of travelers, networks, etc.) and their interactions, thus making it possible to obtain a precise answer to the question of the estimate of the visual impact of an installation.

---

2 ZNIEFF and ZPPAUP are biotic spaces (forest) or abiotic (monument) which, however, benefit from the statutory protection of human design, thus explaining their belonging to an anthropic subsystem.

10.2.2. *Acquiring geographical data: DMG and DMS*

In order to measure certain environmental impacts of landscape planning we need to collect data with respect to, on the one hand, the work itself (possible locations, longitudinal profile, dimensions, etc.) and, in addition, to the space that receives this project (DMG, existing transport system, forests, protected areas etc.)

10.2.3. *The Conceptual Data Model (CDM) starting point of a graph*

The CDM is one of the bases of the MERISE method [MAT 94]; it is used to formalize information systems in general and the data stored in the GIS in particular.

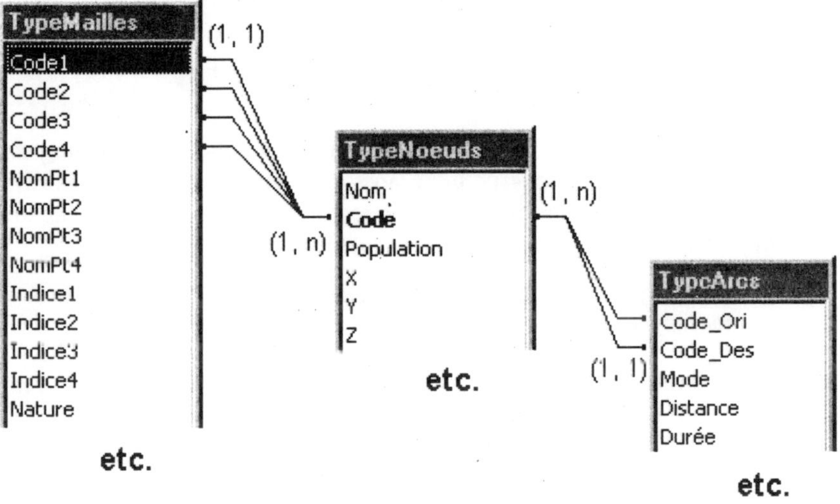

Figure 10.2. *CDM[3] of the three main types of data[4]*

This CDM was implemented in two steps, as follows:

– prior to this research there existed already[5] various structured information lists. They correspond to files: triangular nodes, arcs and facets. These pre-existing files are dealt with as they are with possibilities for improvement (deductive step);

---

3 The CDM rests primarily on the three concepts of *entity*, *relation* and *property*.

4 This CDM was carried out by using Access for the graphic functionalities. The geographical data itself is in text files (ASCII) according to an appropriate format.

5 Laboratoire du centre d'études supérieures d'aménagement EA1373, University of Tours.

– then, for the purposes of this work, a new type of surface data was developed: the square mesh (in vector and not in raster mode). The latter is necessary to restore, among others, the topography of the entrance area of an installation (inductive step);

– in addition, while the typology of data (various files) is based on the nature of the object considered (punctual, linear and zonal), each data type, on the other hand, simultaneously groups semantic and geometrical information specific to this type.

Figure 10.2 summarizes only the three *principal* types of the processed geographical data, their attributes and relations. In this context some clarifications are essential:

– the various tools of the CESA laboratory (D.LOCAT.T, NOD, MAP, RES, FRED, 3D-IMA) generate multiple processing files and intermediate or final results that are not represented in order to avoid overloading Figure 10.2;

– through the codes of the various nodes of the linear or surface elements, the two types of arcs and meshes maintain the relations with the type of nodes. For the cardinalities of these relations we have two eventualities:

- either a cardinality (1, n): meaning that a given node (1) belonging to the *nodes type* can contribute to one or more arcs or meshes (n). In other words, the nodes of the meshes or the arcs necessarily appear in the nodes type. In other words, it means that in the *nodes type* there are no isolated nodes,

- or a cardinality (1, 1): where only one specimen of a node of an arc or a mesh exists in the *nodes type*.

Thus, the *relational* nature of a database management system stems from the preceding relations provided with their cardinalities.

10.2.4. *Principle of multiresolution topography (relations 1 and 2 of the VLP)*

The nature, the quantity and the quality of data necessary depend mainly on the project under study. Since the present work primarily addresses to big planning and housing projects, which need many data, one of the functions of 3D-IMA is, thus, the creation of *specific* DMG. Although it clearly is a function of geographical data *acquisition*, we will see that it is also a function of *analysis*.

Figure 10.3 in the color plates section summarizes for us the principal stages of creation of a DMG referred to as *multiresolution* (often referred to as the *Regular hierarchical terrain model*, [FLO 99, p. 8]). According to this figure it is therefore possible:

– to obtain a surface representation (2.5D modeling: a single z for every pair of (x, y)) of the original terrain on the basis of a simple topographic map with a scale of

1/25,000 with the possibility of overlapping the various resolutions only locally, from where the term of multiresolution originates;

– to create geographical data in various geometrical forms which are specific to various uses: punctual data (a scatter graph or DMG), linear (in the form of segments of straights or arcs) and zonal (data in the form of facets or meshes). All of this information has three coordinates (x, y and z) leading to a better level of abstraction than that of models and 2D GIS.

If the method of multiresolution topography forms part of data acquisition functions, it also implements analysis and calculation functions, and, mainly, interpolation methods [ARN 00]. Interpolation to the nearest neighbor [SER 00, p. 327] enables us to recreate heights on the basis of altitude curves in order to go from a 2D scatter of points (with only two coordinates: x, y) to a 2.5D scatter (a z for per (x, y) pair): *DMG*. Obtaining the 2.5D grid is the result of surface modeling (*DMS*). Lastly, a true 3D representation implies implementing voluminal modeling (several z for certain (x, y) pairs). A 3D vector modeling is used to restore the voluminal aspect of particular geographical information, such as a building, a forest, a listed monument, a nuclear power station, a dam, a viaduct, etc.

In this chapter we will eventually call upon voluminal 3D modeling to represent *noticeable* landscape *elements* after successively implementing a 2D modeling (2D scatter of points) and then a 2.5D modeling (DMG and surfaces) with or without multiresolution.

10.2.5. *Need for overlapping of several spatial resolutions (relation 2 of the VLP)*

This function, which may appear traditional at first, is actually fundamental. Indeed, instead of having a very high definition topography in a generalized manner (a DMG with a step of 10 or 5 meters, for example), we can imbricate various resolutions around or along the processed installation. Our objective is not to introduce a hierarchy in the landscape according to its distance from the observer in the form of plans whose scale is increasingly small as we move away from the latter, as is the case in a recent article by Brossard and Joly [BRO 99]. A ZNIEFF[6] may be located at the edge of the studied terrain, whereas the installation itself is rather more central in this space. In this case a multiresolution topography is implemented simultaneously for the ZNIEFF and for the installation. The goal is to obtain a fine spatial definition only where it is useful, without worrying about the position of the observer or that of the project: "since not all tasks require the same level of detail, the use of high resolution models may affect applications for which many of the

6 Natural zone of ecological, floral and faunal interest.

details are not relevant" [FLO 99, p. 8]. This leads to an unquestionable gain in terms of quantity of geographical data to process.

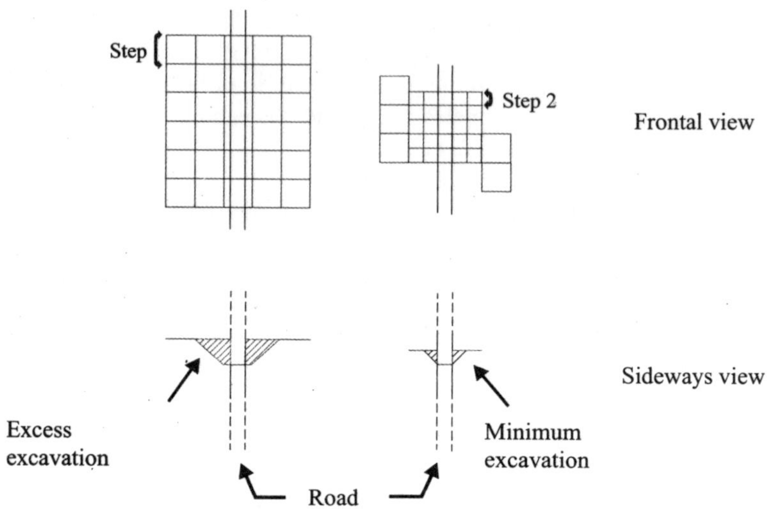

Figure 10.4. *Mounds, excavations and multiresolution*

Which spatial resolutions should be chosen? (50, 25 m, etc.)? And where? Finding exact answers to these questions remains very difficult since each installation and landscape constitute particular cases. This technique is also essential for embankment building and excavation needed to maintain the profile of the works (see Figure 10.4).

10.2.6. *Why a square grid?*

If, potentially, there are various geometrical forms [CHA 91, p. 27] (triangle[7], square, hexagon) to restore any surface (including that of the original terrain), then why did we prefer using the square mesh?

Various reasons may explain this choice.

7 We also have the possibility of plotting a graph containing triangles. This is the case for the planar saturated graph of the center of Tours, where the nodes represent crossroads in the city and the facets are the result of a triangulation using Delaunay's method [SER 97].

On the basis of a regular DMG it is possible to model a surface containing square meshes or triangular facets. In fact, each mesh may easily be transformed into two isosceles triangles with two nodes and the arc connecting them in common. Ultimately, in terms of representation in wire, the same "surface" can be described either through four points connected to each other by four "arcs" (a mesh) or by four points and five "arcs" (two triangles). Several consequences arise from such a transformation:

– a surface composed of triangles (regular geometry) contains twice the number of triangular elements than if it were composed of square meshes. This results in weighing down any possible processing. Moreover, the approach of this research setting often requires processing large quantities of geographical data and the use of a geometry based on triangles does nothing but double the occurrences of possible Triangle objects with respect to the Mesh objects used (see CDM below). For example, performing a simple classification (bubble sort) would have to deal with twice as many triangles as meshes, or even the restitution (the automation of the drawing by AutoLisp) of a topography using AutoCAD® would take practically double the time with triangles than it would with meshes;

– in the case of an DMS formed by triangles the centroid (the dual point) of each triangle has only three close neighbors (the nodes of the considered triangle), whereas linear interpolation usually requires the four closest neighbors, which a mesh structure enables us to easily obtain.

A mesh covers a larger surface than that of the two triangles in this mesh. The definition of topography by facets is unquestionably finer than that offered by meshes. However, this definition is not only more difficult to manage (storage, processing, display), but, moreover, it does not appear to be always useful. To obtain a better spatial resolution, the topography method makes it possible to only overlap various space resolutions where it is useful: an operational multiresolution.

Square mesh topography seems more interesting to us mentally than its equivalent in triangles, since it is closer to the mental representation than we make of the surrounding space (the pieces of land are more often rectangular or even square than triangular and the same applies to houses, buildings and roads). If this observation proves to be founded, then a meshed structure should have a better visual impact on the reader of the cartographic document. Moreover, such topography is graphically lighter (at least one arc per mesh) and, consequently, it would be less complex, more esthetical and thus, most probably, attractive.

10.2.7. *Regular and irregular hierarchical tessellation: fractalization*

In the hierarchical tessellation used for covisibility, when that is necessary to define more precisely the space for which we wish to determine the properties of visibility, a certain number of invariants remain: the form and the limits of the geometric forms (triangle, square, etc.), nodes and arcs. It is, moreover, a whole multiple in terms of surface of the new cube or, conversely, a whole number of new cubes of the same form which occupy the surface of the cube of hierarchical level immediately above. All the elements of auto-similarity and of internal homothety, which is a fundamental property of fractals, are present. The phenomenon can be generalized to statistical homotheties [THI 87].

10.3. Evaluation of the visual impact of an installation: covisibility

10.3.1. *Definitions, properties, vocabulary and some results*

This function, i.e. the evaluation of the visual impact of an installation, corresponds to the determination of the zones of space which are visible from the installation (punctual or linear) and, conversely, the zones of the project which are visible from the surrounding space.

This function is implemented, in the present state of development of this work, with the following elements:

– input data:

- topography of the studied terrain (DMG),

- geographical position and height of the present obstacles (forests and others),

- geographical position and height of the object (profile of installation: embankment, trench, etc.) integrated into its environment;

– parameters of the model:

- geographic position(s) and height of the observer,

- maximum visibility distance,

- population per considered zone (mesh, etc.) of the space,

- Euclidean distance separating the observer (local resident) from the project,

- the direction of movement of the driver (case of a linear transport infrastructure),

- speed of travel of the driver.

Target =	Space	Punctual housing project (like a building)	Linear works
Observer = punctual installation or a single point in space…	Localized visibility = covisibility		

Table 10.1. *Localized visibility*

We distinguish two types of visibility.

A localized visibility (Table 10.1). When we determine the visibility area on the basis of a single point of the object (case of a punctual installation: building [SER 99]), "two points V and W on a topographic surface are said to be mutually visible if and only if interior of the straight-line segment joining them lie strictly above the terrain (…). Given a viewpoint V on a terrain, the viewshed of V is the set of points of the surface which are visible from V" [FLO 99]. In this case the answer is of the "all or nothing" type, i.e. visible or invisible (see Figure 10.5).

Case 1 of Figure 10.5 in the color plates section shows the red zone which is visible to an observer from the green mesh of the works: localized visibility. Moreover, this green mesh of the section of the A28 is visible from all the red meshes in the area. The localized visibility is *symmetric* (the symmetry is *absolute*), hence we have the term *covisibility*. Therefore this term supports two pieces of information at the same time: on the one hand, we have lôcalized visibility and, on the other hand, it is *symmetric*. This fundamental property stems from the fact that the localized visibility is of the *centroid-centroid* type (the observer and the target are each reduced to a single point or centroid with three coordinates x, y and z) [SER 00] (see Table 10.3): "any visibility function can be represented by a visibility graph with arcs that link the nodes corresponding to intervisible entities. The visibility graph for point-point visibility is straightforward because any given point is either visible or invisible from any other point" [NAG 94, p. 764].

The term *covisibility* is new compared to the terms of visibility, intervisibility or such expressions as "field of view", "landscape opening" and "subject to viewing". Indeed, the existing terms and expressions specify neither the nature of the visibility (localized or generalized) nor the property of the symmetry.

Cases 2 and 3 of Figure 10.5 in the color plates section correspond to taking or not taking into account, in the calculation of visibility, a future viaduct which crosses the Choisille valley and which the A28 could use. This possibility (bridge) is presented only as illustration. Indeed, even though the unevenness of the valley is approximately 50 meters, the viaduct solution is not possible due to its cost and the fact that it is already planned that the A28 crosses the valley atop an embankment. Each of the five flagstones of the bridge is a simple three-dimensional parallelepiped containing six facets and two more for the safety fences, so that it really is a three-dimensional modeling. Thus, while the majority of the illustrations presented show only the surface, the visibility model perfectly integrates voluminal information, such as buildings or bridges.

In the two latter cases the implementation of voluminal modeling (3D) makes it possible to take into account the visibility between the piles and under the arch of a bridge or a viaduct: possibility of making the visibility radius "pass" under the bridge and between its piles in the algorithm.

A generalized visibility. Here, we perform the calculation of the visibility area on the basis of many points of the installation in order to simulate, for example, the movement of a driver (case of a linear installation: highway) or of the space. The result obtained is presented in the form of a color scale corresponding to classes of visibility percentages which go from the most visible zones to the least visible or even hidden zones (Figure 10.6 in the color plates section).

Target =	**Space**	**Linear planning (like a road)**
Observer = space or linear installation	Generalized visibility	

Table 10.2. *Generalized visibility*

Figure 10.6 in the color plates section shows six zones of visibility degrees along the works. In order to immediately differentiate the zones that are most frequently visible (generalized visibility) from the section of the A28, the number of classes was limited to six: "*in selective perception it is advisable not to exceed six to seven shades – black and white included*" [BER 67].

For a linear installation visibility is not symmetric. The use of the term covisibility is no longer justified and the statistical results obtained require prudence in their interpretation. For example, for the visibility class [0; 10%] we will say that approximately 9% of the section of the installation is visible from *certain centroids* (or certain "meshes") of the zones of space belonging to this visibility class. Conversely, we will say that *certain centroids* (or some "meshes") of the zones of this same class are visible from approximately 9% of the section of the installation.

Determining the identity (localization) of the centroids of the installation (i.e. approximately 9% of the "meshes" of the installation) which are visible from one of the centroids of the zone of space belonging to the visibility class [0; 10%] requires memorizing all the intermediate results (who do I see? and who sees me? all in all, who sees whom?) of the calculation of generalized visibility.

In conclusion, since for a moving driver the term covisibility is no longer adequate, we will then speak of *generalized visibility*.

Pairs =	Mesh-mesh (Centroid-Centroid)	Mesh-Zone (Centroid-Centroids)	Zone-Mesh (Centroids-Centroid)	Zone-Zone (Centroids-Centroids)
Symmetry of visibility	Absolute (covisibility)	Absolute (covisibility)	Absolute (covisibility)	Unsymmetric (generalized visibility)

Table 10.3. *Symmetry of visibility*

10.3.2. *Operating principles of the covisibility algorithm (relations 3 and 4 of the VLP)*

Having carried out the various stages necessary to create the set of geographical data, the visibility calculation algorithm can be brought into play. The operation of this computer program may be described in three stages.

10.3.2.1. *Loading of geographical data into memory (1st stage)*

This first stage corresponds quite simply to the 3D-IMA model reading and storing in the microcomputer memory of the geographical data files produced previously. There are two main matrices:

– the data matrix concerning the topography of space, including all the information acquired, such as natural obstacles (forests and others). It is "the equivalent" of data that can be found and collected by digitalization and/or vectoring on a topographic map (raster type document) but in digital form: ASCII[8] files of digital data. This data format lends itself well to the restitution of the result in vector form;

– the matrix relating to the installation and its characteristics. This second matrix is at the origin of the opposition between covisibility and generalized visibility. Indeed, the geographical description of each installation consists of at least one object of the zonal type (mesh, facet, etc.). Thus, thanks to the number of objects

8 *American Standard Codes for Information Interchange.*

comprised in the works, the algorithm automatically determines if there is localized visibility (a single zonal entity to describe the object: localized installation) or, on the contrary, generalized visibility (generally linear works comprising many zonal entities).

10.3.2.2. *Preparation of the visibility calculation (2nd stage)*

This last stage preceding the visibility calculation makes it possible to collect interactively certain additional information which are essential for the operation of the algorithm:

– height of the observer: the computer model requires the user to input the height (size) of the observer considered as being located at the installation or in the area. The position of the observer corresponds to the centroid of each facet of the installation or of the space;

– movement speed (strictly positive and less than or equal to 130 km/h), if the observer is a driver;

– maximum visibility distance: if close-up the human eye has a resolution power of about 0.2 mm in diameter, in remote vision this power very quickly loses significance and the distinction of remote objects becomes fuzzy. To integrate this physiological constraint into the algorithm we chose the following solution.

The modeling of an installation is systematically centered on a square topography with sides of approximately 5 to 10 km. In other words, the linear installation will not be modeled as one entity, but section by section (10 km long at the maximum). This gives to the covisibility algorithm its physiological sense with respect to the maximum visibility distance. Indeed, it is estimated that beyond approximately 2.5 km on both sides of the works, the eye distinguishes only the general forms (a forest in the form of a potatoid, a building in the form of a cube, etc.) and very few or no details. Moreover, this solution also makes it possible to implicitly minimize the effect of the curvature of the Earth.

Thus, in the case of a linear installation which extends over hundreds of kilometers it is useless to calculate the visibility for the totality of the works (it is as if we asked a driver located on the A10 highway somewhere around the Parisian region whether we could see Bordeaux located 600 km away).

10.3.2.3. *Calculation of visibility itself (3rd stage)*

At this development stage the model contains all the necessary information and can start evaluating the visibility of the works. Calculating the visibility itself may in turn be subdivided into three stages.

1st sub-stage: systematic calculation of topographic cross-sections

Knowing the location of the observer (*centroid* of the red mesh in Figure 10.7) and the target point, for which we are trying to determine whether it is visible or hidden (*centroid* of the green mesh of the same figure), the algorithm calculates at first the equation of the straight line (line of sight) connecting the observer and the target. Then, it determines all the segments of straight lines (arcs) that cut across this line of sight; this is a limited topographic cut (Figure 10.7 in the color plates section).

Figure 10.7 shows four examples of topographic cross-sections from a vertical perspective (*X, Y plane*). The 3 topographic cross-sections located on the left constitute three particular cases:

– a *horizontal* cross-section parallel to the x-axis;

– a *vertical* cross-section parallel to the y-axis;

– and a final *diagonal* cross-section (angle of 45° to the horizontal);

– the 4th and last cross-section (the one on the right, Figure 10.7) is a random cut.

To determine the areas of space that are most frequently visible to a driver from the installation (Figure 10.7), the topographic cross-sections are calculated from all the meshes of the works to all the meshes of space: we thus have a systematic sweeping of all the potentially visible zones from all of the elements of the structure considered.

Conversely, to determine the sections of the installation which are most frequently visible to an outside observer, the topographic cross-sections are calculated from all the meshes of space to all the meshes of the works by using systematic sweeping. For both events we say that the aiming is multidirectional.

The total number of topographic cross-sections depends simultaneously on the number of entities of the installation and the number of entities of space. Thus, punctual works generate a number of topographic cross-sections which is identical to the number of entities describing space. In the opposite case, the determined number of topographic cross-sections is the product of the number of objects, of which the section of the works is comprised, by the number of objects that form the space of this installation.

2nd sub-stage: visible or hidden?

Once all the arcs cutting through the line of sight have been calculated, the algorithm analyzes the current topographic cross-section. Each arc of the topographic cross-section has a (2D) point in common with the line of sight: it is the intersection point. The latter can have two different altitudes according to whether it is on the line of sight (in 3D) or on the arc of intersection (also in 3D). Consequently, what remains to be done is to calculate (by simple interpolation following the proportionality rule along a line – Thales' theorem) these two possible altitude values of the intersection point and to compare them. If the altitude of the intersection point on the arc of the topographic cross-section is higher than or equal to the attitude, which it has on the line of sight, then the algorithm deduces from this information that the target is hidden for our observer. In order for the model to consider the target as visible, it is necessary that the altitude of all the intersection points on the arcs of the topographic cross-section is lower than that of these same points on the line of sight.

In all cases the final objective is to obtain a *visibility index* per spatial entity. Various types of visibility indices are possible.

Statistical visibility index VI_s. This corresponds to the ratio between the number of times that the spatial (or the works') mesh appears visible from the installation's (or spatial) meshes and the total number of meshes of the installation (or space). This index can be without a unit or expressed as a percentage;

$$VIs = \frac{V}{N} \qquad [10.1]$$

where:

– V: is, for example, the number of times that the mesh of the space appeared visible from the meshes of the installation.

– N: is the total number of meshes (facets) which constitute the works (N = 1: punctual work; N > 1: linear work) or the space.

Consequently, with covisibility, a space mesh which has a $VI = 1$ (100%), is *covisible* (absolute symmetry) from the centroid of the installation mesh. With generalized visibility a mesh of the space which has a $VI = 0.5$ (50%), is covisible (absolute symmetry) from the centroids of a set of meshes corresponding to half of the section of the installation.

$$0 \leq VI \leq 1 \qquad [10.2]$$

The visibility index can be balanced by the Euclidean distance (d) separating the observer (local resident) from the installation which we will call VId (see below).

The visibility index can be also expressed according to the crossing time of the transport infrastructure (see section 10.3.2.5).

Lastly, the visibility index can also be expressed according to the population P, which we will call VIp (see below).

According to the number of installation meshes we can summarize the various scenarios of the visibility index values in Table 10.4.

Number of installation meshes: N	N = 1	N > 1
Visibility index: *VI*	0 or 1	0 <= *VI* <= 1
Visibility type	Covisibility	Generalized visibility
Cartography restitution	2 colors	Color scale (classes of generalized visibilities)

Table 10.4. *Characterization of the visibility index*

3rd sub-stage: balancing with the population

If we consider that the visibility of an installation from a sparsely populated or unpopulated zone (a corn field, a forest, etc.) does not have the same repercussions on planning as that from a populated zone (a borough, a tourist site, etc.), then it seems interesting to estimate visibility according to the population variable.

This balancing is conditioned by the availability of the data regarding the population on a very fine spatial scale (mesh with sides of 100, 50 or 10 meters). Without this data the algorithm omits this last stage. In the opposite case, the visibility index is the sum of the populations of all the meshes from which each element of the installation can be seen.

$$VI_p = \sum_{i=1}^{n} P_i \qquad [10.3]$$

where P_i is the population of the ith space mesh that sees the target (installation).

We will distinguish two sub-types of visibility indices according to the population (Figure 10.8):

– a visibility index calculated only with respect to the real population of the studied terrain that sees the installation; this is the effective population. This effective population is the product of the number of houses per mesh (topo map with a scale of 1/25,000) by the average number of people per household;

– a visibility index calculated simultaneously with respect to the effective and potential population. The potential population relates to space meshes which do not contain houses but from which the works can be seen. In this case the population of these meshes is equal to 1.

In three scenarios of Figure 10.8 in the color plates section (height of the observer: 1.75 m) we voluntarily eliminated the meshes from the topography for better legibility.

The choice of a visibility index type depends on the objectives set, the spatial analysis performed (study of landscape or human impact, planning integration in landscape, etc.) and the availability of the data.

10.3.2.4. *Generalized visibility balanced by the distance to the installation*

Although it is definitely interesting to take into account the relative proximity of the residents in the vicinity of the installation during the determination of the scenarios of road planning, it seems possible to us, however, to go even further by considering not only visibility and the potential and/or present population of the site, but also the Euclidean distance.

The combination of these three constraints (visibility, population and distance from the installation) makes it possible to take into account, for example, all the people who are near, or even very near, to the works but who would not necessarily be inconvenienced by them due to the quite simple fact of the presence of an intermediate visual screen (landscape, woods, etc.) Thus, considering the distance to the works in order to estimate the number of people who are potentially affected by the arrival of a highway infrastructure seems necessary but insufficient.

Covisibility, generalized visibility and anthropic generalized visibility can be balanced by the relative distance of the residents to the infrastructure. The closer the observer is to the installation, the more the balancing of visibility by distance justified. Similarly, the larger the modeled space (more than 10 kilometers long), the more this balancing is essential, especially for taking into account the case(s) of town(s) (concentration of houses, of activities, etc.) that are not in the immediate vicinity of the installation (Figure 10.9 in the color plates section).

Figure 10.9 is an example where generalized visibility of the section of the A28 was balanced by the distance of each space mesh to each installation mesh.

The index of generalized visibility which was balanced by the distance implemented to obtain this map has the following form:

$$VI_{dij} = \sum_{j=1}^{n} \left(\frac{v_j}{d_{ij}} * k \right) \qquad [10.4]$$

where:

– v_j is the result of visibility (according to the traditional topographic constraints) from the ith space mesh to the jth target mesh (the works);

– d_{ij} is the Euclidean distance between the observer (space) and the target (the works);

– k is a visibility adjustment coefficient. It corresponds to the minimal theoretical distance which separates the target (installation) from the observer and for which visibility remains clear.

The result of Figure 10.9 may be regarded as an index of the clarity of the generalized visibility of the A28: the more we move away from the installation, the more details are lost from the visibility. Figure 10.9 shows three zones along the A28 (in the form of *aureoles*) of various colors. The more we move away from the works, the less bright the color becomes: red and magenta are in direct contact with the meshes of the A28, then there is the green, cyan and on the outside we have blue. The latter corresponds to the most remote space meshes, from which parts of the works can be seen. We note the concentric aspect of the space zones of various colors around sections of the A28.

10.3.2.5. *Dynamic generalized visibility*

In addition to the set of constraints considered above (height of the observer, populations, proximity of the residents in the vicinity of the works, etc.), it is also possible to take into account in the 3D-IMA model the *time* and *direction* of movement of a driver on a highway section: space-time directed visibility.

An example simultaneously illustrating *directed static* generalized visibility (see case 1, Figure 10.10) and *directed dynamic* generalized visibility (see case 2, Figure 10.10) is presented in the color plates section.

From an algorithmic point of view, this stage (directed visibility) takes place very early in the covisibility algorithm. Indeed, it is useless to determine the visibility of a target mesh from the site of the observer while knowing that this mesh targets is not in the observer's field of vision (180° for the directed visibility).

The interest of the directed visibility is certain. An observer located on the last floor of the Montparnasse tower does not see the same landscape from its various sides (decision-making aid with respect to the choices of orientation of future large scale development projects). A Parisian elected official would like to know only the landscapes most frequently seen by a driver on the A10 leaving Tours. Conversely, an elected official or a decision maker from Tours would try to get an idea only of the spaces which are generally seen by a tourist on the same A10 highway in the opposite direction. The objective, obvious in terms of management, for a space manager (urban developer, planner or elected official) corresponds to the will to improve the image of a region by using an attractive esthetic framework (entrances to cities or areas with improved visual comfort), thus making it more competitive.

With regard to dynamic visibility, knowing the movement speed of a driver (100 km/h for the example of Figure 10.10) and when the target mesh (pertaining to the landscape) is considered visible by the algorithm (3D-IMA), the latter assigns to it the crossing time of the observer's mesh (located on highway A28). This time is determined by considering a certain speed and network distance of each entity of the installation.

Thus, case 2 of Figure 10.10 represents, according to a color scale, the maximum time (in seconds) of visibility of each space mesh. For example, a magenta space mesh is potentially visible between *53* and *75 seconds* along the section and by considering the southwest-northeast direction of movement.

10.3.3. *Why a covisibility algorithm of the centroid-centroid type?*

Having described the operation of the covisibility algorithm, we now will specify the reasons for choosing an algorithm of the centroid-centroid type.

Let us recall that in the *covisibility algorithm of the centroid-centroid type*, a target mesh is considered visible from the observer's mesh, if and only if its centroid is visible from the centroid of the observer's mesh. Thus, we perform a double generalization of the results of visibility obtained at the level of the two centroids to two corresponding meshes (of space and installation): the *calculations* performed are of the *centroid(s)-centroid(s)* type and the restored *results* are of the *mesh(es)-mesh(es)* type.

With regard to graph theory the algorithm of the centroid-centroid type employs the *dual graph* on the basis of the initial primal graph (DMS).

The generalization of the visibility result obtained at the level of the centroid tested to the entire mesh, to which it belongs, may appear hazardous to some. What are the elements leading us to such a choice?

Continuing on the basis of the point-point type algorithm we performed the calculation of visibility. We did not consider the tested centroid mesh, but its various nodes instead: *1, 2, 3* or *4* nodes of each target mesh. In this case the point-point type visibility algorithm uses the *primal graph* (DMS).

Figure 10.11 in the color plates section exhibits five types of covisibilities:

– *Case 1*: it determines all of the space meshes visible (exactly 330 visible meshes out of 1,024: 20%) from the green mesh of the section of the A28. Here, a space mesh is considered visible, if at least one of its four nodes is visible from the centroid of the green mesh of the installation;

– *Case 2*: it presents the space meshes, at least two nodes out of four of which are visible (275 visible meshes: 17%) from the green mesh of the A28;

– *Case 3*: it shows us only 216 (13%) space meshes, where three nodes out of four are visible from the green mesh of the A28;

– *Case 4*: is shows only 165 (10%) space meshes where four nodes are visible from the green mesh of the future installation simultaneously;

– *Case 5*: it proposes a calculation of the visible field by not considering one, two, three or four nodes of each space mesh, but its centroid. The height of the centroid is calculated using the method of linear interpolation: centroid-centroid localized visibility. Thus, on the whole we obtain 219 (13%) space meshes whose centroid is visible from the green mesh of the A28.

The higher the number of nodes involved in determining the covisibility of each space mesh (passing from case 1 to case 4 of Figure 10.11), the more the visible zone is reduced: a very fine estimate of the landscape visible to the observer.

Thus, in comparison with the first four cases, the last covisibility situation (case 5 of Figure 10.11) seems to us to be a better compromise and a good indicator of the extent of the field of vision.

For specific development ends and in order to better approach the calculation of covisibility, instead of leaning towards the development of particular covisibility calculation algorithms (see below), we chose a solution associating multiresolution topography to a covisibility algorithm of the centroid-centroid type with a result restored in 3D in the shape of a wire mesh (vector data) and with colors, whereas the majority of publications on visibility use raster data.

10.3.4. *Comparisons between the method of covisibility and recent publications*

After having studied in detail the entire method of covisibility, it becomes easier for us to compare and discuss the specificities of this approach with respect to the results of recent and pioneering research.

This comparison encounters two main difficulties:

– a very poor bibliography concerning algorithms of visibility surface calculation of an observer, despite the multiplication of GIS that propose spatial analysis functions, including visibility. "In most documentation of information systems (GIS) it is very rare to find details of the algorithms used in the software" [FIS 93, p. 331]. This first handicap must be a consequence of the GIS market laws which oblige their designers to respond to competition with secrecy surrounding the main analysis functions which they propose. Consequently, the comparison is limited to research publications other than commercial GIS. Moreover, the comparison of visibility functions of certain GIS in the market, mainly through the results obtained, has already been carried out relatively recently by Fisher [FIS 93]. The latter, in order to undertake his comparative study of visibility, had the possibility of experimenting with seven GIS (Idrisi, OSU MAP for PC, PC MAP, Map II, GRASS, Arc/Info and EPPL7), which is not the aim of this chapter;

– besides the GIS there exist, most probably, models of visibility determination which are specific to the requirements and objectives of organizations, such as the military, EDF (EVELYNE software), etc. Unfortunately, we have no hard evidence of this research. The secrecy would be even stronger around military research than around civil research (EDF).

In these conditions the comparison is limited to the recent civil research on visibility. The latter, which also does not clarify all the stages and specificities of the created or exploited visibility algorithm, obliges us to base this comparison on relatively accessible elements.

		Algorithmic characteristics			
	Data	**Visibility type**	**Constraints**	**Sweeping**	**Display**
Recent publications	Rasters [RAN 99], [DU 93] and [BRO 84], rarely vectors [FLO 99] and [NAG 94]	Localized (rarely generalized)	Mainly topological	Partial aiming (rarely complete)	Especially views in the (2D) plane; possibility of overlay
3D-IMA	Vector: (points, arcs and facets: graph theory)	Localized or generalized	Topological and development	Systematic: complete or multidirectional aiming	Different views from the start according to angles and 3D points of view

Table 10.5. *Comparison between various recent publications on visibility*

Thus, if we disregard the potential work done in this field by military laboratories (American, European and others), research that became the subject of publications is not plentiful. We can mainly quote the three most recent working groups:

– English-speaking researchers: Randolph [RAN 94, RAN 99], Nagy [NAG 94] working within the *Rensselaer Polytechnic Institute* (New York), Lee [LEE 92] of the University of Kent (Ohio) and also Fisher [FIS 93] of the University of Leicester;

– the works of Italian researchers: De Floriani and Magillo [FLO 99] of the University of Genoa;

– those of French teams: University of Franche-Comté [BRO 94, BRO 99], COSTEL laboratory of the University of Rennes [DU 93] and the CESA laboratory of the University of Tours [SER 99, SER 00a, SER 00b].

The comparison between these various publications will be summarized as presented in Table 10.5. This reiterates the essential of the characteristics that appear most often in publications dealing with visibility.

Before concluding this chapter we will distinguish in this comparison three primary categories of different elements: data, algorithms and results.

10.3.4.1. *Raster and/or vector data?*

The majority of visibility algorithms consider the studied terrain via a raster grid (pixel matrix coupled to that of corresponding altitudes). In this research we favored the use of vector data created on the basis of an image raster (Figures 10.1 and 10.3 in the color plates section).

The use of the vector mode gives us, on the one hand, a facility in handling graphic entities and, on the other hand, a high degree of accuracy. Thus, it is possible to modify (either via the CAD software used or via the 3D-IMA model itself) the altitude of a set of geometrical primitives (punctual, linear and/or surface data) relating, for example, to a wood, a forest or something else: different scenarios of forest height simulation (10, 15, 20 m, etc.) before the calculation of the visibility.

Contrary to the AMAP (CIRAD – Montpellier) model where the growth of flora in a landscape is simulated according to its age and agronomic constraints, in this research the landscape is modeled as a whole. We will not try, for example, to identify each tree in a forest but rather to consider the forest as a mass: a volume of an average height. Having said that, the use of the vector model enables us to make the grid thicker at any moment (DMG and DMS) and at a very precise site in order to integrate new information on volume(s), thus approaching as much as possible the complete form of this information. This possible return journey in landscape modeling (by addition, suppression or modification of information) is conditioned by the availability of time, means and especially the objectives required. The larger the development project, intended to spread for dozens of kilometers, with a lifespan of around a century and involving colossal expenses, the more justifiable it is to employ this kind of approach (as is done more and more often).

10.3.4.2. *Characteristics of the covisibility algorithm*

The main characteristics of the covisibility algorithm are as follows.

Complete or multidirectional aim. This complete aim is the consequence of the overlap of two main loops in the algorithm. The first makes it possible to traverse the entities corresponding to the set of the possible viewpoints (file of a highway section described by many rectangular meshes). The second loop makes it possible to sweep all the target meshes of the surrounding landscape (file of all the entities describing the landscape around the installation). Thus, this overlap performs a systematic sweep in a precise order of all the entities of the installation and its space simultaneously.

The algorithm of visibility implemented here is of the centroid-centroid type. The exploration is conducted on the basis of a dual observer or origin graph – towards a dual target or destination graph (see above). This algorithm, for very sensible spaces (ZNIEFF, etc.), can be easily made more exact through coupling with the multiresolution topography method. For our purposes this coupling is scientifically satisfactory in comparison with publications that advocate:

– the implementation of a probabilistic model to take into account altitude errors [LEE 93]. "A probabilistic representation of the viewshed is more acceptable than the usual binary product" [MADE 93, p. 331];

– the use of the vague set theory. Knowing that these are relatively complex and sophisticated algorithms, researchers dedicated themselves to the development of parallel visibility algorithms (multiprocessor machines): "because visibility computations are so computer-intensive, they form has natural target for parallelization" [NAG 94, p. 769].

Localized or generalized visibility. It is one of the characteristics of the 3D-IMA model. The latter automatically determines the type of visibility in question. The first (covisibility) is the most frequent, in particular, in commercial GIS [FIS 93]. The second, which is a generalization of the previous one, is much less common because of the quantity of data and calculations that it involves, without forgetting that the researched objectives are often far away from the installation. Figure 10.11 of the submission for review of the article of Brossard and Wieber [BRO 84, p. 10] is the result of the intersection after superposition of various visibility cones obtained with local photographs. It should be stressed that, even if this method was limited to only a few points of view (around 10), the approach itself would remain pioneering in the subject matter. It is one of the first forms of generalized visibility.

Integration of constraints specific to the field of landscape management. Among these we have the maximum visibility distance which was integrated since the start of the method by limiting the surface of the modeled space. In the case of a linear installation, the visibility is studied section by section of approximately 5 to 10 kilometers with overlapping zones between the various sections. Each section corresponds spatially to the segment of the works, which raises interest and concern, first of all, where it crosses towns and affects local residents. The desire to finish a future highway project by using, in particular section by section and not lump visibility simulations could arise during:

– the determination of the various alternatives of the future project taking the landscape [SER 01], as well as the other technical and economic criteria (slope, shortest path, radius of curvature, etc.), into account;

– taking into account the opinion of the future residents within the framework of improving a participative democracy;

– the search for the most sensitive spaces (from a landscape and human point of view: pressure of local residents interest groups) that have priority with respect to "1% landscape and development".

10.3.4.3. *Graphic restitution of the results: color and shape representation*

To discuss, accept or express reservation about a certain visibility calculation algorithm requires either the study of the algorithm itself or the analysis of its results (images, maps, statistics) obtained by taking the stated constraints into consideration. Since the first alternative is generally unavailable, we are left with the second possibility. "Unlike some other GIS functions the viewshed is not actually verifiable in the field nor can it be logically validated, anything more than trivially, by examination of test figures …" [FIS 93, p. 333].

During such a procedure the majority of results regarding visibility reach us in the form of an image raster [BRO 84] with a view in the (2D) plan, even though at a given time the calculations have to involve the matrix of the altitudes (DMG). This form of image or cartographic restitution does not speak in favor of a better visual perception and therefore the understanding or even the appropriation of the results is obtained according to the landscape and the main obstacles. In raster mode it is necessary to resort to the overlay technique to obtain 3D raster visualization. This restitution, although interesting, remains despite everything a typical (2.5D) surface representation.

With a model vector, the change and multiplication of the points and angles of view are one of its intrinsic characteristics. This three-dimensional and colored cartographic representation is more easily perceptible, aesthetically better and seems to better convey the expected message. The user, the resident, the elected official or the decision maker can visually and immediately interpret, at least partially, the results obtained according to the landscape and the main obstacles at the site: the phenomenon of results appropriation by the decision maker. Thus, we estimate that we will obtain better information transmission in the hope of an optimal effect.

The mode of representation of the obtained results is mainly conditioned by the nature of the data used. They are two different ways (raster, vector) to approach modeling and, thus, the representation of a landscape.

10.4. Conclusion

Nowadays the GIS propose to the users multiple analysis functions that are very attractive in many fields, on the other hand, in terms of:

– determining the best sites or routes of an installation;

– estimating the visual impacts of the future installation on the landscape for an optimal and, most of all, equitable implementation of the equipment of “1% landscape and development”.

The majority of GIS do not propose specific functions, such as covisibility and, even less, generalized visibility (simulations of landscapes which are most frequently seen by a driver).

Hence there is the need to have or create tools which are appropriate for the questions and problems treated in landscape management, where the usefulness of cartography and digital imagery is undeniable. Among these models we also find 3D-IMA.

Let us specify that although covisibility and its alternatives give a precise answer to the question of visual impact and the insertion of an installation into the landscape, on the other hand, this function cannot be considered individually when deciding on the future of a potential installation: we need *multicriteria* approaches [SER 01] to help manage decision-making.

Within the framework of implementing a future installation, the visibility algorithm can be used at various stages:

– upstream of project realization: making it possible to determine zones that are most exposed to the view by using calculations of generalized visibility of space to space. Once detected, these zones could possibly be rejected during the search for alternative sites;

– during the search for project variants: the use of visibility helps to evaluate the visual pressure of each variant in order to implement sorting and selection, in particular, in conjunction with other criteria (slope, minimal path, cost, presence or proximity of protected zones, etc.) and taking into consideration the opinion of the residents;

– downstream of the project: for a better implementation of the “1% landscape and development”, the algorithm of visibility (covisibility) may also act very locally to determine with precision the generally affected areas in each town.

The method of visibility, in its current state, integrates certain constraints that make it more humane and operational. Thus, it is possible to determine the visible landscape according to various assumptions on the height of the installation: embankment, ditch or the same altitude as its topography. It is also possible to calculate the visibility by taking into account the intermediate obstacles (constructions, woods, etc.) and the characteristics of the population that sees the installation.

Finally, among possible future developments we wish, on the one hand, to widen the field of constraints of visibility calculation (flow of travelers, effect of proximity of a notable element, such as a monument or a natural attraction) and, on the other hand, to test this tool on cases of installation projects in an urban environment. The second prospect will most probably require significant adaptations of the model to simultaneously respond to the characteristics of the urban environment (high density of geographical information) and the requirements of urban dwellers in terms of quality of life by means of an even partial control of the various sources of potential disruptions (visual, sound, air pollution, etc.) that may be generated during the realization of a new urban development.

10.5. Bibliography

[ARN 00] ARNAUD M., EMERY X., *Estimation et interpolation spatiale*, Hermès, p. 221, 2000.

[BER 67] BERTIN J., *Sémiologie graphique,* Paris-La Haye: Mouton, Paris: Gauthier-Villars, 1967.

[BER 73] BERTALANFFY L., *Théorie générale des systèmes*, Paris, Dunod, 1973.

[BRO 84] BROSSARD TH., WIEBER J-C., "Le paysage trois définitions, un mode d'analyse et de cartographie", *L'espace Géographique*, no. 1, Paris, Doin, p. 5-12, 1984.

[BRO 94] BROSSARD T., JOLY D., LAFFLY D., VUILLOD P., WIEBER J.-C., "Pratique des systèmes d'information géographique et analyse des paysage", *Revue internationale de Géomatique*, vol. 4, no. 3-4, Paris, Hermès, p. 243-251, 1994.

[BRO 99] BROSSARD T., JOLY D., "Représentation du paysage visible et échelles spatiales d'information", *Revue Internationale de Géomatique*, vol. 9, no. 3, Paris, Hermes, p. 359-375, 1999.

[CHA 91] CHASSERY J.-M., MONTANVERT A., *Géométrie discrète en analyse d'images*, Paris, Hermès, p. 358, 1991.

[COL 92] COLLET C., *Systèmes d'information géographique en mode image*, Presses polytechniques et universitaires romandes, collection Gérer l'environnement, p. 186, 1992.

[DU 93] LE DU L., GOUERY P., "Paysage littoral: cartographie des degrés de visibilité", *Mappemonde 2/93*, GIP-RECLUS, p. 9-11, 1993.

[FIS 93] FISHER P.F., "Algorithm and uncertainty in viewshed analysis", *Int. J. Geographical Information Systems*, vol. 7, no. 4, p. 331-347, 1993.

[FLO 99] DE FLORIANI L., MAGILLO P., "Intervisibility on terrains", *Geographic information systems principles techniques management applications*, Maguire Goodchild Lonley, Rhind, p. 543-556, 1999.

[LAA 00] LAARIBI A., *SIG et analyse spatiale*, Hermes, p. 190, 2000.

[LEE 92] LEE J., SNYDER P.K., FISHER P.F., "Modeling the effect of data errors on feature extraction from Digital Elevation Models", *Photogrammetric Engineering & Remote Sensing*, vol. 58, no. 10, p. 1 461-1 467, 1992.

[MAN 95] MANDELBROT B., *Les objets fractals*, Champ Flammarion, p. 212, 1995.

[MAT 94] MATHERON J.-P., *Comprendre MERISE: outils conceptuels et organisationnels*, Paris, Eyrolles, p. 265, 1994.

[NAG 94] NAGY G., "Terrain visibility", *Computers and Graphics*, vol. 18, no. 6, p. 763-773, 1994 (http://www.ecse.rpi.edu/homepages/nagy/doclab.htm).

[RAN 94] RANDOLPH F.W., RAY C., "Higher isn't necessarily better: visibility algorithms and experiments", in Thomas C. Waugh, Richard G. Healey (eds.), *Advances in GIS Research: Sixth International Symposium on Spatial Data Handling*, p. 751-770, Taylor & Francis, Edinburgh, 5-9 September 1994.

[RAN 99] RANDOLPH F. Wm., "Terrain elevation data structure operations", *19e Conférence Internationale de Cartographie*, 1999, Tome 2, p. 1 011-1 020, 1994 (http://www.ecse.rpi.edu/homepages/wrf/research/index.htm).

[SER 97] SERRHINI K., "La métrique du paysage: deux indicateurs spécifiques au relief pour l'aménagement de l'espace", p. 259, *Actes des troisièmes rencontres de Théo Quant*, Besançon 1997.

[SER 99] SERRHINI K., "Evaluation de l'impact et de l'insertion de projets d'aménagement dans le paysage", *19e Conférence Internationale de Cartographie*, vol. 2, p. 1 639-1 646, 1999.

[SER 00a] SERRHINI K., "Le multiéchelles et la covisibilité pour une meilleure insertion des projets d'aménagement dans le paysage", *Mari Europe 2000*, Paris, 18-20 April 2000.

[SER 00b] SERRHINI K., Evaluation spatiale de la covisibilité d'un aménagement. Sémiologie graphique expérimentale et modélisation quantitative, PhD Thesis, Tours, 2000.

[SER 01] SERRHINI K., "Intégration quantitative et multicritère du paysage lors de la détermination de tracés d'un aménagement linéaire", *Mappemonde*, no. 61, GIP-RECLUS, p. 15-18, March 2001.

[THI 87] THIBAULT S., Modélisation morphofonctionnelle des réseaux d'assainissement urbain à l'aide du concept de dimension fractale, PhD Thesis, INSA Lyon 1, p. 144 *et seq.*, 1987, 300 p., see also Part 3: Graph and Fractals.

Chapter 11

Dynamics of Von Thünen's Model: Duality and Multiple Levels

The dynamic model[1] presented below is developed on the basis of the groundbreaking model of spatial analysis developed in 1826 by Johann Heinrich von Thünen [THU 26]. Let us point out that the purpose of the von Thünen model is to determine the location of cultivation in the space of the "isolated State", that is, an isolated plain containing at its center a market town. Von Thünen determines that cultures will form succeeding rings around the town according to the income of the farmer (Figure 11.1). This income is the difference between all the production costs plus the cost of transport and the price on the urban market. The producer attempts to maximize his revenue and the comparison of the potential income from each product according to the price observed determines the production for the following year.

The first issue, which calls into question a generally accepted hypothesis, is to find out if spatial balance can result from a production that is strictly based on the market prices of the previous year, by supposing that they were constant during the season, and be maintained through time. Should this hypothesis not be verified, what are the spatial consequences?

Secondly, generally speaking, what can be the spatial effects of the influence of two or several market town centers whose production zones overlap?

Chapter written by Philippe MATHIS.

1 This chapter follows up on a paper delivered at the "European colloquium on theoretical and quantitative geography", 1997.

The stress is always placed on the spatial characteristics of the phenomenon, i.e. the form of the supply zones of each town and of the production zones of each product.

The considered space is discrete. In our simulations it comprises 600 properties cultivating one of the three types of products (truck farming, silviculture, cereals) according to the prices of the previous year, production cost and distance to the town center which makes it possible to obtain the highest revenue.

Each year, the price in the town center(s) is a function of the ratio between the quantity produced and the quantity required which we will assume to be constant and dependent on the population.

The space is defined by a set of points divided in staged rows connected to each other by a network describes by a graph with triangular faces. The properties are located at the nodes of the graph and the delimitation of field surfaces is defined by the dual of the graph which is formed by the transport network. This surface may or may not be regular; here we chose the simplest solution of the equality of surfaces. However, a different hypothesis may be taken into account.

The method used is simple and consists of a first description of space with the definition of the characteristic points and the relations between these points: the graph. Then the particular nature of the relation is taken into account by using the von Thünen model. In the case of a diachronic comparison the process stops, whereas it continues in the case of a dynamic by successive iterations, whose resulting space in turn becomes the starting space and so on.

The method is perhaps less convincing than a mathematical demonstration, but it is not limited to a few very simplified hypotheses and the systematic exploration of a model, by freely modifying the characteristics of the spaces, makes it possible to clearly delimit its field of validity.

Within this framework the von Thünen model is only a part of a larger group.

11.1. Hypotheses and ambitions at the origin of this dynamic von Thünen model

Two objectives have formed the foundations of this work. The first is to find a dynamic use making it possible, among other things, to test the implicit hypothesis of temporal stability. The second is the use of the von Thünen model in conversational simulation while being able to modify any of its characteristics. We call a simulation conversational when for each iteration it is possible to modify the parameters on screen, to simulate decisions, to use the model practically as a game where we are able to interfere in real-time with a certain number of parameters.

First of all, we wanted to test the hypothesis of temporal stability of the von Thünen model in a dynamic version.

Von Thünen [HUR 94] does not specify the relation between the quantities produced and the prices. However, it appears that this relation potentially contains at least one instability related to the spatial nature of the phenomenon: if we adopt the simplest hypothesis of proportionality, the variation of prices involves a variation of the cost of transport and, thus, of the possible distance to cover between the place of production and the marketplace.

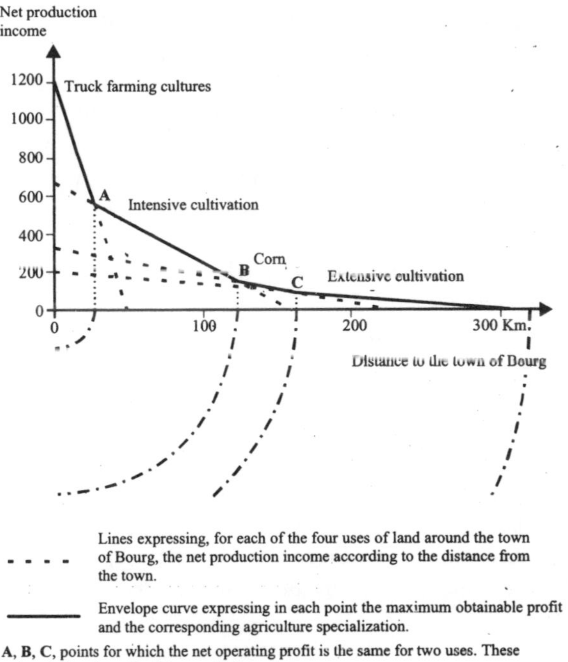

Figure 11.1. *Diagram of the producer's income for each product depending on the price in the market town minus all the production costs and the costs of transport*

Effective production is a function of the cultivated surface since we suppose constant returns, at least initially, and in the simplest von Thünen model this surface, which can generally vary according to the case, is a ring around the town.

Any variation in price according to the difference between the quantity required and the quantity produced involves a modification of the producer's income which is defined as the difference between the price and all the production costs plus the cost of transport. However, this is valid for each property. Since the production costs are fixed, the income depends on the costs of transport. However, the total variation of production is a function of the difference of the squares of the distances that these prices make it possible to cover, when all other parameters are identical for each producer.

Consequently, the stronger the variation in production will be, the greater the distance from production and the smaller the cost of transport. It will also depend on the product considered: the remotest production will be able to extend without obstacles, whereas the most central production must shift the others and thus generate an income that justifies the change of culture in the surface considered.

Thus, *a priori*, there is an amplification and instability process, all the more so since, in particular for foodstuffs with a constant demand, the elasticity of the quantity with respect to the demand is very low and, consequently, the elasticity of the prices with respect to the quantities is normally much higher than 1. This further amplifies the phenomenon and, thus, logically, its spatial consequences.

In addition to the development dynamics when we introduce cultivation rotation, the development of yield remains to be determined. Similarly, it is necessary to be able to introduce risks, differential developments in the productivity of the modes of cultures, work, transport, soils, in the consumptions, in populations, in a global, spatial, sector manner, etc.

The second ambition is to relax the hypotheses [HUR 94], to remove the most constraining, which are mainly those related to the mathematical tool, in order to be able to explore the capacities of this fundamental model in less simple environments, less theoretical than those used by the majority of economists. This implies, in particular, rejecting a uniform continuous space and the Euclidean distance.

We wish to be able to:

– simulate in real-time and interactively, and to immediately have the figures of production locations;

– freely change the characteristics of space: regular spaces of Christallerian type, irregular spaces, random spaces generated by a model, etc.;

– change from the transport function to demand, whether proportional or not, including load ruptures, but also the existence of several means of transport;

– vary the number of towns on request;

– analyze exactly the problems of production zones limits in the case of several market towns;

– freely change the production costs, introduce qualities or specific "vocations" of the soils;

– explore the influence of spatial scales.

Finally, it seems necessary to be able to explore the consequences of behavioral inertia, instantaneous adaptation, training and anticipation. Therefore, this implies having a basic dynamic structure, into which we can gradually integrate new dimensions.

11.2. The current state of research

Due to the extent of the combination of the various hypotheses that have to be explored, only initial results are presented here. The main objectives, such as the simplification of the hypotheses and real-time simulation have been achieved. The dynamization of the model made it possible to confirm the hypotheses of instability. The introduction of risks is possible, but has not yet been analyzed because the consequences of dynamization appeared significant. Culture rotations and behaviors still need to be introduced: inertia, training, anticipation, instantaneous adaptation being the selected hypotheses.

11.3. The structure of the program

The dynamization of the von Thünen model that we carried out a follow up on the models of space description that constitute part of the D.LOCA.T.[2] "software" and uses a visual BASIC routine to implement the parameters, MAP[3] software for the "maps" and Excel for graphs.

2 D. LOCA.T is a patented software developed by members of the CESA laboratory.

3 MAP is a patented software developed by members of the CESA laboratory.

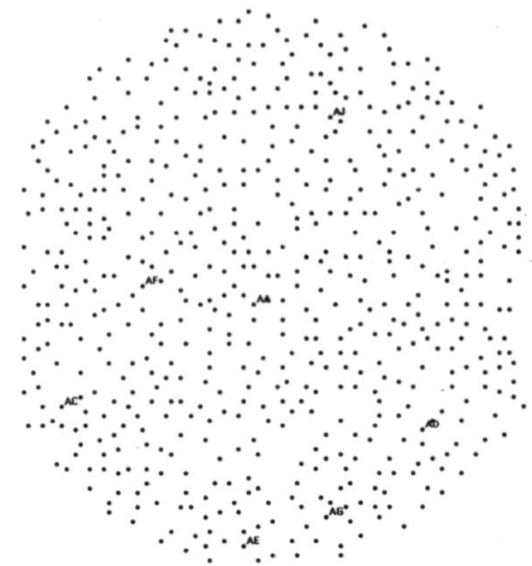

Figure 11.2. *Random space with 600 fields locations and 1 to N possible towns. Seven towns were kept in this space*

It is implemented in 32 bit FORTRAN, uses the Power Station Microsoft compiler and runs on a desktop computer. The space presently used is either a random space (600 – Figure 11.2 – or 1,225 zones or fields: both spaces look very similar), or a space of the Christaller type (1,225 zones or fields – Figure 11.3) which is generated by the RES[4] program, which makes it possible to obtain networks with planar graphs.

The current limit on the number of points is due mainly to the calculation time, in particular, that of minimal distances because Floyd's algorithm has a complexity of O(n3). This doubling of the number of nodes involves a multiplication of the calculation time by eight, which means a quadrupling of the size of the matrices and thus of the necessary memory. However, the programs are conceived to be usable regardless of the number of points used.

The program of the dynamized von Thünen model imports a file of properties (or producers) locations, a file of markets towns and a file of distances from the properties to the market towns. The necessary parameters are fixed by using an interface, i.e. a window, in visual BASIC, which makes it possible to define on screen the desired values for these various parameters, either at the beginning of the

4 RES is a program developed by Hervé Baptiste, which makes it possible to generate a space where random locations are connected by a saturated planar network. Towns or central squares can be assigned to these locations with possible population constraints. The model can generate more than a million different spaces. See, by the same author, [BAP 99].

period, or during each iteration. The fields or the market towns are located at the nodes and the arcs represent the circulation network (Figure 11.3).

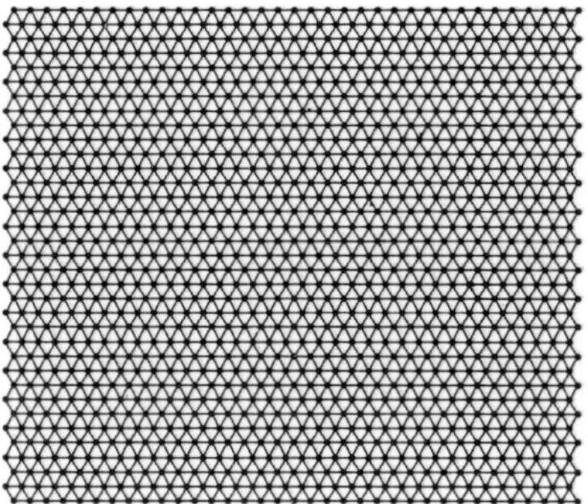

Figure 11.3. *Christallerian circulation network*

The dynamic simulation space could be defined as: 1,225 locations, seven kilometers between each and two central sites with 200,000 inhabitants each[5]. In this simple and regular case the limits of the properties are defined by the dual of the circulation network, which is a regular, planar and saturated graph. This network has a graph with hexagonal faces as a dual.

On the basis of production and transport costs, distances and prices of products in the various towns, the program determines the potential income from each production for each property, and all the farms specialize in the production that maximizes the private income.

The model respects the hypotheses 1 to 5 described by Huriot [HUR 94, p. 265 *et seq.*]. Currently, the model exhibits myopic behavior, since it does not anticipate and it maximizes the expected profit only for this period.

At the beginning the conditions are as follows:

– no soil use;

– no predetermined zones;

– by simplification all the farms are identical, but this is not necessary; specific characteristics may be input for each farm (surface, quality of land for each product, technical profitability, costs, etc.);

– three products are taken into account, but we can use more, if so desired;

---

5 See section 11.4.3.

– no *a priori* capital constraints, but they could be easily introduced for each farm;

– the market is not unique: there are as many of them as there are towns which are taken into account. Thus, for the same product there is a specific price for each point of sale, which is a function of the relation between local supply and demand;

– there is no exchange of products between various markets and demand is confined to each one of them;

– the choice of the production for each farm is a function of the maximum revenue expected income by considering the revenue, i.e. the prices of the previous period minus production costs and the related transport costs. In the long term it is envisaged that preserving inertia, learning and anticipation will be integrated.

Let us observe that the decision is made traditionally: each property chooses the product that generates the highest income for the most interesting market in the case of several market towns. Thus, there is a comparison of income for each product and market. Naturally, there is no production at a loss. The choice is therefore more significant for the producer when there are several markets, since by assuming that there are three products and five towns, he has to take into account 15 potential incomes! That renders the determination of the production zones of each town much more interesting since these zones become all the more dynamic.

Determining the total quantities produced is simple when the product chosen for each farm, the production surface, the output and the place where it will be marketed are known. We can sum up for each market town the quantities of each product that will be offered for sale there.

Thus, we have a cellular model where each farm is a "cellular agent" with a behavior pattern that is the same for all in this case. Each farm individually decides its production and the market where it will be sold according to its particular interest: maximization of revenue.

The cellular model defining the nature and quantity offered for each market makes it possible to move on to the global level of pricing for each market which is obtained by comparison between local supply and demand. The difference in price depends on the difference in quantities.

The prices in year T, which are determined overall for each market, will influence the choices of production of the agents during year T+1 and thus the production distribution space, but this will depend on all the prices in each market.

Consequently, the supplying zone of a town does not depend only on the prices established there but also on prices in other markets that are a function of their populations and distances.

The interrelationships are thus essentially of a spatial nature, since they make it possible to pass from the level of the agent to the global level. The reasoning is not the same: it is necessary to know all the production to define the price, whereas the latter enables each agent to define his production in the next period, while remaining completely unaware of the reaction of other agents: the agent's rationality is limited.

11.4. Simulations carried out

The exploration of the model mainly related to the effects of the space scale factor, the effects of transport costs, price trends, the effect of price/quantities elasticities, the effect of the number of towns, the shifting of the borders of production zones, which implied a significant number of simulations from 20 to 150 iterations.

Four simulations are presented here: the first two refer to a random space with 600 farms and with a single central town as in the original model, the difference being in the value of the damper coefficient of the price/quantity relation. These two simulations emphasize the extreme sensitivity of the model.

The third simulation refers to a regular Christaller space with two identical towns. This "academic case" makes it possible to highlight the variations in the spatial location of the border between the two provisioning surfaces and the strong variations of production surfaces of the various products.

The last simulation again uses a random space of 600 fields where we find six market towns of unequal importance, which, in a more complex case, makes it possible to predict drive and spatial domination effects.

Each simulation comprises the various parameters used, one or more graphs of the product's price variations according to time and a simplified representation of the production areas of the various products and the provisioning zones of the various towns, if necessary.

Each simulation begins with specific characteristics: the space used, the various files, the number of towns, products, iterations to be made, the useful agricultural surface, a technical factor which is a coefficient of scale, the type of output, the type of transport costs, the growth rate, the type of output of the land followed by transport costs per product, the unitary per capita consumption, the production costs per product and the absorber or amplification coefficient of the price/quantity ratio.

The produced price/quantity ratio used in simulations has a very simple, linear form:

pppactkj = pppantkj x (100 + epqkj x (-dpjk)/100) if dpjk < 0 [11.1]

pppactkj = pppantkj/(100 + epqkj x (dpjk)/100) if dpjk > 0 [11.2]

where pppactkj is the current price of the product k for the town J, pppantkj is the former price of the product k for the town J, epqkj is the price/quantity absorber coefficient and dpjk is the difference between the quantity produced and quantity demanded expressed as a percentage.

Let us recall that the choice of production in year n is individual for each farm. It is determined by the highest revenue that the prices of year n-1 generate. The variation in production volume is the sum of all the individual choices.

11.4.1. *The first simulation: a strong instability in the isolated state with only one market town*

The windows, through which the various parameters are implemented, are preserved in the form of files. That makes it possible to define a tree structure of simulations and to explore systematically.

The first simulations immediately showed the great sensitivity of the model and its instability in time, contrary to the traditional assumptions of the existence of a stationary balance [HUR 94]. The simulation presented below and illustrated by Figure 11.4 is a typical example.

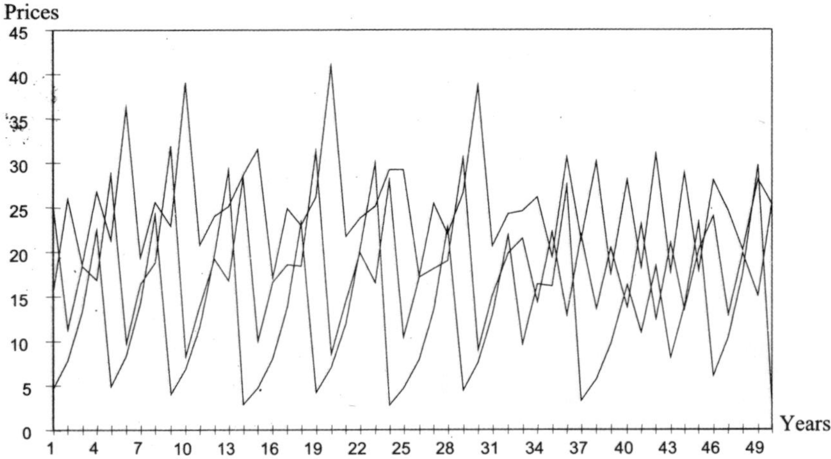

Figure 11.4. *Price of the three products to the nearest constant in the market town, with a price/quantity elasticity of 0.7*

The first simulation uses a price/quantity elasticity of 0.7, which could have been replaced by elasticity in its proper sense. The surprise comes from the fact that traditionally the price/quantity elasticity is considered to be greater than 1 for foodstuffs, whereas we systematically had to use a reverse multiplier between prices and quantities.

Despite that, the fluctuations are very strong from one year to the next, and the curve obtained exhibits a quasi-periodicity and a behavior that is unforeseeable in the short term. This quasi-periodicity, which evokes a chaotic phenomenon, is the result of the interaction of the three products chosen with individual rationality according to the income of 600 agents that are independent, but localized spatially.

Here, the sum of the individual choices does not lead to the overall optimum. The hypothesis that we can formulate on this subject is that any increase in price involves the possibility of cultivating a product at an additional distance Δ with respect to the prior distance d. However, this additional distance Δ from the market center enables an additional production on a surface that increases proportionately to d and is equal to:

$$\delta S = [\pi(d+\Delta)^2 - \pi d^2] \quad [11.3]$$

that is, $\delta S = \Delta\pi(\Delta+2d)$ [11.4]

and production will thus be of type $P = f(\pi\Delta(\Delta+2d))$ [11.5]

Secondly, it is clear that the modification of one or several production surfaces will consequently involve an adaptation of the other surfaces. Indeed, only the growth of the third ring will have very little influence on the two preceding ones, whereas a strong increase in the prices of this product will lead to an increase in income and, thus, a reduction of the previous product in the space that has become less profitable "to the space margin".

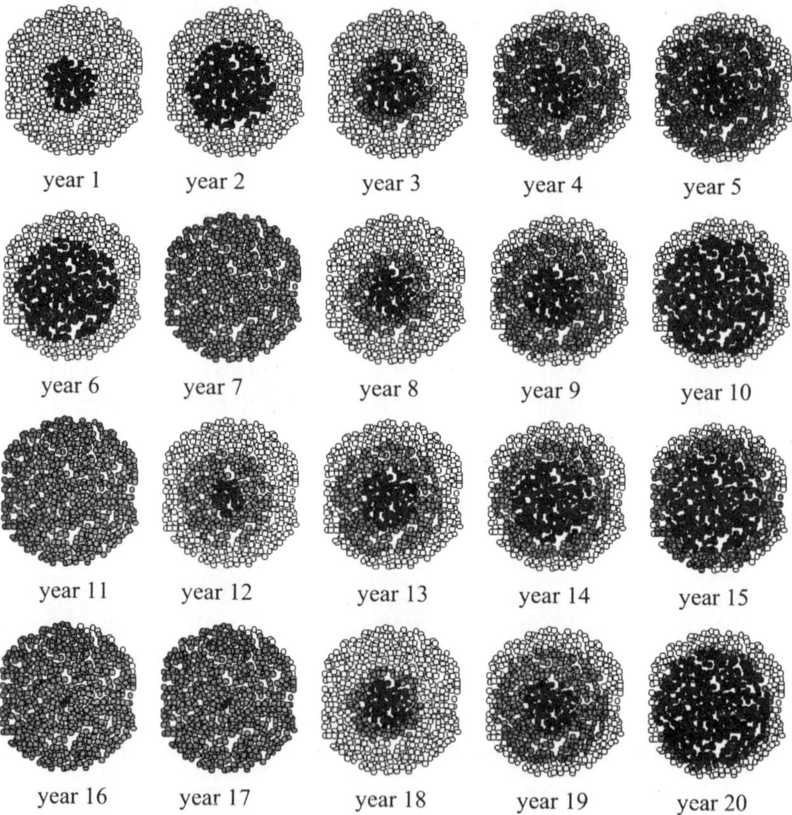

Figure 11.5. *Variation in the space of the cultivation zones of the three products*

On the other hand, a variation in the price of the product cultivated in the second and, all the more so, in the first crown will generate an adaptation of all the products that can be reflected in time. Indeed, *a priori*, nothing makes it possible to consider that the variation in the price of one or several products related to the inadequacy between production and consumption will make it possible to solve the problem the following year spontaneously and exactly. There are multiple reasons for this. The main one stems from the spatialized character of production: an increase or reduction in price involves an increase or reduction which is a function of the square of the distance variation to the market.

We have here a mechanism of amplification where the disturbances will not necessarily diminish, as observed over the first 20 periods of the simulation (Figure 11.4).

The effect on space is considerable, since we note a great instability not only of the cultivation zones, but also of the cultures themselves, which will appear and disappear according to market prices (Figure 11.5).

The maps indicate the zones of cultivation in the first years, but taking into consideration the price graph clearly indicates that the phenomenon persists in time with perhaps a quasi-cycle long superposition.

11.4.2. *The second simulation: reducing instability*

The second simulation differs from the first only by the value of elasticity which is 0.3 here. We can immediately note the effect on the price curves (Figure 11.6), curves whose variation amplitude is restricted, but always preserves this quasi-periodic character with a duration of eight "years", which evokes the Juglar cycle[6], during which the fluctuations seem to be limited in order to regain their amplitude of the start.

In this simulation, space "competition" develops primarily between products 1 and 2, as we can observe from the graphs, which we have increased to this end.

The product whose fluctuation is the strongest is the one produced at the edge of the town, i.e. truck farming, then comes silviculture and, finally, cereals on the outside.

We note that only a considerable fluctuation in the price of product 1 would enable an extension of the surfaces thus triggering a weaker fluctuation of product 2, which will push back product 3 by generating a price hike the following year. This one-year shift between the price maxima of each product is explained by the increase in production surfaces following the increase in income (Figure 11.7),

6 A cycle highlighted by the French economist Clément Juglar in 1862 that has a duration of seven to eight years and is an intermediary between the short and long Kondratiev cycles.

which involves a reduction of the surface dedicated to the next product, whose income has changed only a little.

We note a quasi-periodicity of change during the first 30 years which is followed by a period of low amplitude fluctuation of only two years and then the phenomenon restarts with strong amplitudes.

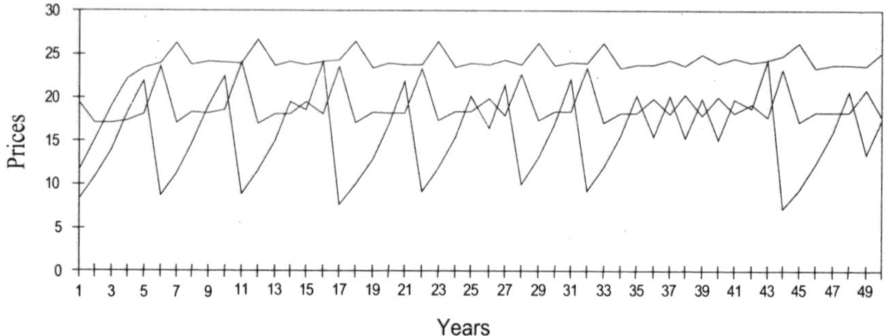

Figure 11.6. *Price of the three products to the nearest constant in the market town, with a price/quantity elasticity of 0.3*

It is clear that such fluctuations could not happen so radically in reality, although there are known cases where certain productions were partially supplanted by others, such as, for example, the development of kiwis at the expense of apricots, or completely as witnessed by the disappearance or quasi-disappearance of certain types of fruit or domestic animal races. It would be necessary, however, to introduce production cycles into the model, because certain cultures, such as silviculture, take many years to develop and due to this fact the fluctuations are damped by a "delay constraint".

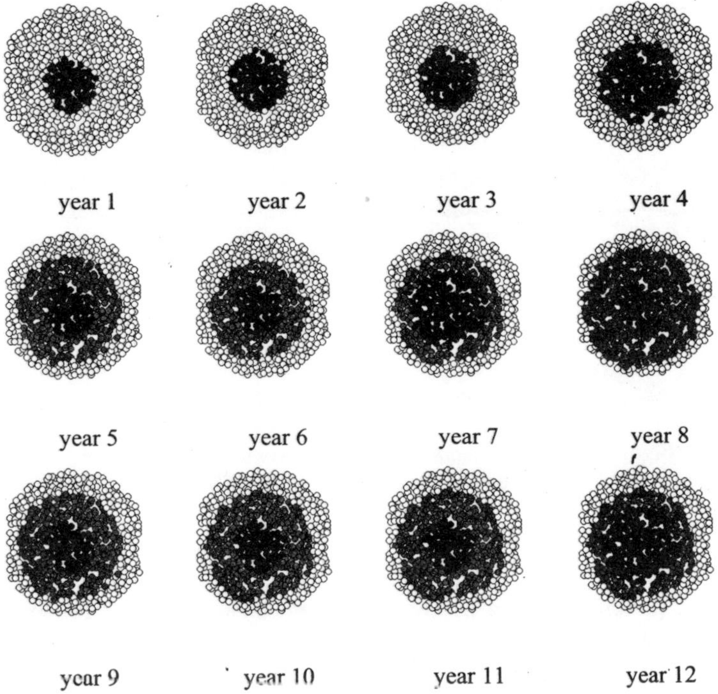

Figure 11.7. *Variation in the space of the cultivation zones of the three products with a damping effect*

11.4.3. *The third simulation: the competition of two towns*

We used a Christaller space (Figure 11.3) conforming to the one described above, with two equally important market towns, which are symmetricly located in this homogenous space.

The first observation is that, even in this case where all the factors seem identical, a very fast differentiation between the two towns appears, immediately for one product and after two and four years for the two others (Figure 11.8).

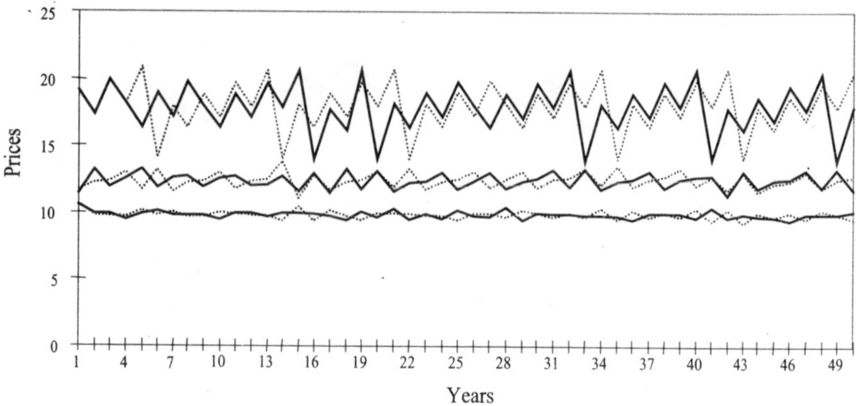

Figure 11.8. *Price trends for the three products in both towns. The prices are evaluated to the nearest constant. There are 1,225 sites*

It should be noted that the quasi-periodicity remains a constant and that, during the simulation period, the order of the fluctuations changes: the prices for products in the same town are not the first to vary. That is particularly visible for product 1 (top curve) where the dotted lines represent the prices in the first town and continuous lines represent those in the second. We observe a quasi-periodicity, but with shifts between the towns.

We establish the origin of the fluctuation by carefully observing the production zones: the number of fields between the two towns is odd and this is a phenomenon to which we had not paid attention at the beginning also in programming, which is necessarily discrete, the town on the left is selected first and therefore its production zone is the largest, which triggers the fluctuation amplified by the appropriation of the most distant quasi-median field for cereal production, due to the systematic rounding-off. The town on the left benefits from three additional fields compared to that on the right, that is 0.5%, from a total of 627 really productive fields for the two towns out of a potential 1,225, when taking into account the selected parameters. Space is intentionally oversized, so that the cultures are not restricted

This inequality of production surfaces triggers a differentiation in prices, which in turn sustains the phenomenon, even more so since the layout of the fields in a quincunx in turn introduces an inequality which we observe starting from the second year: five extreme zones for town A and four for B. The inequality persists when the zones of cultivation arrive at the limit of the expanse taken into account.

It seems that we are witnessing a "butterfly effect" at the start, which is later amplified as it is clearly shown by the sometimes complex fluctuation of the zones by product and of the border of the production zones of the two towns which can be seen below.

The graph (Figure 11.9 in the color plates section), since it would be inappropriate to call this a map, on the left represents the production zones of each product for both towns: product one in blue, the second in green and the third in red. The fields that do not provide any production to the urban market were left blank. The graph on the right represents the production zones of each town: the fields whose produce is sold in the town market on the left are in blue and those in the town market on the right are in green.

The apparent symmetry of product zones disappears from the second year for products 2 and 3 and the phenomenon is amplified throughout the later years until the near disappearance of product 1 (truck farming) for the town on the right during the 7th year, and the total disappearance during the 15th year for the same town. The cereal zones also fluctuate alternatively between the 14th and the 17th year.

Moreover, we can note that the expansions and reductions of the surfaces occupied by each culture do not fluctuate simultaneously, as it is clearly shown in year 15 for the town on the right where we have the disappearance of truck farming at the same time as a strong expansion of product 2 and a reduction of the cereal zone. We could provide multiple examples of this type.

With respect to the border between the provisioning zones of the two towns, the results here are very close to those of a Palander model between two companies, of which we have here almost a partial simulation. Indeed, the shape of the border in is the majority of cases a discrete hyperbole and the costs of transport are evaluated from the fields to the markets, whereas for Palander they are estimated from the production facilities to the market, which in both models symbolize the slope of the costs of transport.

More generally, the graph of product's prices for both towns alone does not make it possible to measure the spatial complexity of the development of the production zones, which in turn will modify local supply and therefore prices, thus leading to new developments of the production zones.

11.4.4. *The fourth simulation: the competition between five towns of different sizes*

This simulation has the same parameters and the same space as the first two, but takes into account five towns of different sizes: 132,000, 53,608, 31,645, 21,771 and 16,289 inhabitants respectively, which are distributed according to Zipf's law.

This inequality of populations leads to the inequality of the quantities necessary for subsistence and, consequently, of the prices in each market.

This more complex set approaches reality and, although it is more difficult to interpret, it makes it possible to analyze the possible phenomena of drive, domination, etc.

Considering the various graphs makes it possible to detect strong tendencies (Figure 11.10).

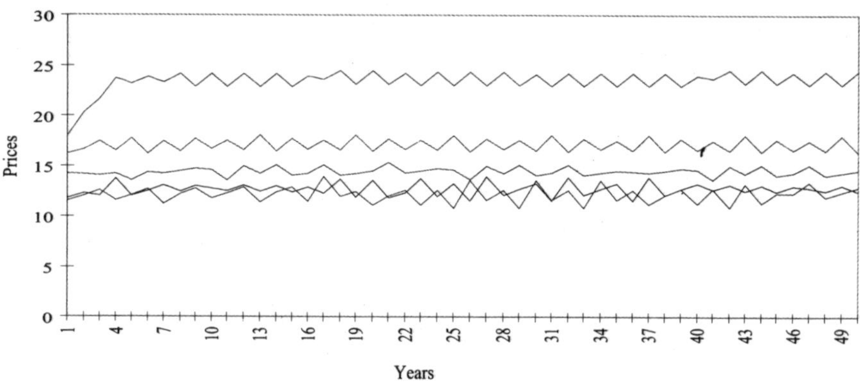

Figure 11.10. *Price of each product for the five towns*

At the beginning no surfaces are used and the prices per product is identical for all towns are arbitrary for year 0.

The first observation is of a very different nature from the curves of product 1 for the five towns, whose location is clearly visible in the graphs (see Figures 11.11 and 11.12 in the color plates section). While for the largest one the curve is relatively regular in its quasi-periodicity, the others, on the other hand, seem to be subjected to an influence that makes each of them specific, which means that the least populated towns have the most irregular curves, i.e. the fifth town. We find the same type of phenomenon attenuated for 'the second product, with a clearly marked irregularity for the last three towns.

For the third product, although the amplitude of variations is rather low as a whole, nonetheless, it appears stronger for the smallest, which may be explained by the fact that its territory is the most constrained by the three most populated towns that surround it. Its production zone is limited because the producers prefer to market their produce in other markets where it can be sold at a higher price, which repeats the development of the production zones. Everything occurs, as if the smallest towns were, consequently, much more sensitive than the largest.

Indeed, a rapid analysis of these makes it possible to note that the first five, six or seven years are dedicated to the growth of the production zones before an almost regular fluctuation of the borders between the zones dominated by each town appears.

We may also observe a fluctuation of the center-periphery type, which is confirmed by the analysis of the price graph of product 3 for the five towns: to each minimum of the most populated town center there corresponds a maximum for the other four towns, which is not true for product 2, for which we have three maxima for two minima and vice versa.

With regard to product 1 whose fluctuations are the most considerable and whose role is essential, the periods are different and shifted in time.

This periodicity, as if the four peripheral towns did not reach their maximum market price simultaneously, triggers a sort of a circular motion at the periphery, which seems to be corroborated by taking into consideration the production surfaces per product (Figures 11.11 and 11.12 in the color plates section).

On the one hand, product 1 does not reach its maximum expansion simultaneously in the center and at the periphery, but, on the other hand, it does not reach it simultaneously for all the towns at the periphery either. That seems to occur as if we had a rotation of maxima for this production, which followed the trigonometric direction.

11.5. Conclusion

The first conclusion is the corroboration of the initial hypotheses:

– on the one hand, the von Thünen model appears unstable in its present configuration and with the simplest proportional price/quantity function. This seems to stem from the fact that an increase in price and therefore in the distance at which the goods can be produced involves an increase in cultivable surfaces as a function of the square of this distance;

– on the other hand, not only is it perfectly possible to generalize the model, but also the more finely a space is described, the more interesting it appears since the complexity of data enriches the model and reveals other types of phenomena;

– the importance of the spatial dimension (location of production, analysis of the limits of the production zones) is evident and it makes it possible to find and generalize the results and to join the Palander and Devletoglou models [DEV. 65, p. 140-160].

The conclusion on instability must be refined because a certain number of existing "dampers" have not yet been taken into account, such as, for example, the fact that in silviculture once a piece of land has been planted its use is not modified the following year and the necessary time is left for the trees to grow before cutting them down and the fact that there are constraints in the succession of the cultures, etc. Nonetheless, even by taking these significant nuances into account, it seems that the invisible hand has limited efficiency. Due to production spatiality, the choice of each producer according to the potential income is not enough to ensure a balance between supply and demand in this case.

The second major conclusion that makes it possible to highlight dynamic analysis is the chaotic behavior of the model: fluctuation, pseudo-periodicity, butterfly effect.

A significant number of fields where production is determined individually and rationally according to the price of the previous year and by maximizing their income according to two or several markets in no way ensures either price stability or production localizations stability.

The individual and immediate rationality of the behavior seems to increase the total instability. However, the fluctuation mechanism generates its clear borders itself: in general, the chaotic system is permanent, although certain parameter constellations produced systems with "explosive" prices.

The total stability of the prices seems inversely proportional to the spatial stability of the cultures and it also seems that the relative stability of cereal prices is the consequence of the greatest capacity of this product to fluctuate in space. This element appears fundamental: spatial disposition constitutes an absorber for fluctuation, which is corroborated by "explosive" simulations that we have carried out because in these specific cases, due to the inability of supply to adapt to demand, prices constantly continue to increase. In this case, the resolution to the crisis lies in other means than those described in the model.

The simulation makes it possible to show that if production is directly a function of the income, that triggers fluctuations in prices and production zones. Indeed, the variation of income implies a variation of the possible distances from production, which involves a variation of the surface of production, which is a function of the square of the radius, thus leading to a systematic amplification of production margins and therefore of price and so forth. This triggers pseudo-periodic chaotic development.

In the model, the spatial dimension comprising all the fields or cells makes it possible to connect the agents who individually decide on their production according to the revenue that enabled the prices of the previous period, thus exhibiting a short-sighted behavior and a market mechanism which fixes the prices with respect to supply and demand.

The simulation can be considerably enriched by taking into account various types of constraints that make it possible to approximate real situations. We can thus reproduce complex developments of production surfaces by type of good or market town, or follow the developments of the borders of productions zones, etc.

Having said that, many developments are necessary to exploit all the possibilities of the model, in particular the risks, the specific qualities and the introduction of stabilizers, such as the constraints on production. It is necessary to explore other price/quantity functions, but especially to introduce a price memory and a learning capacity, to open the model to the outside, to really introduce multimodality of transport, in particular mass transport between the various centers or market towns with means of transport of great capacity, etc.

In conclusion, we have developed a traditional cellular model on the basis of the traditional von Thünen model by using a description of space by graphs, which clearly showed the concrete possibility of passage between a graph and a cellular diagram, once the dual is used, as it is here, to take the surface into account. In this case we only used the dual to delimit the borders of the fields by considering them all as being equal in order to simplify matters. However, duality makes it possible to vary these sizes[7] considerably and, moreover, we can incorporate fields together to simulate phenomena of land concentration or agreements.

In the present case, the use of multi-levels is limited to individual choices when taking into account the effect of the overall balance or imbalance by evaluating the local offer and during the spatial analysis of the structures generated by the cellular representation.

Despite all the extremely strong limitations of this model, and beyond its immediate objectives, we believe that its chief interest seems to lie in the beginning of the corroboration of our fundamental hypothesis which was presented and expressed throughout the third and the fourth simulation, for which passing from the individual to the global level can only be achieved through the introduction of space.

7 See Chapter 10.

11.6. Bibliography

[BAP 99] BAPTISTE H., Interactions entre le système de transport et les systèmes de villes: perspective historique pour une modélisation dynamique spatialisée, Thesis, Tours, 1999.

[DEV 65] DEVLETOGLOU N.E., "A Dissenting View of Duopoly and Spatial Competition", *Economica*, vol. 32, p. 140-160, May 1965.

[HUR 94] HURIOT J.-M., *Von Thünen, économie et espace*, Economica, Paris, 1994.

[PUM 01] PUMAIN D., SAINT-JULIEN T., *Les interactions spatiales*, Collection Cursus, Armand Colin, Paris, 2001

[THU 26] VON THÜNEN J.H., Der isolierte Staat in Beziehung auf Landwirtschaft und nationalökonomie, vol I Hambourg, Perthes, 1826.

Chapter 12

The Representation of Graphs: A Specific Domain of Graph Theory

12.1. Introduction

A certain number of recent publications by the various teams of the GDR "LiberGeo" have all simultaneously demonstrated the efficiency of graph theory in describing networks and their various characteristics, such as speed [LHO 97], the characteristics of transport supply and optimization [CHA 97], mass and value flows [ROB 00], structural dynamics [BAP 99], but also its deficiencies in terms of morphological indicators, which can be the effectively mitigated by using the properties of fractals [GEN 00].

Upon reflection these various publications do not seem to be contradictory since, although each attempts to develop its own tool in a relatively exclusive manner, they are actually deeply complementary and close. It results from this observation that fractals could be regarded as representations of graphs under certain conditions.

This hypothesis was implicitly reinforced by the fact that the representation or drawing of a graph is an aspect that, curiously, has been completely left out in graph theory because, apparently, it is not an important problem within the mathematical research dynamics, whereas it appears essential for users and, in particular, for network modeling.

Chapter written by Philippe MATHIS.

12.1.1. *The freedom of drawing a graph or the absence of representation rules*

The graphic representation of G is extremely simple: the location of the nodes in the figure, i.e. drawing of the graphing, does not count, as opposed to the representation of the relations between nodes. That obviously supposes that nodes are simple points without declared attributes of the coordinate type. This is interesting because it gives great freedom of graph representation. However, in order to represent, for example, a transport network, and if we wish the result to "resemble" the observation, i.e. to be similar to a map, it would be appropriate to make this representation more precise by supplementing it with additional properties or constraints so that the process of developing the representation can be repetitive and the result reproducible.

However, the usual representations are done on paper or on a computer screen and are therefore plane representations. This type of projection, whose nature is implicit in the majority of publications on the subject, is actually neither obvious nor traditional. Indeed, if we go back in time, the founding fathers of the theory showed more diversity than what has been preserved through usage.

12.2. Graphs and fractals

It is not mentioned that the representation may be "projected" onto a plane or a three-dimensional space, as done by Euler [EUL 1758], or be the projection of one onto the other, as in the famous example of Augustin Louis Cauchy [CAU 76] shown in Figure 12.1.

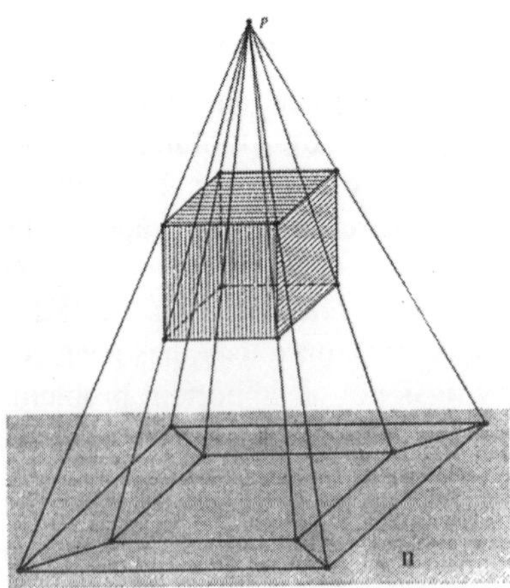

Figure 12.1. *Cauchy projection (Figure 5.5 in [BIG 76])*

Poinsot will develop work on polyhedra but, most of all, it is Lhuilier and Headwood who will develop projections on a sphere and a torus (see Figure 12.2) [HEA 86, LHU 86, POI 86], which is particularly interesting for the construction of a dual.

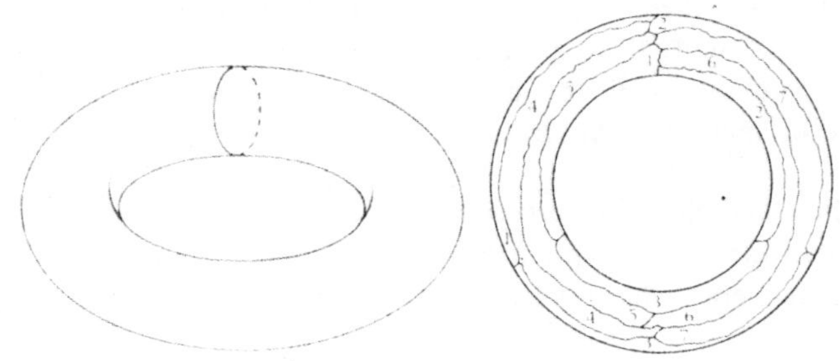

Figure 12.2. *Headwood projection (Figures 5.8 and 7.6 in [BIG 76])*

It thus clearly appears that a graph can be drawing on a plane which can be considered as a sphere of infinite radius. Therefore, on a sphere whose radius is not infinite and on a transformed torus that is topological of the sphere, the draw must specify it by taking into account the consequences that it may have. This is the particular case for the definition and the drawing of the dual, which presupposes the existence of an infinite face, which is obviously not always the case.

The implicit hypothesis of Berge therefore cannot be sufficient and it is not only important to know how the nodes are connected, but also on which type of surface the graphic representations of the graphs are drawing.

Let us consider the simplest case, i.e. that of Cauchy, where it is obvious that the representation suggested is also a particular case: a projection onto a plane parallel to one of the faces of the cube, whose projection center is located on the median of the face considered.

Many other projections of a cube onto a plane are conceivable such as, for example, the particular one where the projection center is located on the extension of a diagonal of the cube and of course without counting all the intermediate situations which generally produce regular forms. Thus, the number of possibilities grows considerably without necessarily modifying the number of nodes and arcs, and all may be the projection onto a plane of the same graph with three, four, or N dimensions.

It is obvious that if the representation of the graph is automated and repetitive, it is necessarily so by means of an algorithm, which must be accessible in its principle and its source code, lest the whole programming of the software used should be questioned. However, the current algorithmic [BAN 01, COR 94] and discrete geometry publications do not deal with the problem, so to speak.

The problem that consists of supplementing the traditional methods of producing graphs so as to make them reproducible and therefore verifiable is, thus, relatively obvious. This is essentially a problem of rendering objective a process whose algorithms probably already exist in several software, the access to the source code of which is unfortunately not yet free.

On the other hand, the initial problem of whether we can describe fractals in graph form and create their graphic reproductions with the same properties of self-similarity and internal homothety as the simple Mandelbrot fractals is much richer.

12.2.1. *Mandelbrot's graphs and fractals*

Benoît Mandelbrot himself described in his work [MAN 77, p. 46 *et seq.*] the generators of his fractals as linear graphs, without, however, naming or individualizing the points that he very clearly makes appear at the ends of the segments of straights as shown in Figure 12.3, which have been selected among many examples.

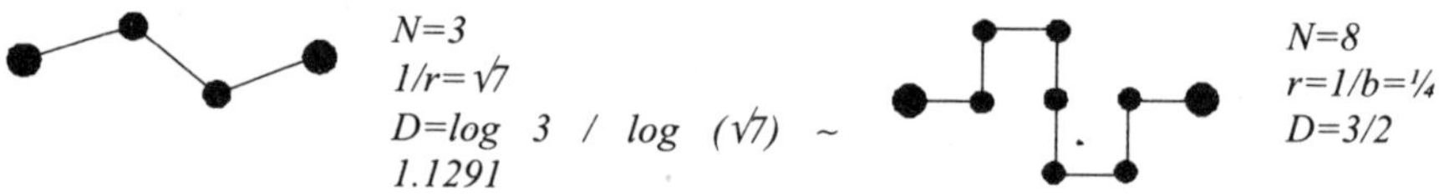

Figure 12.3. *Mandelbrot's fractal generators where N is the number of arcs, R is the increase rate and D the fractal dimension [MAN 77]*

We may consider without any difficulty that the generator is a simple graph, whose drawing, however, has particular characteristics, such as flatness, the number of elementary segments, their sizes and their successive geometrical positions (angle to the previous, area of space, etc.).

Figure 12.4. *A first stage graph (here a simple arc) and the adjacency matrix associated with the graph*

We can then individualize the nodes by naming them and this generator could thus be implemented on multiple scales by studying the size of the elementary segment.

Let us consider the first graph (Figure 12.4): we have the arc x, y: 1,2 which is defined by the adjacency matrix. The graph drawing is member of a plane, the arc has a length = l and it is horizontal.

Let us consider the adjacency matrix of the second graph (see Figure 12.5): we proceeded to subdivide the graph which meant creating two intermediate nodes on the arc of the previous graph, since the length of the new arcs is equal to 1/3 of the previous arc, that is r = l/3.

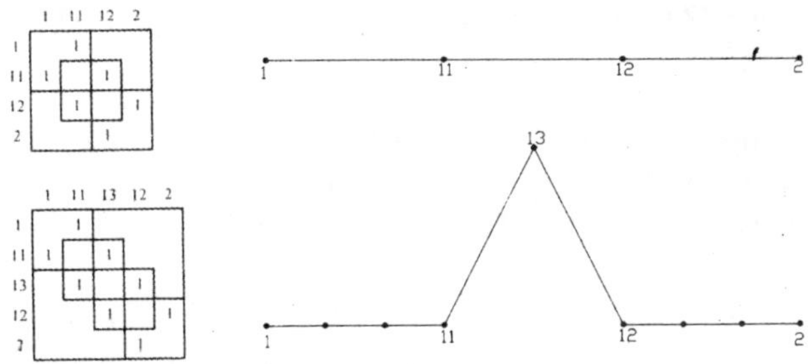

Figure 12.5. *Second and third stage: extension of the graph and replacement of the central arc by two arcs of the same length in the positive part of the plane and associated adjacency matrices*

It has to be said that in the case of an extension of the graph by intrapolation, the created nodes 11 and 12 must be inserted between the previous nodes. If this simple constraint is followed, we note that the second matrix is self-similar to the previous one and that, if we divide this matrix into five 2*2 submatrices, the three diagonal sectors are identical to the first level matrix.

We will note that the graph drawing must follow the order of the adjacency matrix.

Let us replace the central arc by two arcs of the same length, that is l/3, as illustrated in the third diagram. We have created a new node 13, which is inserted between nodes 11 and 12 of the previous matrix (Figure 12.6). The new matrix still possesses the property of self-similarity.

In order for the drawing to be reproducible, we specify the location of the new node with respect to the previous segment which must be situated in the positive (or negative) part of the plane.

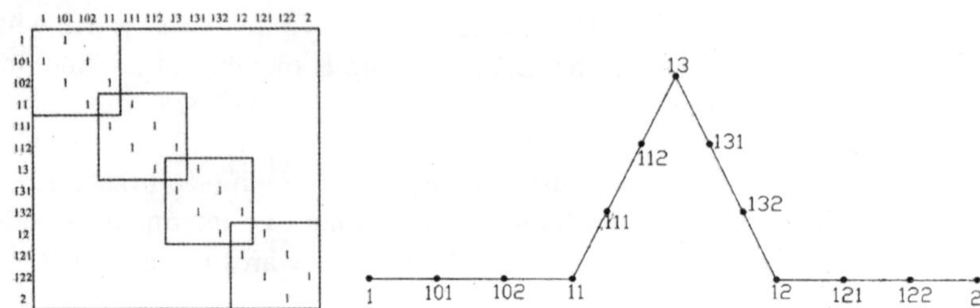

Figure 12.6. *First iteration of the graph of the von Koch curve*

We have the first level of the von Koch curve and therefore we only have to repeat the same sequence of operations for each arc to obtain the second level of the von Koch curve, which is illustrated by the following diagram (Figure 12.7).

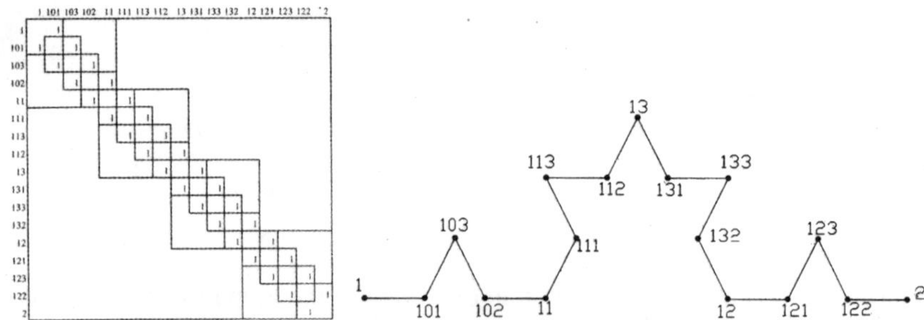

Figure 12.7. *Second iteration of the graph of the von Koch curve and self-similarity of the adjacency matrix*

We can conclude from this first example that:

– the von Koch curve can be described in graph form;

– as long as the new nodes are inserted between the previous ones, the adjacency matrix possesses the property of internal homothety and develops identically to itself, since each indicated submatrix reproduces the adjacency matrices of the previous level;

– the graph drawing must obey specific constraints in order to reproduce the fractal in its traditional form:

- the drawing must be plane,

- the arcs are segments of a straight line,

- the length of the inserted arc has a ratio r with the original arc,

- the middle arc is replaced by two arcs of the same length,

- the decimal notation of the nodes created on an arc adds a figure to the notation of the node on the left. In the case of replacing the initial arc by two arcs of a length identical to that of the original arc and the creation of a node of the second degree (subdivision), the notation is on the same level as that of the two nodes of the replaced arc. This notation makes it possible to know the level of each element and the length of each arc considered.

12.2.2. *Graph and a tree-structured fractal: Mandelbrot's H-fractal*

The von Koch curve is a typical example of graph extension by intrapolation. Rightly, the question arises about the possibility of graph extension by extrapolation. Wc will tackle this question by using Mandelbrot's H-fractal (Figure 12.8) [MAN 77, p.164].

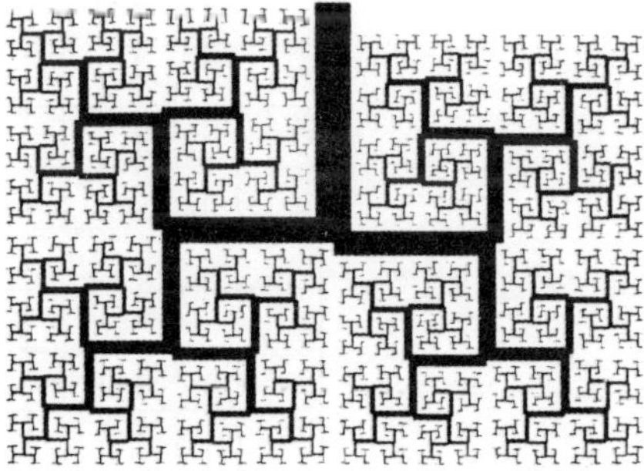

Figure 12.8. *Mandelbrot's H-Fractal [MAN 77]*

Let us consider Figure 12.9 starting from node O of the plane, where we construct two co-linear arcs on both sides of this node. The adjacency matrix is obvious: a symmetric (3,3) matrix.

We perform the extension of the initial graph by adding an orthogonal arc (segment of a straight) at each of its ends, whose length is half of the previous one and whose number is that of the original node increased by a figure or a number (classification of position).

The newly created nodes are listed in order but here, following the nodes of the previous level, it is an extension by extrapolation.

As previously, it suffices to restart the procedure, i.e. the number of times needed to develop the fractal graph and its adjacency matrix. Figure 12.9 shows us the result of four successive iterations, which produce a fractal graph with 31 nodes on the basis of the first graph with three nodes and two arcs. The adjacency matrix is self-similar and the drawing under the stated conditions is identical to Mandelbrot's H-fractal.

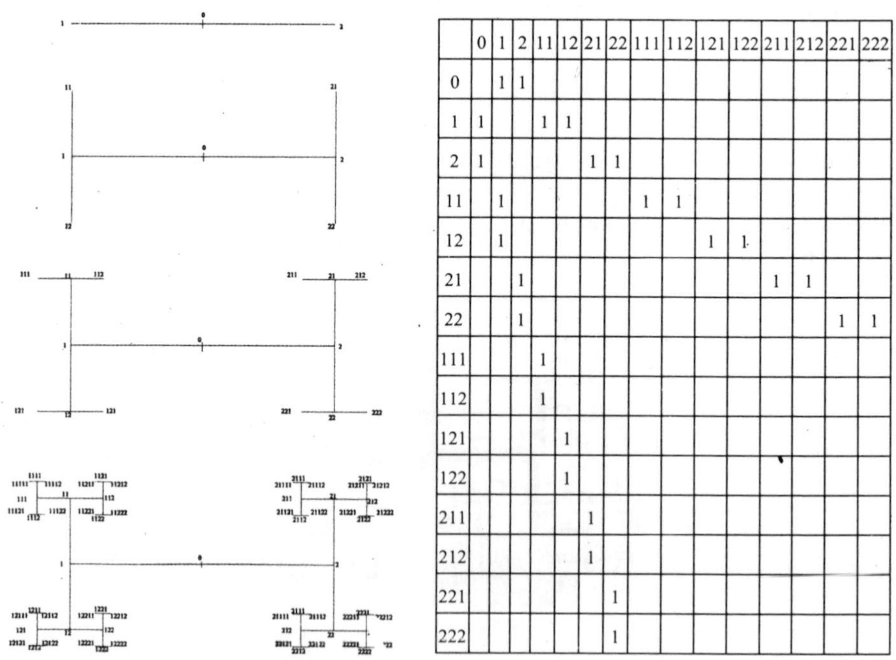

	0	1	2	11	12	21	22	111	112	121	122	211	212	221	222
0		1	1												
1	1			1	1										
2	1					1	1								
11		1						1	1						
12		1								1	1				
21			1									1	1		
22			1											1	1
111				1											
112				1											
121					1										
122					1										
211						1									
212						1									
221							1								
222							1								

Figure 12.9. *The first four stages of the process of constructing the fractal*

Let us note, however, that if we do not observe the condition of orthogonality of the two new arcs and replace it by co-linearity for one and orthogonality for the other by reversing the conditions according to the branch, we obtain a perfectly regular tree structure which naturally is also a particular graph: a binary tree or a complete m-ary tree with m = 2 in this case (Figure 12.10).

This example shows the obvious need for a precise and complete algorithmic definition of the implementation in order for it to be reproducible.

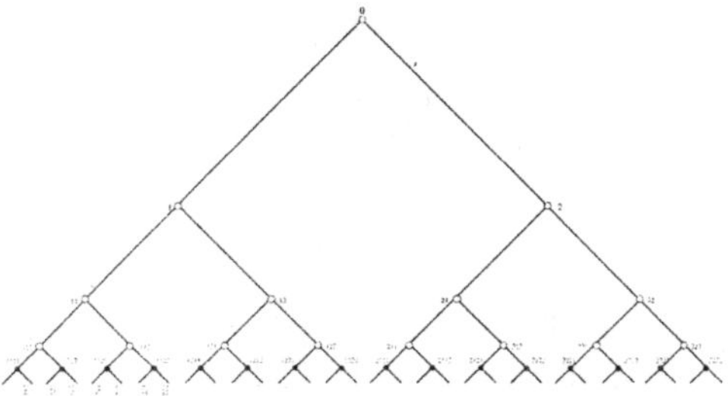

Figure 12.10. *The fractal is identical to the previous one; only the implementation is different in one of its constraints: the orthogonality of the branches issued from the same node instead of co-linearity*

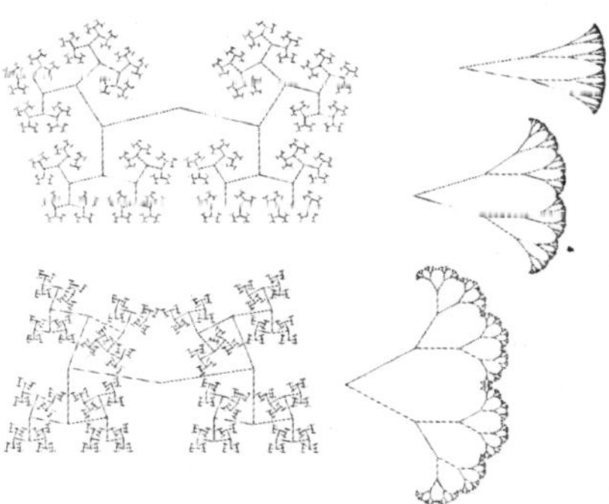

Figure 12.11. *Mandelbrot's fractal: graphs that are isomorphic to the two previous figures. The adjacency matrix is the same and only changes the geometrical constraints of implementation [MAN 77]*

The graph drawing must follow specific constraints in order to reproduce the fractal in its traditional form:

– the drawing must be flat;

– the arcs are segments of a straight line;

– the length of the extrapolated arcs has a ratio r with the original arc;

– the arc is extended by two arcs of the same length, which form a given angle with the previous arc;

– the decimal notation of the nodes created on an arc adds a figure or a number to the notation of the original node.

This notation makes it possible to know the level of each element and the length of each arc considered. If the arc is of level n and the initial arc has a length l, then the rate of increase r is:

$$L_n = l_1 * r^{n-1}$$

These conditions are valid only for a symmetric drawing, but it is possible to have different ratios r for each branch, as well as unequal angles, as demonstrated by the representation of a Pythagoras tree below.

12.2.3. *The Pythagoras tree*

The example of the Pythagoras tree [LAU 91] enables us to consider a more complete use of the graphs by taking into account duality. The method of construction of the tree is based on the first demonstration of the famous theorem. We associate a square with each side of a right-angled triangle, which is placed with the hypotenuse oriented downwards and we continue in the same way by assigning to the opposite sides of the smallest squares a right-angled triangle, whose hypotenuse corresponds to the length of the square considered and so on (Figure 12.12). In this figure overlaps are acceptable.

We then obtain the symmetric or asymmetric Pythagoras tree according to the choice of the right-angled triangle; a choice that we may consider fixed at first.

Let us consider the quasi-dual[1] of this figure, which we can describe like a graph. The quasi-dual is defined by the centers of each figure or unit face, which here are squares or isosceles right-angled triangles and the arcs joining them.

1 The quasi-dual is a dual graph whose arcs connecting the node located on the possible infinite face are not taken into account (see section 12.3.1).

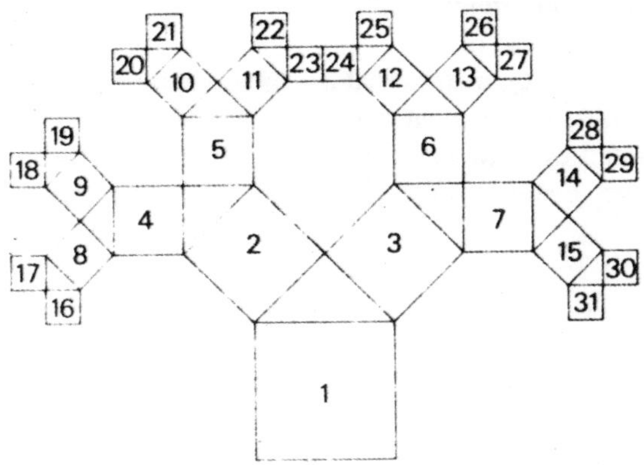

Figure 12.12. *Pythagoras tree: construction of the first iterations [LAU 91]*

We thus obtain Figure 12.13, which is a form of H-fractal obtained on the basis of the same adjacency matrix but whose new arcs form an angle of $\pm \pi/4$ with the arc of the previous level.

The two manifestations of the Pythagoras tree are dual and the drawing is much simpler and an esthetic by using the quasi-dual that corresponds to the "dual graph of the map" defined by Rosen [ROS 91, p. 488].

The Pythagoras tree can just as easily be produced in an asymmetric fashion on the basis of the same geometrical construction by removing the condition that the right-angled triangle is isosceles and by fixing a value, or even by admitting a random variable, either on the basis of a description by an adjacency matrix angle and different or randomly variable angles.

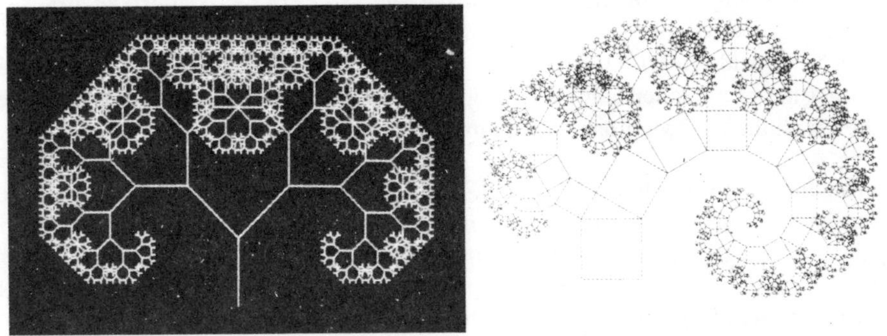

Figure 12.13. *The dual Pythagoras tree and a non-symmetric tree [LAU 91]*

12.2.4. *An example of multiplane plotting*

The previous adjacency matrix enabled us to show that the number of possible graphs drawing is not limited, even if all are plane. However, this characteristic may well be disputed and even cancelled. We can, for example, consider the H-fractal whose two branches could be implemented on two orthogonal planes as shown in Figure 12.14.

It suffices to state as a replacement for condition 1 that: the representation is biplanar, branch 1... belongs to plane P1, branch 2... belongs to plane P2 orthogonal to P1, since the two planes have an intersection following the straight supporting arc 12.

We see here a biplanar fractal. We can just as easily consider that each new arc is orthogonal to the plane of the two arcs of the previous levels. We would then have a multiplane fractal.

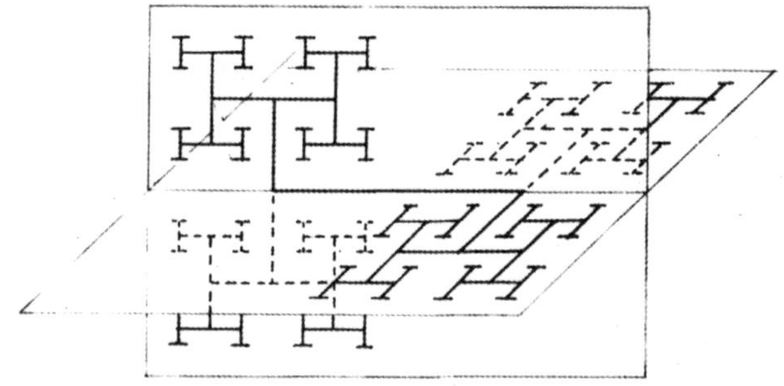

Figure 12.14. *Biplanar fractal with orthogonal branches*

12.2.5. *The example of the Sierpinski carpet and its use in Christaller's theory*

Let us consider a ternary tree [LAU 91, p. 11-12]. On the basis of a node we construct the graph extension by extrapolation, with each branch forming an angle of $2\pi/3$ with the following one with a ratio of r (Figure 12.15).

This tree enables us to define a quasi-dual, such that each of the four nodes of the primal is the center of a dual face formed by an equilateral triangle, whose four equilateral triangles also form an equilateral triangle. This figure itself gives us its reduction ratio: each face of the dual is equal to ¼ of the face of the previous level and each arc will be equal to the $1/x$ of the preceding arc.

Let us add a constraint to this drawing: we regard as full only a face with a terminal node, i.e. a leaf.

By analogy, the face of the unique node will be regarded as also being an equilateral triangle of identical size to that determined by four of the immediately lower level. Since this graph has only one node, this will be *ipso facto* final and the triangle full.

At the immediately lower level, only the faces associated with the three final nodes will be considered as full and the central face will be empty.

During the following phase where the process is reproduced, each node of the primal leads to a ternary extension and each full triangular face is divided into four triangular faces, of which one is empty, one is the central face, etc. (Figure 12.16).

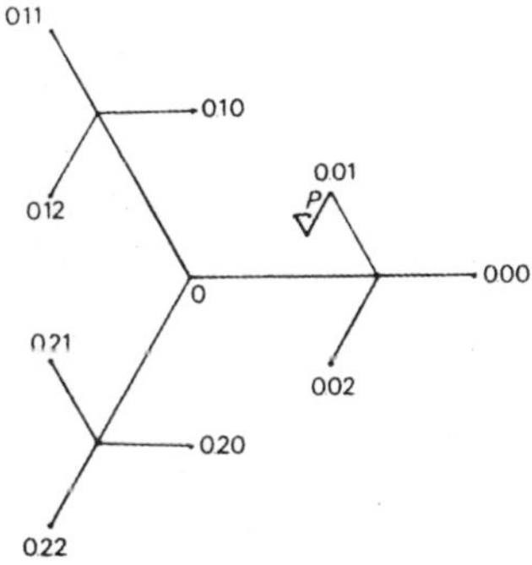

Figure 12.15. *Ternary tree as a generator*

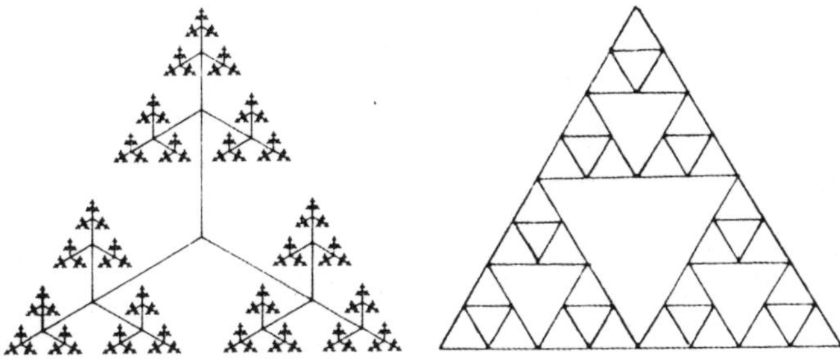

Figure 12.16. *The Sierpinski carpet: ternary tree third iteration and the quasi-dual: the Sierpinski screen or carpet; full surfaces correspond to leaves of the tree [LAU 91]*

We thus obtain a Sierpinski carpet on the basis of a graph and his extension by extrapolation (Figure 12.16).

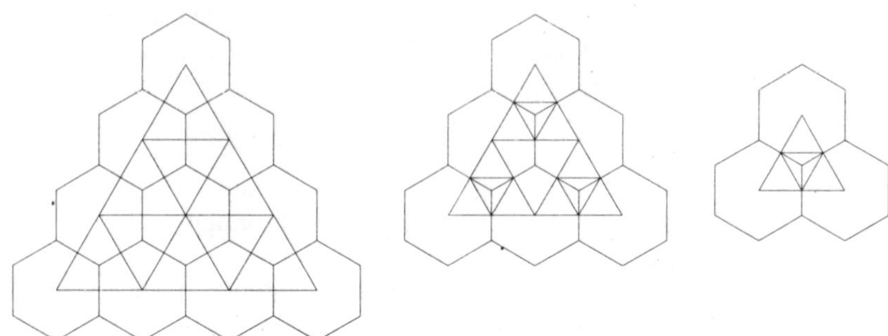

Figure 12.17. *Christaller's hexagons obtained on the basis of a ternary generator*

If the condition that only the final nodes correspond to full triangles is removed, we obtain Christaller's hexagons, which in the case (Figure 12.17) are of the market principle K = 3. The hexagons are the dual of the dual of the ternary trees when the space is tiled.

12.2.6. *Development of networks and fractals in extension*

Until now the fractals considered were convergent fractals due to a reduction ratio less than 1. In this case these fractals have a higher limit and the internal homothety applies only from an origin. If, on the other hand, we consider a magnification ratio greater than 1 (Figure 12.18), we then have a tool to account for the extension of networks, the growth of cities, etc. and we return to the percolation-type phenomena.

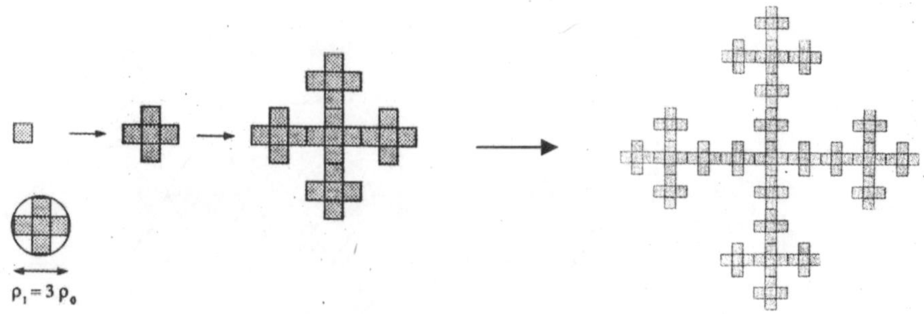

Figures 12.18 and 12.19. *Fractal in extension (Frankhauser [FRA 94, p. 257])*

12.2.7. *Grid of networks: borderline case between extension and reduction*

The growth of networks is not achieved solely by extension and creation of new external arcs, it is also carried out, in particular when the density increases, by applying an increasingly fine grid. The problem of the fractal grid is in fact simple: it is the borderline case between a tree-structured extension and a convergence towards a limit. If we consider a reduction factor $r \leq 1$ and a magnification factor $a \geq 1$, the borderline case, obviously, has the value 1, at which the fractal reproduces identically to itself. The problem of the increasingly fine grid (possibly until a limit) is solved by a reduction coefficient of the generator size with a possible delay. The example of the H-fractal makes it possible to show it.

In this example we use Mandelbrot's H-fractal whose reduction factor takes the limit value 1. The first fractal has a generator size fixed at two. The second generator with the same reduction factor has a size fixed at 1, that is, twice smaller. The second process develops with a delay that is arbitrarily fixed here at 2 iterations. When the first process arrives at the seventh iteration, the second process is only at the fifth (Figure 12.20).

However, since this fractal is twice smaller in size, it only concerns the center of the previous figure. We can thus interpret a first wave of urban allotment in a city and then after two periods of time x a second denser grid develops giving, for example, access to the center of the previous pieces of land in order to enable their constructibility.

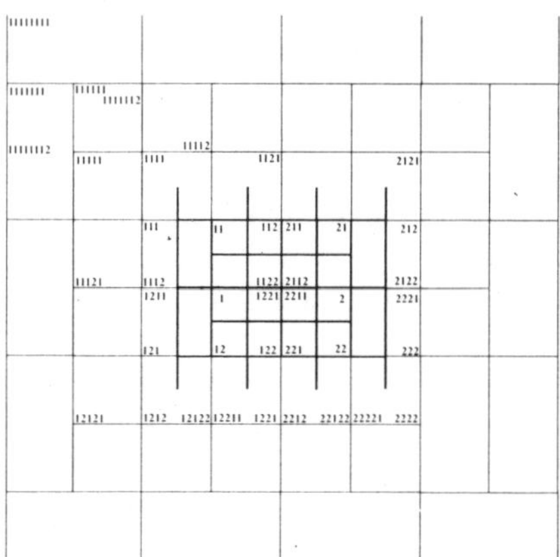

Figure 12.20. *Example of a grid with delay by using a borderline reduction factor =1 and a delay in the reduction of the generator size*

We could just as easily reverse the process: the first dense urbanization followed by the second peripheral one with a less tight network, etc. However, this type of fractal by construction has double points (mentioned in the figure above), which should be eliminated from the adjacency matrix. Similarly, when two arcs are superimposed over a single precedent, the latter two should be kept in the adjacency matrix, since the intermediate node is identical to a subdivision of the previous graph. This solution is preferable due to a finer definition of the network than described by the graph, even though in terms of the incidence matrix this choice is heavier, since it implies an additional row and a column. We could also use a multigraph by making an expressway with fewer crossings coexist superimposed with a normal road.

12.2.8. *Application examples of fractals to transport networks*

The example is very clear here: the networks of trunk and departmental roads are, broadly speaking, saturated planar graphs, with a triangular mesh and without possibility of adding an arc without it cutting through another. The internal homothety is thus perfectly real, only the position of the nodes changes and overall that implies a quasi-random variation. The succession on various levels of planar graphs (department, region, France and Europe) thus constitutes a true fractal structure, since not only are the graphs of comparable nature, but also the main cities are the same on four levels. On the other hand, when we go down a level and the scale increases, there appear new cities and graph nodes, as well as new arcs.

We have the case of an extension of the graph by intrapolation of the von Koch type for a much less linear road diagram and of the double triangular grid type.

When we descend to sufficiently fine levels, we are obliged to take into account the natural geophysical constraints, which are imposed on the network infrastructure. The last two levels are disturbed, for example in the case of the Indre-et-Loire department, by the presence of the Loire and the lack of bridges which on this scale renders the planar network unsaturated.

In the agglomeration of Tours and especially in Tours center, the old rhomboidal grid imposes itself and developed from the old town in urbanized areas starting from the second half of the 19th century. Hence, this result in the unsaturated nature of the graph and an illustration of the grid as a borderline situation in a dense space (Figure 12.21 in the color plates section). We may stress that we have a perfect illustration of several types of grid in this zone, which corresponds exactly to the fractal grid with the use of a borderline reduction factor and a delay for the grid with the largest generator, as in the second possibility expressed above.

The second example drawn from the same hypothesis is also clear: we have an irregular tree structure, which is apparently close to a tree structure of the stochastic type linked to the definition of the shortest paths from Perpignan to the rest of France (Figure 12.22 in the color plates section).

The irregularity of the tree structures results from geographical obstacles: Massif Central, Rhone Valley, etc. In the case of this simulation, we only had matrix indices at our disposal to compare the two solutions (number of nodes connected by the tree structure, lengths of the paths, etc.), but no morphological indices.

The obvious fractal nature of this tree structure then makes it possible to use the calculation of the fractal dimension and spread [GEN 00] of each branch.

12.3. Nodal graph

Graph theory in its current state does not enable a sufficiently precise description of networks, in particular with respect to the nodes, vertices of the graph.

Indeed, the nodes are "elements" which are connected by relationships. In the description of networks by graphs, the nodes represent places connected by circulation network elements: roads, railroads, canals and, more generally, elements of the natural or artificial infrastructure which enable the circulation of one or several means of transport. It is what Chapelon refers to as the functional binomial [CHA 97] by association of the infrastructure and means of transport.

Conversely, we may say that the node is what makes it possible to "connect" several relations by describing the various interrelationships between the elements and thus their independence, which is illustrated very precisely by the transfer or fluence graphs [CHO 64, p. 39]. With respect to the description of networks, in the same inversed manner, a node is what makes it possible to pass from an arc to another, i.e. from one infrastructure element to another. Here the node is an interconnection, an interface, or even a correspondence for the same or for different means of transport.

However, in graph theory nothing defines this passage: we can implicitly pass from an arc to any other incident arc to the node without any constraints apart from those that are explicitly specified or those that can be easily deduced from the definition of the characteristics of the nodes and/or the arcs, if they exist. These cases are simple and the most important ones are determined by considerations, arc capacities, optimal paths, etc.

If the graph is directed, that imposes a constraint, that is, the existence of arcs leaving so that the considered node, will not only be a dead end. However, one of the conditions of existence of a circulation network is precisely strongly connected, i.e. the possibility of going to any point and returning from there. However, certain crossroads cannot be described by a node of an undirected graph: it is in particular the case of the following semi-interchanges (Figure 12.23).

Indeed in this case, according to the direction of movement on the highway arc, we cannot either leave or return on the highway arc from the crossroads considered, which otherwise enables all the other possibilities.

It is necessary to descend to a large scale description in order to be able to describe it. We must be able to pass automatically from one scale to another by using an algorithm, so that the passage may be represented according to the needs.

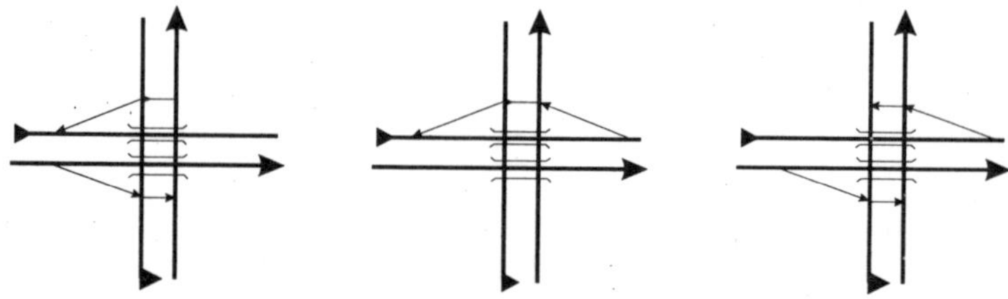

Figure 12.23. *Semi-interchanges*

Moreover, the problem of maximum capacity in a graph describing a physical network does not appear related to the maximum capacity arcs but more precisely to that the nodes. Indeed, let us consider the simplest case: traveling along two axes against the flow of traffic. If the maximum physical capacity of a car lane is 2,000 vehicles per hour, and we suppose that our arcs have only one lane and that the level of saturation has almost been reached, it is clear that the problem will immediately revert to a problem of traffic flow and that it is impossible to make the hypothesis of a realistic flow capacity of 4,000 v/h, which would be already double the allowed maximum for a lane on an arc with fluid flow. In the case considered, the capacity of the crossroad – maximum theoretical capacity – would be, at the most, equal to half the sum of two flows if the overlap was carried out without deceleration, which is still an unrealistic hypothesis.

Such reasoning can be generalized to less simple crossroads than the chosen example and, then, in order to determine a maximum theoretical capacity, it becomes necessary to break up the flows into divergent and convergent flows. Therefore, we will attempt to demonstrate that it is possible to mitigate this present deficiency of graph theory, while remaining within the theoretical framework defined beforehand, by using graph extension. This is referred as nodal zoom by Chapelon, although here it would not be used to change the scale of representation, i.e. to describe a city previously represented by just one node on an interurban relation scale, but simply to study with precision how a specific node of the graph functions or can function. This level of analysis is known as nodal.

Let us consider a simple node S of an undirected graph of the fourth degree (Figure 12.24). We will define a subdivision of this graph that associates with it four nodes of the second degree near to the considered node, i.e., S1, S2, S3 and S4 respectively. It is possible to go from S1 to S2, S3, S4, from S2 to S1, S3, S4, from S3 to S1, S2, S4 and from S4 to S1, S2, S3 passing by S since these nodes are 2-connected. These possibilities are described by the adjacency matrix of the second order (Figure 12.24).

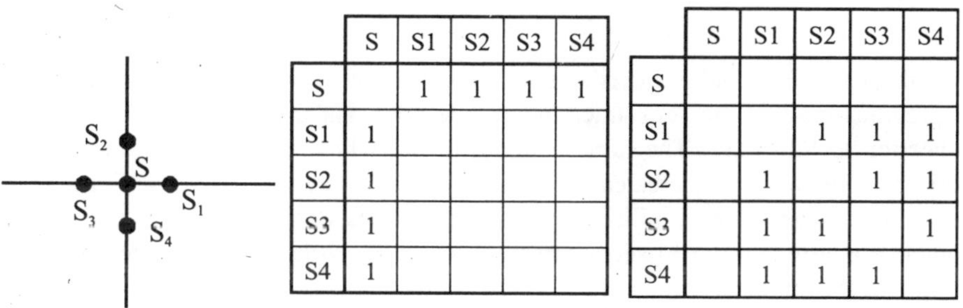

	S	S1	S2	S3	S4
S		1	1	1	1
S1	1				
S2	1				
S3	1				
S4	1				

	S	S1	S2	S3	S4
S					
S1			1	1	1
S2		1		1	1
S3		1	1		1
S4		1	1	1	

Figure 12.24. *Nodes and adjacency matrices of the first and second order*

We reduce the graph by removing the intermediate node S and in order to ensure the same connectivity we extend it by creating the arcs S1S2, S1S3, S1S4, S2S3, S3S4 and S2S4. We thus transform the 2-connected relations into 1-related relations by reduction of the previous graph. Thus, we obtain a nodal graph consisting of four nodes of the fourth degree which enable the same passages from one arc to another as in the original graph (Figure 12.25). The nodal adjacency matrix of the first order has the same relations as the preceding adjacency matrix of the second order, but its size has decreased by an empty row and a column.

It should be noted that this operation is merely the opposite of simplification of flow graphs which were first used in France by Ponsard [PON 69] and then by Lantner [LAN 74] in Dijon. It partially corresponds to the derivation of a transformation by the method of nodes duplication [CHO 64, p. 39].

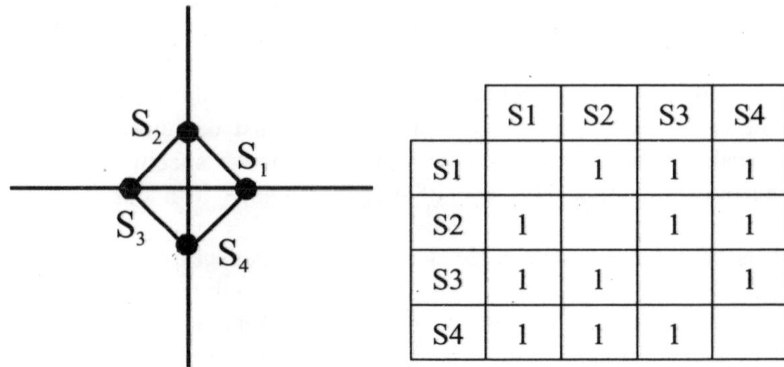

	S1	S2	S3	S4
S1		1	1	1
S2	1		1	1
S3	1	1		1
S4	1	1	1	

Figure 12.25. *Node reduced by the deletion of S and the resulting adjacency matrix of the first order*

In this case the nodal graph is a complete subgraph. If we apply this technique to the previous case of the semi-interchange, we note that the S1 node has only one connection to S2: S2S1. The same happens to S6, whose only connection to S5 is S6S5. The nodal graph created in this way proves capable of describing with precision the various possibilities and impossibilities of passing from one arc to another. We note that the adjacency matrix cannot be made triangular: there exists a circuit S5S4S3S2S5, as shown in the figure below.

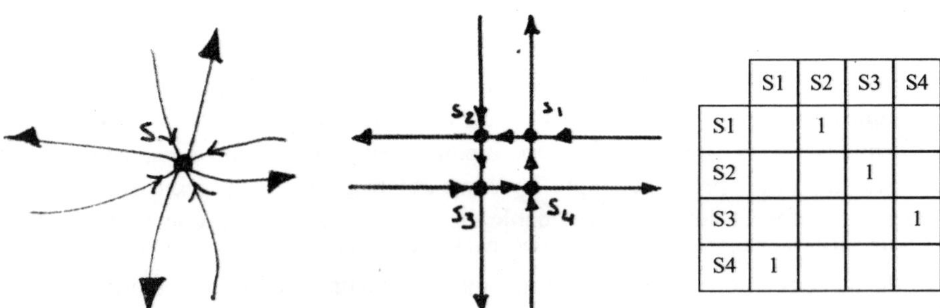

	S1	S2	S3	S4
S1		1		
S2			1	
S3				1
S4	1			

Figure 12.26. *Directed nodal graph and matrix*

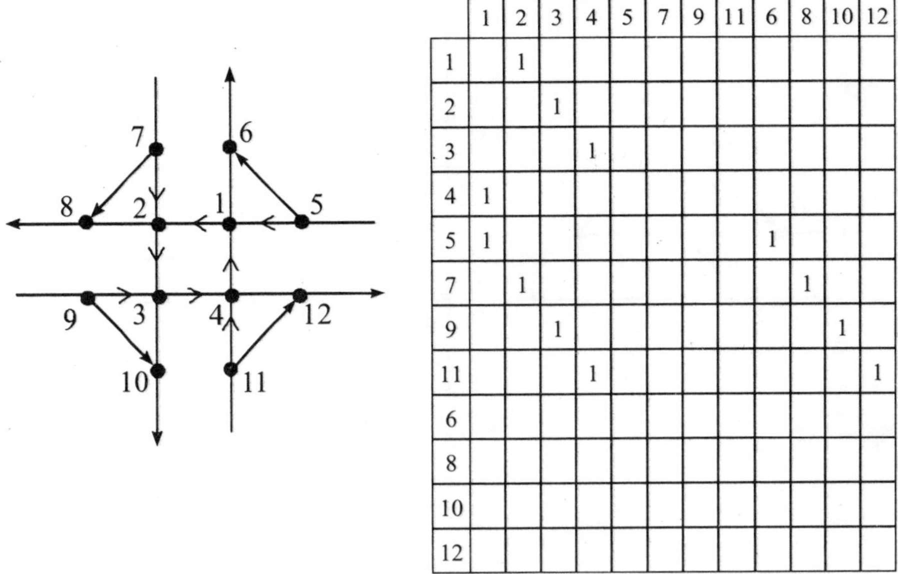

	1	2	3	4	5	7	9	11	6	8	10	12
1		1										
2			1									
3				1								
4	1											
5	1								1			
7		1								1		
9			1								1	
11				1								1
6												
8												
10												
12												

Figure 12.27. *Extended G3 nodal graph and adjacency matrix with the nodes according to the functional order: first the crossings then the divergences and convergences for each arc*

Let us consider the same example but with a directed graph with four internal semi-degrees and four external semi-degrees G1. The solution that consists of breaking up this G1 graph into a subgraph with four nodes, like the G2 graph (Figure 12.26) with two internal semi-degrees and two external semi-degrees, is not satisfactory in terms of describing the flows and it certainly does not make it possible to describe the semi-interchange considered above.

Nodal decomposition can be achieved in different ways according to at least three solutions. The second solution is a variation of the first variation mentioned above. The same method is thus reused: extending the graph by creating a subdivision and a list of possibilities of passing from one to the other, i.e. determination of the connectedness of the second order and a list of adjacency of the second order (see Figure 12.28).

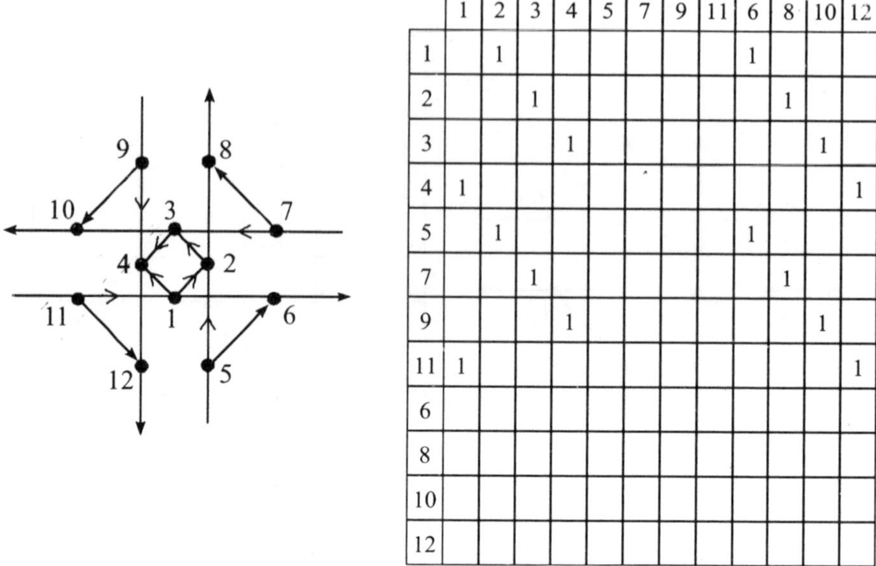

	1	2	3	4	5	7	9	11	6	8	10	12
1		1							1			
2			1							1		
3				1							1	
4	1											1
5		1							1			
7			1							1		
9				1							1	
11	1											1
6												
8												
10												
12												

Figure 12.28. *G4 graph of a complete simple crossing and adjacency matrix: second solution*

However, this second solution has a disadvantage that can limit its use: it modifies the nature of the graph. Indeed, while remaining plane the graph is no longer planar, since certain arcs intersect outside a node. There is an intersection in the shape of an interchange, which transforms the planar graph into GRS due to the absence of bijection between the graph itself and its projection onto a plane.

The third is a little bulkier but has the advantage of making the nodes specialized. For a better analysis of the flows we introduce an additional constraint: we consider only convergent or divergent nodes but not mixed nodes. We will call a node divergent if from it a flow can split into two, and convergent if two separate flows meet at this node. We will draw the graph as shown in Figure 12.29. However, it should be mentioned that this graph is no longer planar. As before, its projection onto the plane is only a projection of a GRS and there is not bijection between the two. Using this presentation of the nodal graph, it is easy to describe any type of crossroads.

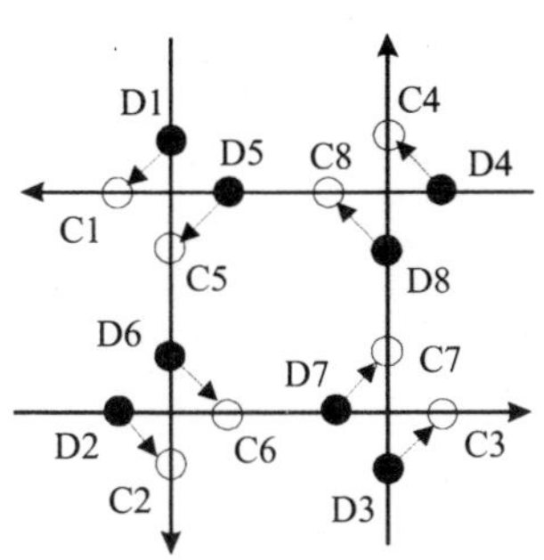

Black nodes are divergent
White nodes are convergent

	D1	D2	D3	D4	D5	D6	D7	D8	C1	C2	C3	C4	C5	C6	C7	C8
D1									1				1			
D2										1				1		
D3											1				1	
D4												1				1
D5									1				1			
D6										1				1		
D7											1				1	
D8												1				1
C1																
C2																
C3																
C4																
C5						1										
C6							1									
C7								1								
C8									1							

Figure 12.29. *G5 graph with eight divergent nodes and eight convergent nodes*

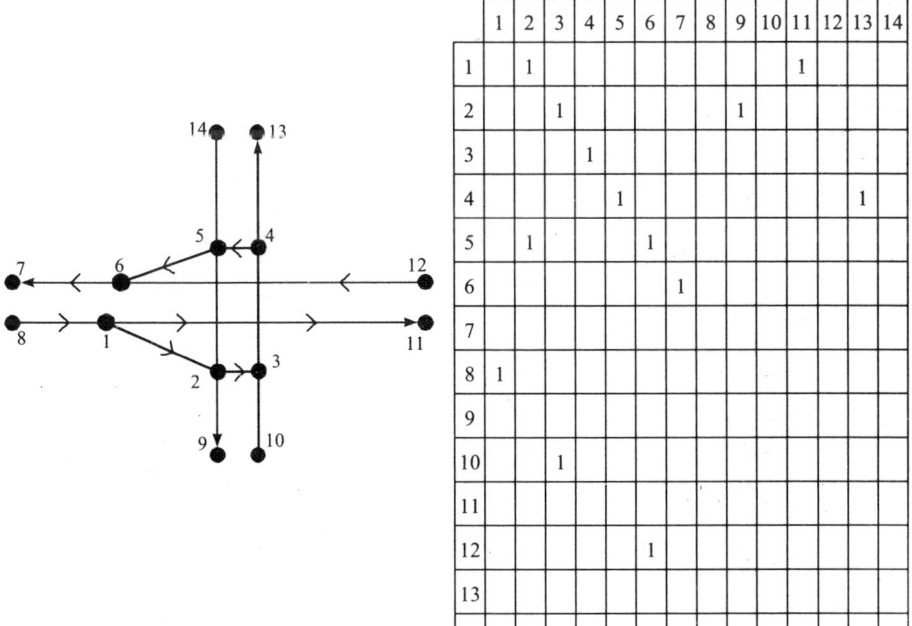

	1	2	3	4	5	6	7	8	9	10	11	12	13	14
1		1									1			
2			1						1					
3				1										
4					1								1	
5		1				1								
6							1							
7														
8	1													
9														
10			1											
11														
12						1								
13														
14					1									

Figure 12.30. *Nodal graph of the semi-interchange*

Generalization of the nodal decomposition

The nodal decomposition with divergent and convergent nodes is automatic and can be carried out on a part of the graph or on its entirety at our discretion.

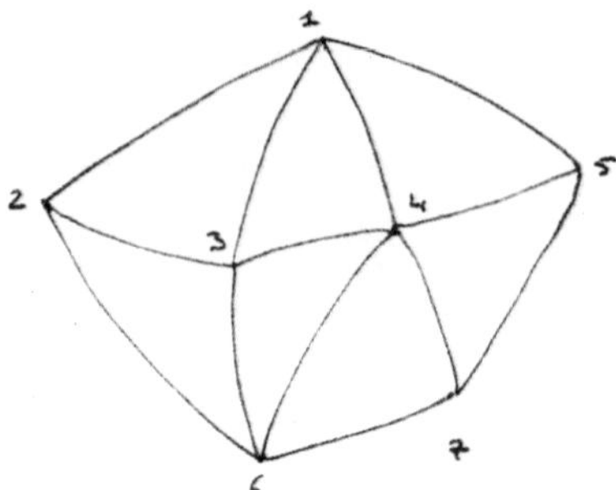

Figure 12.31. *Any undirected graph G*

Let us consider the graph G above (Figure 12.31) and the adjacency matrix below.

	1	2	3	4	5	6	7
1		1	1	1	1		
2	1		1			1	
3	1	1		1		1	
4	1		1		1	1	1
5	1			1			1
6		1	1	1			1
7				1	1	1	

For each node we can use the zoom to reveal the nodal graph by using the technique described above, with specialized or unspecialized nodes (Figure 12.32).

The graph then appears in its extended version, but its properties are identical to the previous one in terms of path layout. If the adjacency between two arcs was achieved via a node in the first graph, the adjacency in the second graph is achieved via a new arc and two nodes and the 1-adjacency has been transformed into 2-adjacency. We must specify once again that the projection of the graph is not planar.

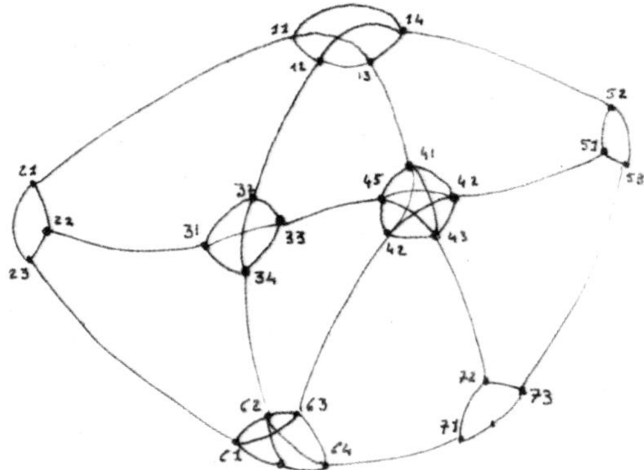

Figure 12.32. *Graph with GDS nodal decomposition of the undirected graph G*

Property: each node created to replace a primitive node in an undirected graph has the same degree and so forth: this property is preserved by transformation.

The nodal matrix has the property of self-similarity with respect to the previous matrix if we consider the sectors identified by the first figure of the node of the nodal graph identical to that of the node which it replaces. On the other hand, the diagonal of the previous matrix is replaced by the nodal matrix of each node.

•

The transformation of a node could be described by a complete m-ary tree, where the number of wires is given by the degree of each root node.

The "generator" here is thus the transformation of the node into a nodal graph. Mathematically, this generator could be used successively several times. Both graphs may therefore be considered as fractals. The size of the diagonal submatrix, which replaces the node of the simple graph, is the degree of the node in the case of an undirected graph and the sum of external and internal semi-degrees in the case of a digraph. The nodal transformation is therefore fractal.

The graph then appears in its extended version, but its properties are identical to the previous one in terms of path layout. If only the adjacency between two arcs was achieved via a node in the first graph, in the second graph the adjacency was achieved via a new arc and two nodes. Moreover, an algorithm can easily be written for this transformation.

	11	12	13	14	21	22	23	31	32	33	34	41	42	43	44	45	51	52	53	61	62	63	64	71	72	73
11		1	1	1	1																					
12	1		1	1					1																	
13	1	1		1								1														
14	1	1	1															1								
21	1					1	1																			
22					1		1	1																		
23					1	1														1						
31						1			1	1	1															
32		1						1		1	1															
33								1	1		1					1										
34								1	1	1											1					
41			1										1	1	1	1										
42												1		1	1	1						1				
43												1	1		1	1									1	
44												1	1	1		1						1				
45										1		1	1	1	1											
51													1					1	1							
52				1													1		1							
53																	1	1								1
61							1														1	1	1			
62											1									1		1	1			
63															1					1	1		1			
64																				1	1	1		1		
71																							1		1	1
72														1										1		1
73																			1					1	1	

12.3.1. *Planarity and duality*

The property of duality of graphs is very interesting, but polysemous. Applied to a particular draw, it allows us to make a network correspond to the faces that it generates, which is fundamental and absolutely necessary for a space analyst.

Among various accepted variations we will specify the meaning of duality that we employ, which is a sufficiently general definition for our purposes and operational, so that the creation of a dual could be algorithmic. The objective of taking duality into account is to become able to transform a graph into a cellular network.

A graph is referred to as plane when it is drawn on a simple surface: a plane or a sphere or a torus, as defined by Lhuillier [LHU 76], Heawood [HEA 76]. Let us recall that a plane graph is known as planar, if none of its arcs are intersected outside the nodes and that a planar graph is known as saturated or complete when it is impossible to add an arc to it without it intersecting another arc outside the node.

All the faces of a saturated planar graph are triangular. Indeed, if one of its faces was quadrangular, it would be possible to add to it a new diagonal arc and thus the planar graph would not be planar saturated.

The problem of projection onto a plane of a three-dimensional graph. Let us return to the historical example of Cauchy [CAU 76]. Each node is projected onto one and only one point, and similarly each arc has one and only one projection. However, it is not the same for the faces of a cube; indeed, the projection of the lower face of a cube partially covers the projection of the upper face.

Let us consider point M belonging to ABCD, which is the upper face of the cube. It is projected onto the plane in M', just as point N located on the bottom face EFGH is projected onto N', points M, N, M' are aligned, N' merges with M' and the point is double. The same applies to any point of the cube projection where even the stops of the projection squares contain double points, since they belong simultaneously to two faces.

If we do not consider the cube formed only by nodes and arcs, the projection does not contain a double point unless a particular choice has been made by the projection center. If it is located on the extension of a diagonal there will be a double node, if it is located on a diagonal plane passing through two sides, these will be superimposed in the projection and, more generally, if it is located on a plane passing through two sides, these may have a common projection. On the other hand, if we take into account the faces of the cube, all the points of the projection are doubled.

However, the case of the cube of Cauchy is not unique and many graph plottings can be regarded as plane projections of three-dimensional graphs or more.

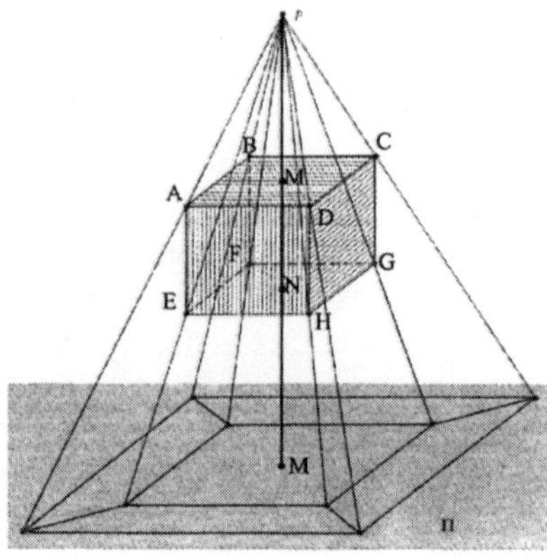

Figure 12.33. *Cauchy projection in [BIG 76]*

The condition of existence of a 3D volume is four adjacent nodes of degree d° ≥ 3. Let us consider Plato's solids below (Figure 12.34) [ALD 00]: we note that the simplest of them is the tetrahedron or pyramid with a triangular base whose nodes are of the third degree, as those of the cube and the dodecahedron. On the other hand, the octahedron and the icosahedron have nodes of the 4th and 5th degree respectively.

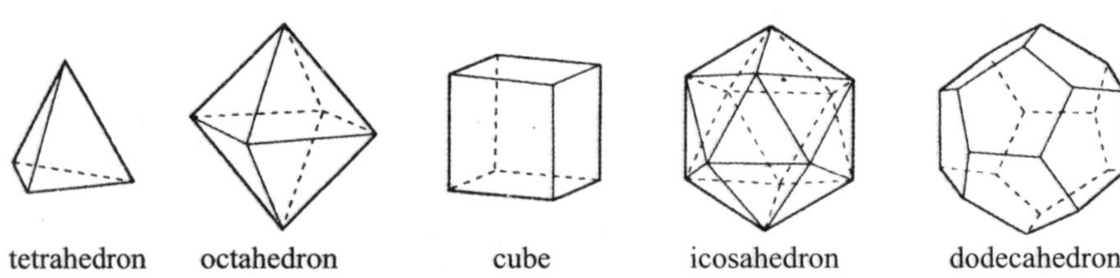

Figure 12.34. *Plato's solids, three-dimensional graphs [ALD 00]*

We can deduce from this that any graph with at least four adjacent nodes four by four and thus of at least the third degree would contain a tetrahedron. We can also deduce that irregular polyhedrons have similar properties.

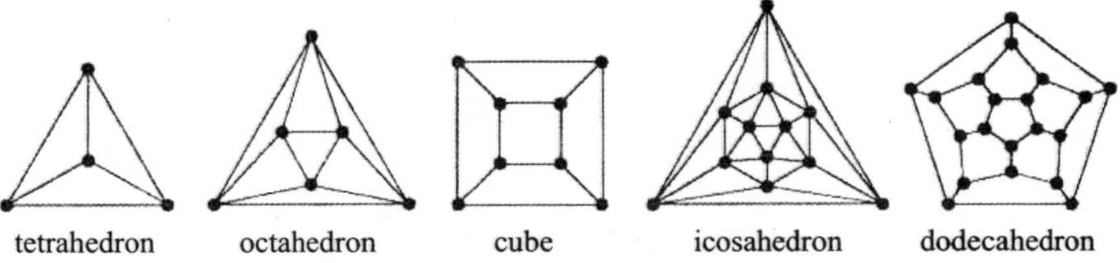

Figure 12.35. *Planar graphs of Plato's solids [ALD 00]*

Nothing forces us to only consider a three-dimensional space. Let us consider the hypercube below (Figure 12.36).

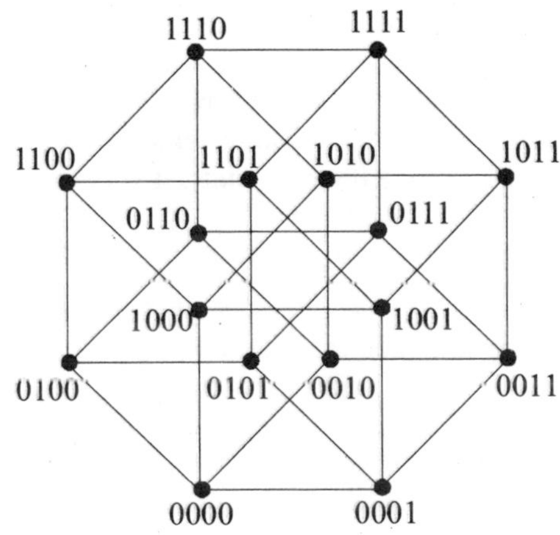

Figure 12.36. *Projection of a hypercube onto a plane*

This graph shows us that all the nodes are of the 4th degree. Its projection onto a plane is possible. In particular cases we may obtain a projection of the hypercube which repeats Cauchy's projection. The projection of a hypercube with a central projection source yields two overlapping cubes connected by edges, which is incidentally the plan of the Grande Arche in La Defense, Paris (Figure 12.37) [DOL 01].

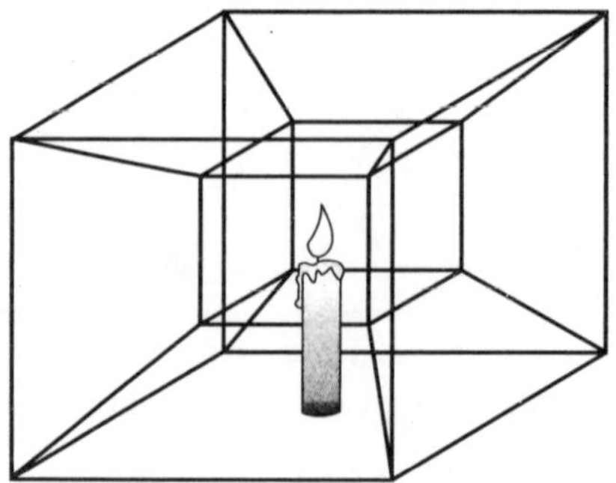

Figure 12.37. *Central projection of a hypercube in a three-dimensional space – (Science Avenir)*

The 3D graph thus obtained yields a plane graph with four overlapping squares in a straight projection. If we have a volume, at least one face will be at least a double face, even if the nodes and arcs taken separately the framework of the figure are not. This is apparent for all of Plato's solids and for their transforms; indeed, in order for the projection not to have double points, it is necessary that the projection center is internal to the volume considered.

12.3.1.1. *Duality*

The duality of the graphs is polysemous. For example, Beauquier, Berstel and Chrétienne [BEA 92, p. 273 *et seq.*] use the concept of duality within the framework of linear programs related to the determination of the minimum cost flow defining a dual linear program… Band-Jansen and Gutin [BAN 01, p.150 and 667] use it in the same sense and also in that of a *dual of a matroid.* Xuong uses it to define a transposed binary relation and also within the framework "of what is referred to as the principle of duality of directed graphs which is obtained by reversing the direction of the arcs… [XUO 92, p. 69 and 329]".

On the other hand, many other authors, for example, Cormen, Leiserson and Rivest [COR 94], as well as Minoux and Bartnik [MIN 86] do not even mention it in works on algorithms. Further on we will use and discuss a definition of duality in line with that of Withney [WIT 32], Berge [BER 70, p. 20], Kaufmann [KAU 64, volume 2 p. 358], Balakrishnan and Ranganathan [BAL 00].

Although Euler used a concept very close to duality, if not just a brilliant anticipation thereof, for the bridges of Königsberg, Withney seems to be the first to explicitly define the dual graph, which he calls the *Geometric dual of G*. However, the idea goes back to Kemp in 1879 and Heffter in 1891, and, it seems, to Maxwell even earlier…

The duality of a graph consists of associating the faces of the graph called primal with the nodes of a graph called dual. Berge [BER 87] [BER 70] provides the following definition for it. "Let us consider a planar graph G, connected and without isolated vertex; we will make it correspond to a planar graph G* in the following manner: inside every face s of G we places a vertex x* of G* and to every edge e of G, we will make correspond an edge e* of G*. This will connect the z vertices x* and y* corresponding to the faces s and t which are located on both sides of the edge e. The graph G* thus defined is planar, connected and does not have isolated nodes: it is called the dual graph of G".

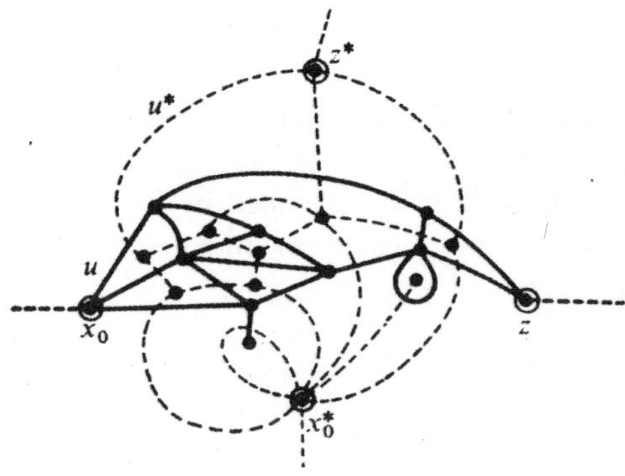

Figure 12.38. *In Berge [BER 70]*

The example of construction of the dual provided here by Berge (Figure 12.38) is a particular case in the sense that the dual has the same nature as the primal: a 1-graph. The diagrams below (Figure 12.39) seems to make it possible to show a property of the dual which has not yet been underlined.

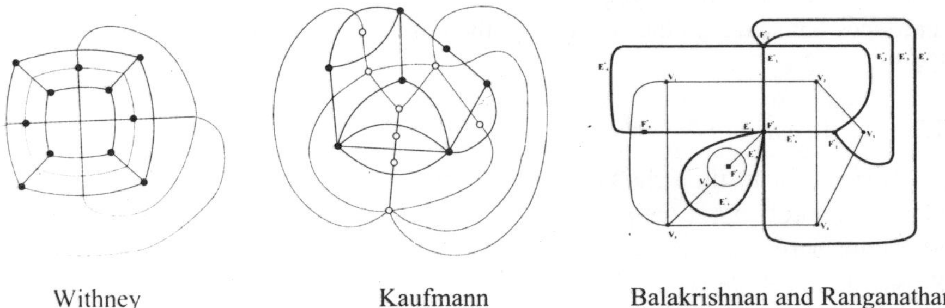

Figure 12.39. *Property of the dual*

The Withney diagram [WIT 32, p. 148-149] is very simple and at the same time paradoxical, as we will see below. The Kaufmann example [KAU 64, volume 2, p. 358] is much more interesting because he shows that the dual in this case is a multigraph, but does not mention it [KAU 64, volume 2, p. 357-358]. The construction of the dual by Balakrishnan and Ranganathan [BAL 00, p. 165] confirms Kaufmann's.

The two latter examples enable us to deduce the following theorem:

– the dual of a 1-graph is a multigraph if and only if the primal has at least one node of the second degree.

Indeed, by construction, the nodes of the dual belonging to faces which are located on both sides of the two arcs connected by a node of the second degree will be connected by at least two arcs. This is the case of the last two graphs. It is in this sense that the example of construction provided by Berge is just a particular case, like that of Withney, which is a particular case of a graph whose connected faces are only separated by single arcs.

This enables us to specify that if the subdivision of the graph (by insertion of second degree nodes) does not affect the planarity or non-planarity of the graph [ALD 00, p. 260], on the other hand, the nature of the dual is changed depending on whether this insertion is carried out between a peripheral face and the infinite face or between any two adjacent faces.

This particularity poses a problem of algorithmic definition of the dual, since if it is carried out by using an adjacency matrix, for example, the latter must be at least three-dimensional and, thus, possibly very concave. In the case of a definition by pointer, the case must be envisaged in such a way that the first value is not crushed by the second, thus making a calculation for a line of optimization of the shortest path or maximum flow type impossible. The algorithmic description of networks and faces conceals another difficulty: the description of the infinite face of a plane graph.

12.3.1.2. *A return to the origins*

This return to the origins is justified by an ambiguity in the definition of the planar: the problem of the possible infinite face. Let us consider the famous problem of Königsberg's bridges (Figure 12.40). The population of this town in Eastern Prussia found it amusing to try and find a path that would make it possible to cross each bridge of the city by using every bridge just once. Some of them believed that this was impossible but no proof of this impossibility had been given.

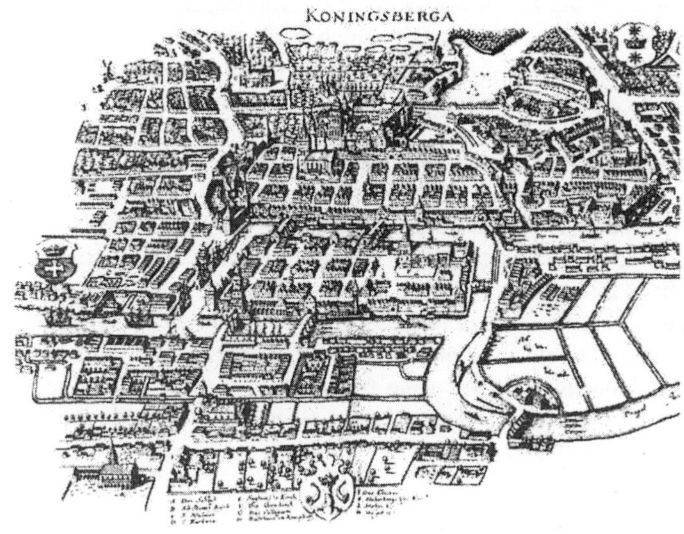

Figure 12.40. *The bridges of Königsberg*

In 1736, Euler presented the problem in its most general mathematical form [EUL 1736], from where we draw the following chart (Figure 12.41).

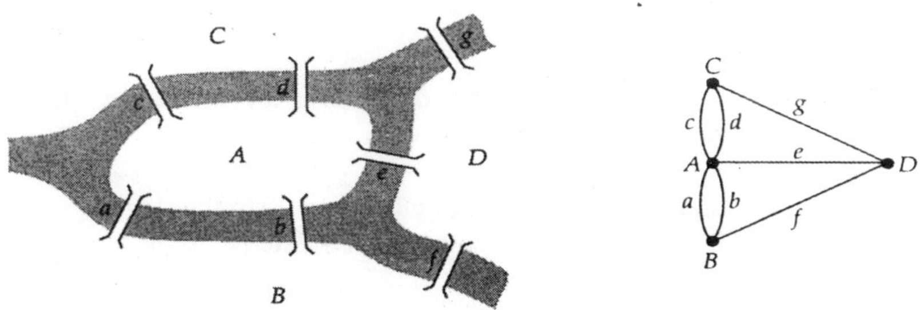

Figure 12.41. *Euler's graphic representation*

This presentation is most interesting partly because the map of the city is replaced by a simple diagram showing its principal characteristics and it formulates the problem in such a manner that the diagram becomes superfluous. However, the diagram itself is very interesting since it poses, in fact and for the first time, the problem of duality. Indeed, it associates each district of the city with a node and connects these nodes by an arc which passes by one of the city bridges, thus defining a multigraph. By using this it passes from the primal, which is the map of the city and its districts (finite), to a dual[2], which is a simplification of the circulation network. The fundamental problem of duality is thus posed: the comparison of a network and the associated surfaces, which is our most important problem.

Both representations above are dual and planar. The graph describing the communication network makes it possible to specify the dual of the sectors and, in particular, its nodes. However, that immediately presents a fundamental problem: that of existence or non-existence of the infinite face. If it exists, the dual of the network defines three faces and an infinite, which may be chosen between C, B or D. Nonetheless, that does not define four districts, as in a city. If we take four finite delimited districts into account, the dual of the dual is not the primal and there is no bijective correspondence between the two, since the dual of the dual according to Berge should have an infinite face with an associated dual node and thus, it would then have an additional node.

Let us stress here that the starting point seems to be the various districts, to which we initially associate a node and we then connect by arcs passing through each bridge. Moreover, the districts are finite, although the ambiguity is there, in the problem.

The dual of the Cauchy cube

The problem presents itself differently when the graph is a three-dimensional volume.

Definition of the dual of the cube

On the basis of the planar graph [ALD 00, p. 264], it is identical to the Withney diagram mentioned below.

The determination of this dual corresponds perfectly to the definition of Berge. We note, however, that this dual has a node which is external to the larges square, as if face 6 was the outside face!

---

2 We find this procedure implicitly in Kaufmann [KAU 64, volume 2, p. 283] and in Rosen [ROS 91, p. 488].

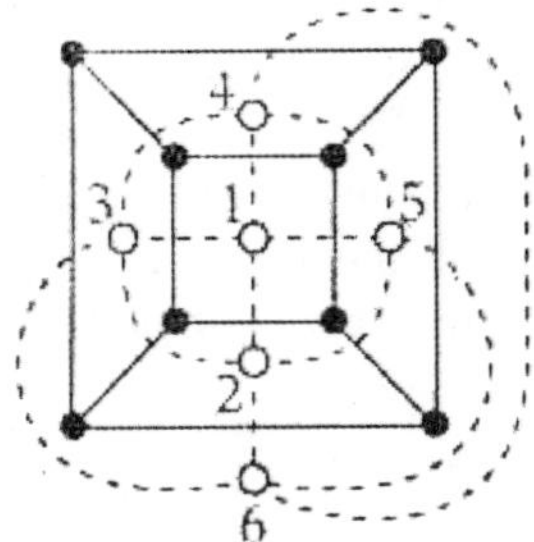

Figure 12.42. *The dual of Cauchy projection*

The definition of the dual on the basis of a cube in the three-dimensional space and, more generally, of a convex polyhedron is as follows: "the dual of a convex polyhedron can be constructed by placing a vertex at the center of each face of original polyhedron, and joining a pair of vertices with a line-segment whenever the corresponding faces of the original polyhedron are adjacent along an edge" [ALD 00, p. 273].

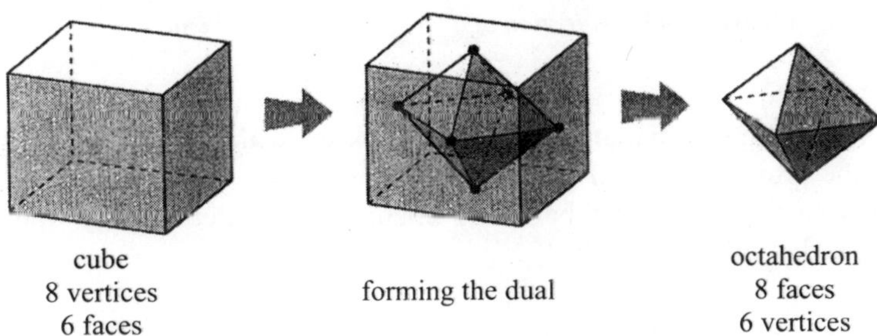

Figure 12.43. *Cube and dual of the cube: the octahedron*

Let us stress in this case that, if the dual nodes are located on the faces of the primal, the arcs joining the nodes of the dual, on the other hand, do not intersect with the arcs of the primal and instead only the projections intersect. That clearly shows that the definition of Berge only applies to a plane and, moreover, planar graph. It should also be added that the dual represented in this manner is a particular case where the dual nodes are located in the middle of the faces of the primal. The dual of a regular Plato's solid is not necessarily a regular Plato's solid.

Actually, if we specify that an arc connects the dual nodes of two adjacent primal faces, the intersection of the arcs on the plane is simply a consequence of the planarity of the graph in one case and of the projection of the graph, and thus of its flatness, in the other case.

A more difficult problem is that of the possibly infinite face, i.e. of the external node (node 6 of Figure 12.42). Let us use again Cauchy's projection but this time by including the dual (Figure 12.44). We note the absence of an infinite face, but the existence of a double node. The lower and upper nodes of the octahedron are projected onto a single point of the plane in orthogonal projection, but if the projection on the plane was oblique, these two nodes would be distinct.

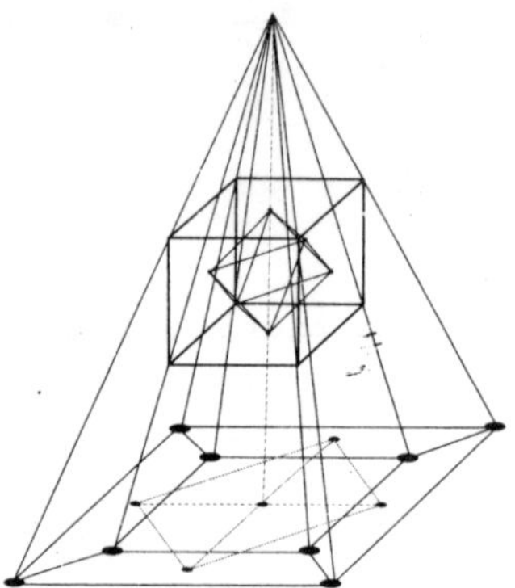

Figure 12.44. *Projection of a cube and of its dual onto a plane*

Similarly, each external arc of the dual intersects with two arcs; this is due to the fact that the projection center is too far away from the upper surface of the cube, which is something that can be easily changed by increasing the upper square. However, is this still a projection?

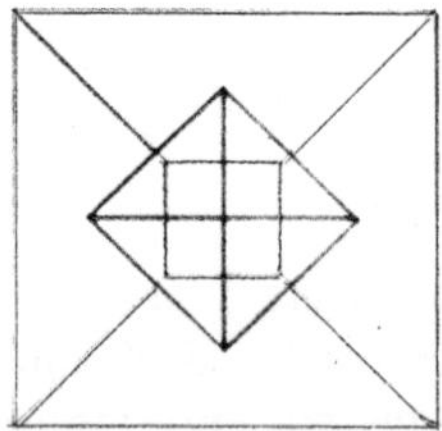

Figure 12.45. *Straight projection of the cube and its dual by enlarging the upper square*

In fact, if we project the cube and its dual orthogonally onto a plane which is perpendicular to a diagonal plane of the cube, the central edge of the octahedron is projected according to a point located in the middle of the projection of the face. Therefore, in order for this edge to be projected beyond the basic node of the cube face, the projection center has to be internal to the cube. Then, the projection cannot be plane and it can be done onto a sphere whose center belongs to the space delimited by the cube.

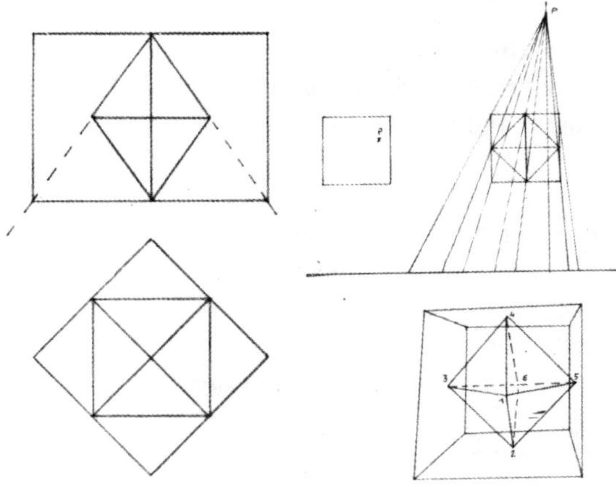

Figure 12.46. *Projection of an octahedron and a square (left) and any projection of the Cauchy cube (right)*

Consequently, the projection of the dual of a cube onto a plane does not correspond to the usual definition of the duality of planar graphs. The projection of dual arcs intersects two projections of primal arcs and the node that is double for a normal central projection is divided for any oblique projection, as above (the cubic graph is represented in profile, the vertical of the projection center is materialized on the square and the projected figure is seen from above to make it possible to observe the deformations).

This projection with a double face satisfies the Euler relation between nodes, arcs and faces without there being an infinite face:

– let G be a connected graph with e arcs and v nodes. Let r be the number of areas in a representation of g;

– then, r = e-v+2.

In this definition we have removed the planar term because the projection of a convex volume onto a plane is not planar, as shown in Figure 12.46. Taking into account an external face would be against Euler's relation in this case.

The non-orthogonal projections of the Cauchy cube and of its dual transform the projection of the cube by giving it a trapezoidal aspect if the projection center is located on a symmetric or quadrilateral plane, which is more irregular in other cases.

However, the most visible result of this is mainly the separation of the node and double arcs, which distinguishes the two upper and lower nodes of the octahedron, as above. The projection is then planar and non-planar.

We thus obtain two different results on the basis of the same graph. The question is then to find out whether there is possible isomorphism between the two dual graphs (nodes and arcs) and whether there is isomorphism between the faces of the two dual graphs.

Let us consider the dual obtained on the basis of the graph projected onto a plane and that obtained by its projection onto a plane (Figure 12.47). We can make a node of the dual D' correspond to each node of the dual D. A pair of adjacent nodes of D' corresponds to each pair of adjacent nodes of D and thus there is bijective correspondence between the two graphs. Furthermore, if we move node 6 of D' towards the exterior, will there still be bijection between the faces of D and D' (Figure 12.48)? Is this operation possible?

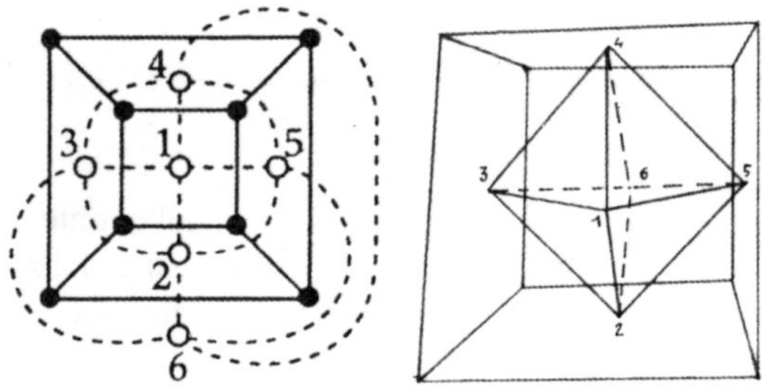

Figure 12.47. *D, dual of a plane projection (left); D', dual of the cube projected onto a plane (right)*

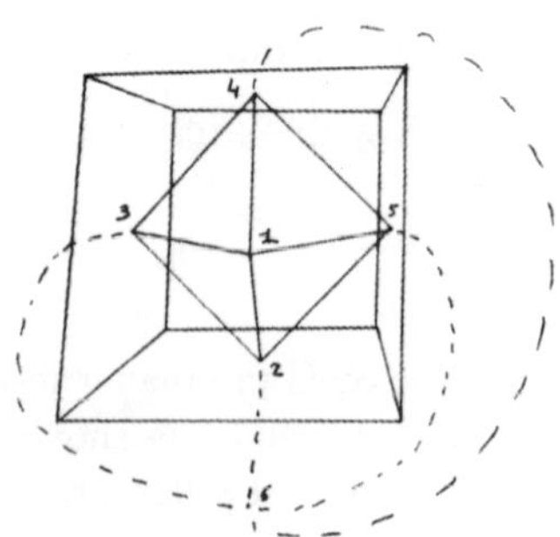

Figure 12.48. *Movement of node 6 of D'*

Indeed, the surface of the cube is convex and the external part of the plane does not belong to the projection. Due to this fact node 6' of D' no longer belongs to its definition face, as we demonstrated above. Consequently, D can only be the dual of the planar graph G, if and only if G is not a projection of a cube.

General definition of the dual

If we refer back to the two definitions, we note that the second is more general than the first because it specifies "joining a pair of nodes with a segment of a straight line each time the corresponding faces are adjacent along an arc". This applies to the three-dimensional graphs, whereas Berge's definition specifies that "*to any edge e of G* (the italics are our addition) will correspond an edge e* of G* that will connect the z nodes x* and y* corresponding to the faces s and t which are located on both sides of the edge e".

Specifying *"to any edge e of G"* is not necessary since, on the one hand, it is a simple projected consequence of the adjacency along an edge of the two faces connected to a node of the dual and, on the other hand, it introduces a restriction, which means that this definition of duality is valid only for planar graphs.

The second definition is more general because it applies to plane graphs and to planar graphs, for which primal and dual arcs necessarily intersect and, of course, it also applies to three-dimensional graphs, for which primal and dual arcs do not intersect in space, but may do so once projected onto a plane.

12.3.1.3. *Duality: a periodic fractal generator*

Let us consider the following theorem.

If G is the drawing of a connected graph with n nodes, m arcs and f faces, then G*, which is the dual of G, has f nodes, m arcs and n faces.

A tetrahedron has four nodes, four arcs and four faces. Its dual will also have four nodes, four arcs and four faces and it will be a reversed tetrahedron written into the original one and reciprocally.

We saw that dual of a cube, i.e. 8 nodes, 12 arcs and 6 faces, was an octahedron with 6 nodes, 12 arcs and 8 faces, and vice versa.

The icosahedron has 12 nodes, 30 arcs and 20 faces. Its dual is the dodecahedron, which has 20 nodes, 30 arcs and 12 faces.

More generally, the dual of a convex graph with n nodes, m arcs, f faces in a three-dimensional space has a dual, which has f nodes, m arcs and n faces and which in turn has a dual with n nodes, m arcs, f faces.

Corollary: the graph and the dual of its dual have the same number of nodes, arcs and faces.

Consequently:

THEOREM 12.1.

*If the graph G is Plato's solid and if the nodes of the subsequent duals are defined by using faces, the dual G** of the dual G* will also be Plato's solid of a comparable nature to graph G.*

The dual of a tetrahedron is a reversed tetrahedron written into the original, whose dual is a reversed tetrahedron written into it, and so on...

The dual of a cube is an octahedron written into the original, whose dual is a cube written into it, and so on...

The dual of an icosahedron is a dodecahedron written into the original, whose dual is also an icosahedron written into the original, and so on...

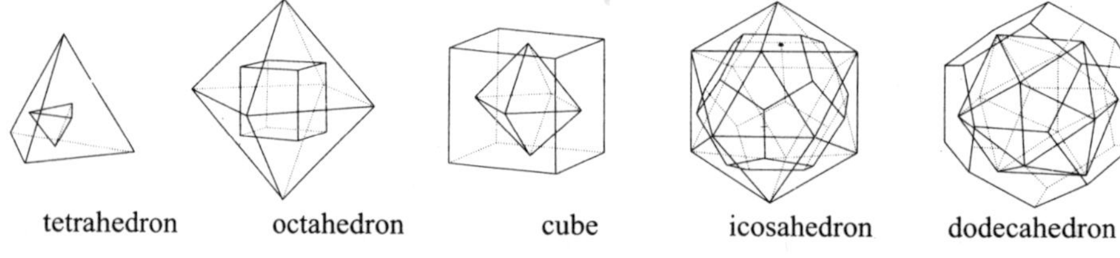

Figure 12.49. *Plato's solids and their duals (according to [ALD 00], modified by us)*

THEOREM 12.2.

Duality constitutes a periodic fractal generator for Plato's solids with internal homothety whose ratio r is constant, every two transformations (Figure 12.51).

The determination of the dual of Plato's solid along with the definition adopted until now leads to defining a dual is inscribed into the primal and, thus, of a reduced volume. However, by taking into account the symmetry of the transformation, we can reverse the process and look for the primal of Plato's solid which is itself considered as a dual.

Since the dual is inscribed, the primal will necessarily be ex-inscribed and therefore have a larger volume.

THEOREM 12.3.

A primal Plato's solid has Plato's solid inscribed into as a dual, and is itself the dual one of Plato's solid inscribed around it with the same ratio.

The theorem above corresponds to the traditional definition of decreasing fractality with a constant reduction ratio, i.e. 1/3 for the sides and 1/27 for the volume, for a cube, on the basis of an original solid, with duality as generator. We can reverse the process by adopting primality defined as above as generator on the basis of this same solid. We will then have a fractal transformation with a constant growth factor. By taking into account the perfect symmetry of the process, the starting point of the fractal inclusion series can be found "in fine" at any level.

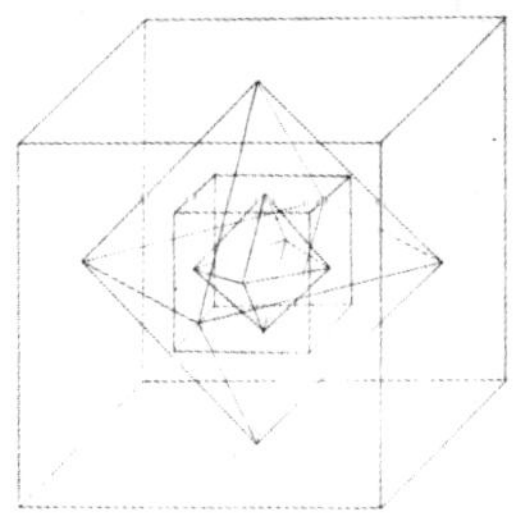

Figure 12.50. *Periodic fractal by generation of duals*

THEOREM 12.4.– *for a volume whose faces constitute a closed convex surface, the dual of a dual is never the primal. We will call ante primal the graph whose dual is the primal considered. Any Plato's solid can be regarded as the dual of an ante primal. Thus, a primal would always have a dual inscribed into it and an ante primal inscribed around it, and there is homothety between the two: the ante primal and the dual, are therefore self-similar.*

We can generalize the latter property to plane graphs (Figure 12.51):

– let us consider any planar graph G. We will call "quasi-dual" the dual G' whose possible vertex located in the infinite exterior is not taken into account (case of the projection of a graph with a non-convex surface);

– the quasi-dual G' of G, which thus has no external vertices, will be the dual inscribed inside for the considered graph;

– G' corresponds to the "dual graph of the map" defined by Rosen [ROS 91, p. 488], or to the graph of the map of the 10 departments quoted by Kaufmann [KAU 64, volume 2, p. 283];

– we will define by symmetry an ante primal, whose considered primal will be the quasi-dual;

– the planar quasi-dual written inside is a subgraph of the ante primal.

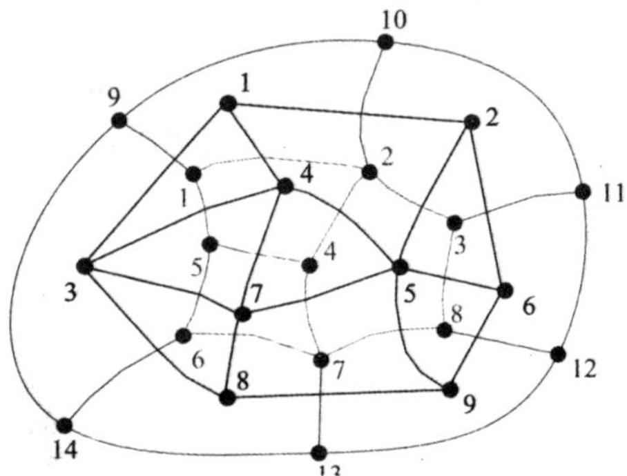

Figure 12.51. *Any quasi-dual graph inscribed and ex-inscribed*

We will call graph periphery the chain of arcs delimiting the interior and the exterior of the graph.

The quasi-dual has as many vertices as the primal has finite faces.

The quasi-dual has as many edges as the primal has internal edges.

The ante primal has the same number of internal arcs as the primal plus twice the number of external arcs or the same number of arcs as the graph plus the number of external arcs.

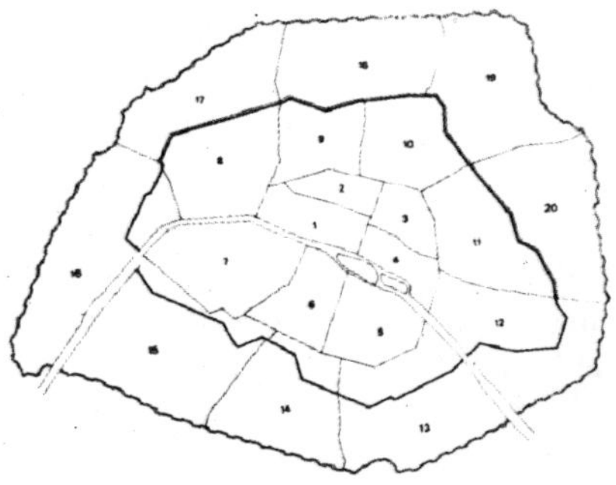

Figure 12.52. *The division of Paris into 20 districts and the old city limits*

The ante primal will have as many external faces as the dual written inside it. The growth will be of comparable nature by addition of a constant number of faces for each successive ante primal.

This process recalls the growth of cities as the plan (Figure 12.53) of the growth of Paris under Haussmann shows.

Let G(1-12) be the quasi-dual exterior of G (1', 2', 3', 4', 5'), the ante primal.

The graph G(1'-5') has G(1-12) as ante primal. The ante primal is the external dual ex-inscribed of graph G. The graph G(1', 2', 3', 4', 5') is a subdivision of the graph G(1', 5').

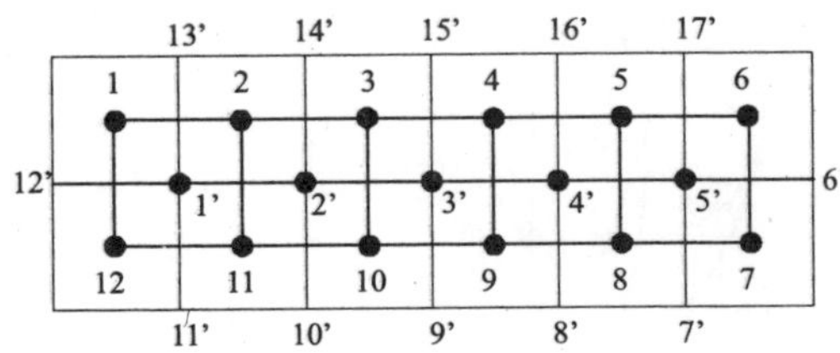

Figure 12.53. *Ante primal graph and the dual written inside it*

Let us consider the annular graph and its projection onto a sphere: there is no infinite surface but two possibly symmetric external points belonging to the two opposite faces of the graph.

Plato's solid is a figure written inside of or around a sphere and the graph of this solid can be projected onto this sphere.

Let us consider, for example, a regular graph projected onto a sphere and covering it entirely, which is the case of a central projection of a cube onto a sphere, as shown in Figure 12.54.

In this case the dual, which is a spherical plot of the central projection of the cube, exists but cannot have an infinite face. In this case again Berge's definition does not apply. The dual of this figure will also be the central projection onto the sphere of the dual tetrahedron of the cube.

The dual can be drawing either by using the spherical projection of the cube or the projection of the dual tetrahedron of the cube. More generally, we can show that the graph, which is the central projection onto the sphere of a convex volume, does not necessarily consist of identical faces, as demonstrated by a soccer ball, or even regular ones. This graph will have a dual projected onto the aforementioned sphere; which is itself a projection of the dual of the convex volume obtained as indicated by Labelle [LAB 81, p.98]: "let P be a convex polyhedron. Its dual is obtained by taking as nodes the centers of the faces of P and by joining the centers of the adjacent faces with segments". This is still a particular definition in the sense that the nodes of a dual are not necessarily in the middle of the faces but, on the contrary, may be located anywhere on the face as long as they are mathematically distinct from the nodes and arcs of the primal.

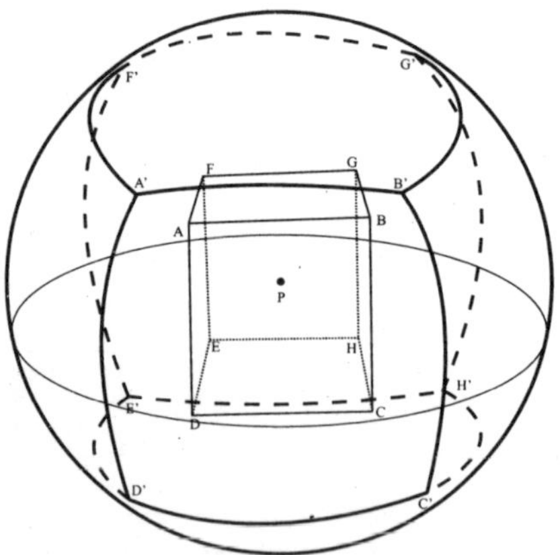

Figure 12.54. *Central projection onto a sphere*

This problem is not purely theoretical; it is an extremely concrete problem for any geographer, for whom the ground is spherical in the first approximation. At a certain scale a network will have a spherical projection and thus no infinite faces.

We arrive at the definition provided by Rosen [ROS 91, p. 488] of the dual graph of the map: "any map can be represented by a graph. To establish this correspondence each area of the map is represented by a node. Arcs will connect two nodes, if the areas which they represent have a common border. Two areas which only touch in a point are not regarded as adjacent areas. The resulting graph is the *dual graph* of the map. From the manner in which we construct dual graphs, it is clear that any map has a planar dual graph", such as, for example, the projections presented by Rosen (Figure 12.55).

However, this definition of the dual graph used by Headwood for the problem of five colors and then of four colors of a map does not take into account the possibility of an infinite face. Similarly, the number of arcs does not correspond between the primal and the dual: external arcs do not have corresponding dual arcs. Let us add that this dual is identical to the Eulerian tradition of the bridges of Königsberg. Due to this fact this definition is different from that of the geometrical dual of Withney consecrated later in other works.

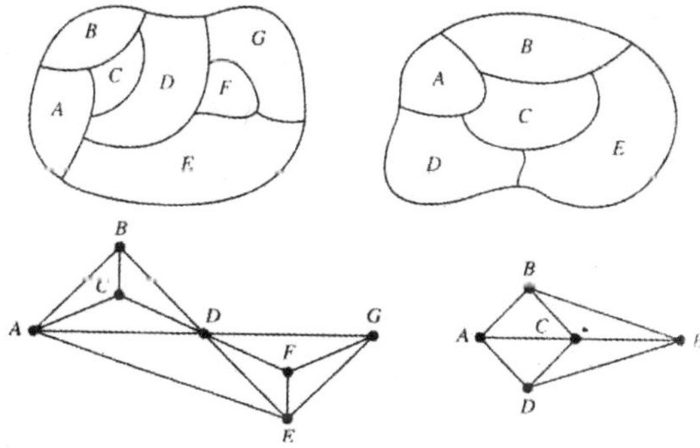

Figure 12.55. *Dual graph of the map, which is itself a primal graph according to Rosen [ROS 91]*

The same problem arises when we project the dual onto a torus as Headwood did [HEA 76]. In this case an external node can be created, but it is not necessary.

Let us consider the projection onto a torus: in this case the dual will be very specific. The external arc actually constitutes a loop, which is the same arc which intersects the two primal arcs.

The problem of the torus grid is an old one, since it dates from the works on perspective by Paolo Uccello in the 15th century (Figure 12.56).

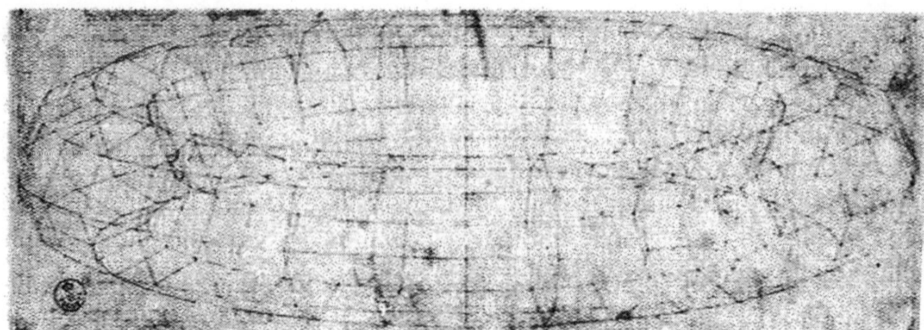

Figure 12.56. *Torus grid by Paolo Ucello (Flammarion 1972)*

The dual of this graph will be a possibly regular and identical graph obtained by using either Withney's or Rosen's methods, which in this case both give the same result.

We can conclude for the time being that duality remains a polysemous concept whose definition remains to be specified in graph theory where two definitions prevail, but where apparently only one of them makes it possible to obtain a cellular graph.

12.4. The cellular graph

Cellular graph and its uses

The following discussion aims at enabling the passage from a traditional graph to a cellular network thus making it possible to use a traditional cellular model, on the one hand, and of a multiagent system, on the other hand, which may contain reactive agents or artificial intelligence to simulate the behavior of network users.

Traditional transport system modeling by using a description in graph terms of its physical network is based, as we saw, on precise assumptions, such as complete rationality and total prior knowledge of the network, which is a necessary condition for any search for any optimum in a network, be they the shortest paths in distance, time, cost, maximum flow, the search for a Hamiltonian circuit, etc.

However, we have observed that the maximum capacity of a graph does not depend, in general, on the capacity of the arcs, but on the capacity of the nodes. This is a fundamental reason for the representation problems of the explanation of the decomposition technique of nodes into nodal graphs.

We could observe, nevertheless, that the technique of nodal graphs led only to their decomposition into convergent and divergent nodes, which is still insufficient to determine the real capacity of a node because it does not take into account the priority rules of various flows, their real overlap and interpenetration capacities, and thus the variations in speed of flows, and, consequently, the capacity of the lanes, which is itself a function of this speed, as well as the storage capacity of the arc upstream of the node, which is traditionally defined as a queuing problem.

In order to be able to explore the range of possibilities with respect to the circulation of traffic flows through a network, it seems necessary to go down to the level of the agent, whose individual behavior will be determined, among other things, by the rules of the highway code.

The chosen hypothesis is that the total state of the system results from the behavior of all the agents, their capacities of training, adaptation and anticipation.

To simulate an MAS (multiagent system) [IRON 97] in a transport network, one of the solutions is to have the definition of the network in the form of a set of cells which make it possible to apply the rules of percolation-type behavior.

It is obvious that a method which makes it possible to pass from the network modeled in the form of a graph to a cellular graph would constitute a powerful tool employing few means and making it possible, among other things, to globally simulate the transport system and simultaneously all its users through simple behavior rules, and thus to dynamically study their interaction.

Our intention is to show that on the basis of the previously developed considerations on duality and nodal decomposition, it is possible to define a cellular graph supporting an MAS, and that this transformation can be expressed as an algorithm.

Transformation of any arc into a cellular arc

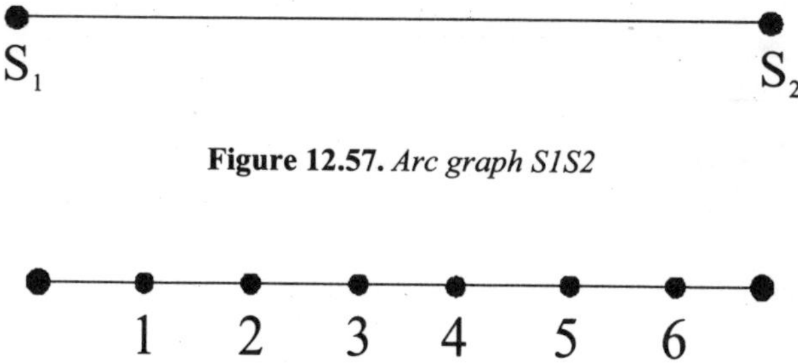

Figure 12.57. *Arc graph S1S2*

Figure 12.58. *Subdivision of the arc graph S1S2*

Let us consider any arc AB (see Figure 12.57). Arc S1S2 is an arc possibly belonging to a plane graph but not necessarily planar. The coordinates of S1 and S2 are known: they are attributes of the nodes and, naturally, fixed. The length l of the arc is therefore known. Let us perform a subdivision of the arc S1S2 with n nodes of the second degree, so that they have between them an arc l/n+1=u. The length "u" will be variable depending on the user, as we will see below.

Let us consider the ante primal of S1S2 with the preceding subdivision of n intermediate second degree nodes (Figure 12.59).

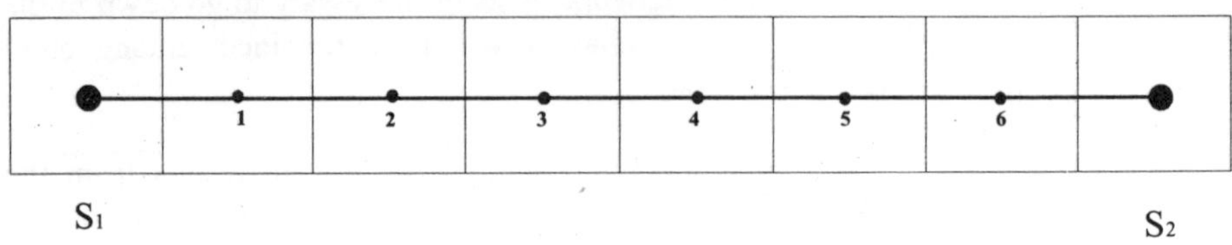

Figure 12.59. *Cellular arc*

We have transformed the arc S1S2 into a cellular arc with n+2 cells of a length u, which can correspond, for example, to the average statistical space occupied by a vehicle or a passer-by.

The cells [LAN 97] which are created in this manner will have a statute that can be specified at leisure and may or may not be occupied. However, let us immediately remove an ambiguity in the case of a road arc, where a cell of a cellular arc can only be occupied by one element at any time and in the majority of the cases it would then be necessary to use a directed graph, digraph or directed multigraph (Figure 12.60). The cellular arc could then be used as support for MAS micro-modeling, for example, in the case of movement against the traffic flow. In the more general case of a symmetric directed arc, the construction principle is identical.

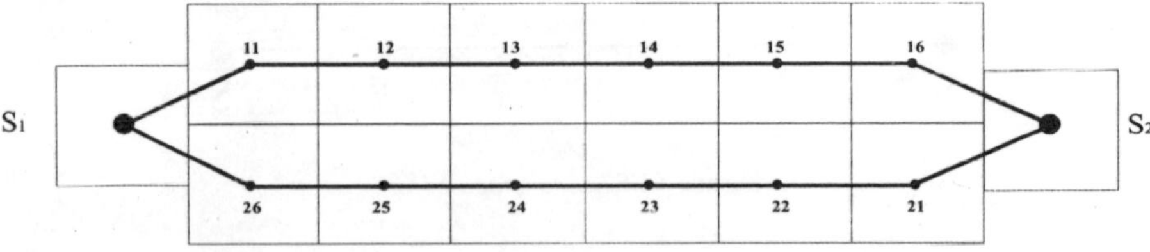

Figure 12.60. *Symmetric directed cellular arc or cellular digraph*

The notation principle is simple: the first set of numbers indicates the node towards which the arc is directed, the second indicates the node of origin of the arc, the third possibly indicates the number of the arc in the case of a p-graph (or of the lane, in the case of an arc whose capacity is defined by several lanes starting from the lane farthest on the right) and the fourth indicates the number of the cell starting from the destination node. This orientation of the cell number makes it possible to automatically determine the distance from the destination node, which is necessary for the behavior of user agents. We will associate various values with each cell, such as occupation by a car, truck and public transport, which in this case can simultaneously occupy several cells according to their sizes.

However, we are then obviously presented with the problem of the cellular definition of the node because the previous pattern is not satisfactory at this level. We will use the nodal graph developed into a cellular nodal graph: we transform the node S (Figure 12.61) into four nodes S1, S2, S3, S4 (Figure 12.62), thus adding four arcs of length u according to a rule opposite to that of simplification of flow graphs, since here we do not simplify but rather break up [CHO 64, p. 39]. We take as a starting point the derivation method of transformation by using the method of duplication of nodes for the nodal graph.

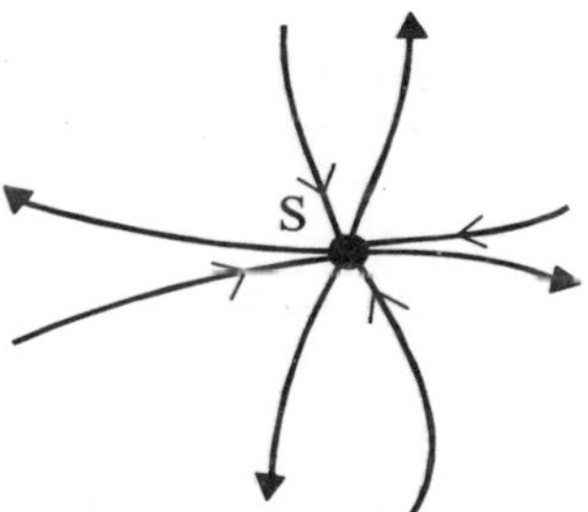

Figure 12.61. *Node of a digraph*

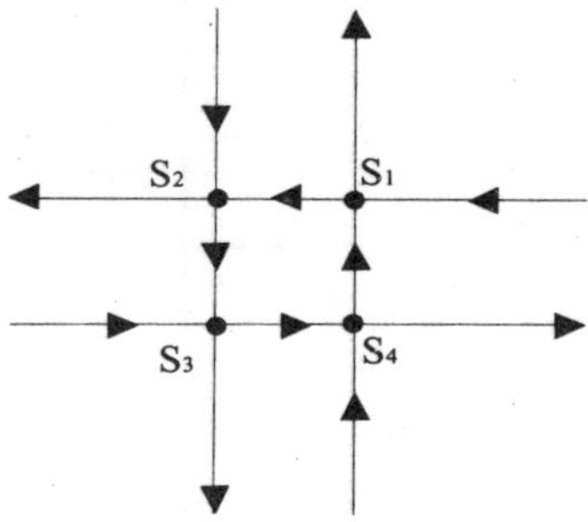

Figure 12.62. *Simplified decomposition of the digraph node*

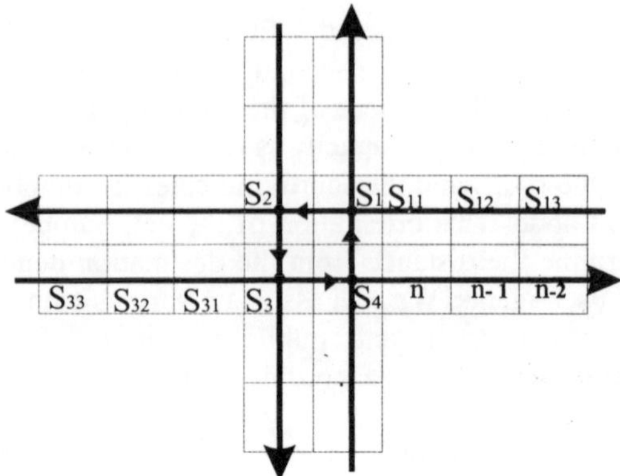

Figure 12.63. *Node of the cellular graph of a simple crossroads: one lane in each direction*

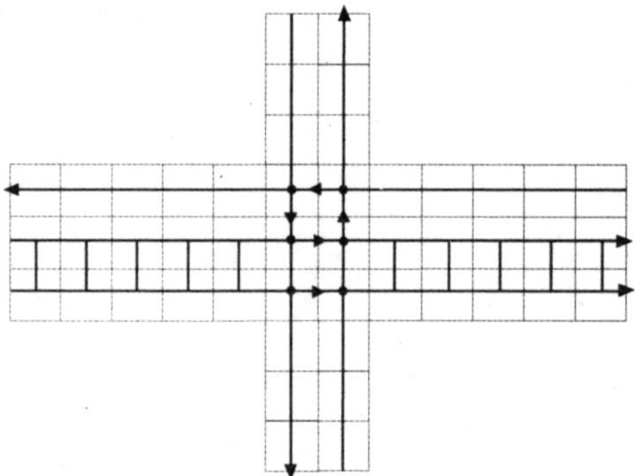

Figure 12.64. *Crossroads with two lanes in one direction*

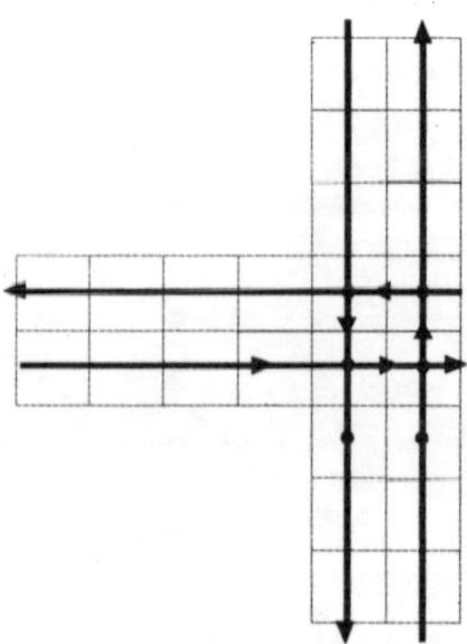

Figure 12.65. *"T" or three-way junction*

The size of the junction is automatically determined by the width of the cells and the number of incidental arcs (entering + leaving) (Figure 12.65).

The same process is used for directed multigraph describing three lanes (Figure 12.64), four lanes, two in each direction (Figure 12.66), crossroads with three lanes (Figure 12.65) and more, and the roundabouts, for which we may or may not use convergent/divergent nodes.

However, it is preferable to use the simplest forms that are closest to the material reality, with limited decomposition in order to introduce as little bias as possible into the behavior of the simulated agents[3], which is defined in the case of the ordinary land vehicles by the rules of the Highway Code[4]. The objective here is different to the nodal decomposition which we described previously. Indeed, the present objective is to make it possible to simulate, under conditions close to reality, the crossing of crossroads by reactive agents whose rules of behavior comply (or possibly do not comply) with the rules of the Highway Code and the main driving constraints. In the case of nodal decomposition with convergent and divergent nodes, the objective is to determine not the capacity of each arc, but the deceleration induced by flow divergence or convergence and, thus, the capacity of the node according to speeds under the assumption of fluid flow.

With the cellular MAS modeling, the random simulation hypothesis of non-compliance with rules or constraints and of the occurrence of an accident is taken into account.

In this case, we no longer have to deal with flows and the determination of an optimum under constraints, but with behaviors of possibly diverse, erratic or even criminal agents.

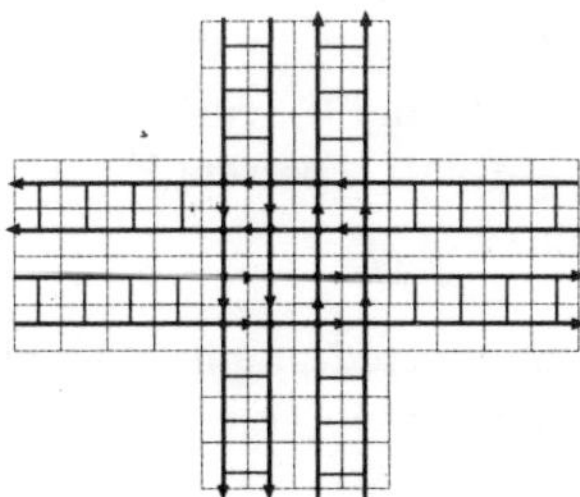

Figure 12.66. *Crossroads between two streets or roads with twice two lanes*

3 Mathis, Philippe and Maupuis, Gregoire: Programming an automat VST-CESA, September 2000.

4 Decoupigny, Christophe, PhD "Modélisation fine des émissions de polluants issues du traffic urbain" Tours, March 2006.

One of the great advantages of this solution with respect to that of the nodal graph is that the projection of the cellular graph remains planar and thus its computerized description is largely facilitated: to each arc there corresponds a vector of cells and each of them is clearly identified.

On the other hand, in the case of interchanges, i.e. of a very specific non-planar graph, we also have to remain faithful to reality, which is more complex, and the use of divergent and convergent cellular nodes makes it possible to describe such interchanges with precision.

Nonetheless, we saw with nodal graphs that the decomposition follows a defined process, an algorithm which is generating a fractal representation. This process can be reversed on demand and according to the selected scale of the representation and the user is free to consider the node traditionally, decomposed or in cellular form.

The application of this transformation of a network represented by a traditional graph into a cellular graph makes it possible to take into account another level of simulation: that of the user himself with his own constraints, his limited rationality, his imperfect knowledge of the network, etc. The example of the graph of the center of Tours clearly shows that the capacity and saturation problems are obviously due to the morphology of the network itself: there are multiple intersections, the roads are often one-way, with one or two lanes with problems of parking, obstruction of lanes, etc.

12.5. The faces of the graph: from network to space

A graph is defined as a set of points and relations between these points. The graph drawing, i.e. its representation on a surface, is in general a plane and therefore it takes into account only such elements as the crossroads of a network, which is represented by a node of the graph, and the relations between them, the roads or the railroads, for example, which we will represent by an arc in the shape of a straight line segment, a polyline or a curve.

Initially we only consider undirected graphs. The planar projection of the graph G divides the plane into areas, of which one is open. These areas, which certain authors call faces of the graph, have a status and rarely or poorly specified properties. Mainly, these areas or faces appear, on the one hand, in the statement of Euler's formula, which is applied either to a planar graph with an infinite face or to a projection of a convex volume without an infinite face:

– let G be a connected simple graph with e arcs and v nodes. Let r be the number of areas in a representation of g;

– then, r = e–v+2;

– in addition, right from the start authors have been concerned with the problem of colors. How can we apply colors, if the "faces" are not surfaces?

Finally, in the definition of the dual graph to each area or face we associate a node of the dual graph, which can be localized not just in the middle of the faces but anywhere in the face, since it is mathematically distinct from the nodes and the arcs of the primal. This condition enables the dual node to cover, to the nearest ε, the entire surface of the area considered.

This definition is valid regardless of the graph considered: planar or three-dimensional polyhedron. However, this causes some difficulties due to the flatness of a face, if it is in general supposed and real in the case of Plato's solids, which is not automatic once the number of arcs, when they consist of straight line segments, is more than three. Indeed, a triangle is always flat, which is not the case of a quadrangle or more, since the nodes do not necessarily belong to the same plane.

In the case of Plato's solids, the dual is particular, since its nodes are implicitly located at the center of the faces of the primal. The dual whose nodes are not at the center of the faces of the primal would be irregular and would in turn have an irregular dual, which would then not be isomorphous with the first, but rather its transform, still with the same number of nodes, arcs and faces.

The face is possibly plane and can be traversed by a point. It is used as a support for a possible dual node and has the role of separating the arcs and nodes from an elementary cycle of the primal.

At first we make the assumption that the surface on which the graph drawing is plane and that it is not the projection of a convex three- or more dimensional surface and, thus, there are no double points.

When we look at the map of a city (Tours, Paris, Chicago, Barcelona), we note that small islands are simple geometrical forms, a generally regular space paving: very regular and formed by right-angled or square parallelepipeds for the town of Millet where Hippodamos (5th century BC) had invented, according to legend, the orthogonal plane, which we now find in Chicago, Manhattan or Tours; octagonal for Barcelona with the plane of Cerda; square or triangular for Paris with the works of Haussmann, etc. It is the same for the majority of pieces of land in the rural world where Roman centuriations left generally clear traces.

Thus, we have an self-similar form of sectioning where fractality has a limit in urban planning: the size of the piece of land, which varies according to the times and construction methods used, is perfectly visible in a large number of cities, such as Tours, where the medieval fragmentation of Bourg Saint Martin is definitely smaller than that of districts urbanized later on. Even when the pieces of land were large due to their origin, i.e. gardens and parks of religious congregations, they are allotted according to the contemporary socio-economic and technical considerations. Frankhauser [FRA 94] and his team [GEN 00] showed how fractality could be used, on the one hand, to describe and, on the other hand, to characterize the city.

On the basis of these faces and by using a fractal transformation it is possible to define a cellular in order to approach the fragmentation as much as possible. The cells or pieces of land thus defined will have attributes of various natures: built-up, courtyard, park, private or public space, etc.

By considering cells and attributes we are in the domain of application of cellular models. However, the two-dimensional graph then becomes insufficient to describe a building or the city itself, which is not flat.

We then have the case of a GRS, which is the graph with spatial referencing defined at the start, since in the case of a built-up cell or piece of land it is quite obvious that the land use coefficient can be high, thus reflecting the existence of buildings with many or fewer floors, and there again the simple graph with geographical referencing is no longer sufficient.

This justifies in another way the taking into account of three-dimensional graphs beyond the purely mathematical problem of the dual. Indeed, graph theory simply brings together a set of elements with a collection of relations between these elements. The dimension of the graph is neither specified, nor necessary for the majority of algorithms. It is simply that the practice of representing graphs on a plane used to prevail.

Three-dimensional analysis becomes necessary to simulate a city, since it has three dimensions. A cellular graph lends itself admirably to this use as Serrhini had suggested in his three-dimensional visibility model: the GRS is a must.

This method makes it possible to pass from a graph and the faces of a graph to a set of cells which are clearly defined for arcs and nodes, and others for the faces of the graph. It is then possible to use MAS and cellular modeling. The former enables modeling and simulation of flows, whereas the latter enables the simulation of stocks by taking into account the attributes of cells, simultaneously with a classical four stages simulation on a more global scale.

12.6. Bibliography

[ALD 00] ALDOUS J.M., WILSON R.J., *Graphs and Applications, an Introductory Approach*, Springer-Verlag, London, Berlin, Heidelberg, 2000.

[BAL 00] BALAKRISHNAN R., RANGANATHAN K., *A Textbook of Graph Theory*, Springer, New York, 2000.

[BAN 01] BANG-JENSEN J., GUTIN G., *Digraphs: Theory, Algorithms and Applications*, Springer-Verlag, London, Berlin, 2001.

[BAP 99] BAPTISTE H., Interactions entre le système de transport et les systèmes de villes : perspective historique pour une modélisation dynamique spatialisée, Thesis, Tours, 1999.

[BEA 92] BEAUQUIER D., BERSTEL J., CHRETIENNE P., *Eléments d'algorithmique*, Masson, Paris, 1992.

[BER 70] BERGE C., *Graphes*, Dunod, Paris, 1970.

[BER 87] BERGE C., *Hyper graphes*, Bordas, Paris, 1987.

[BIG 76] BIGGS N.L., LLOYD E.K., WILSON R.J., *Graph Theory*, Oxford University Press, 1976 (1986).

[CAU 76] CAUCHY A.L., "Recherches sur les polyèdres", Premier mémoire, *J. Ecole Polytechnique 9* (Cah.16) (1813), 68-86., in *Graph Theory* 1736-1936 Biggs N.L., Lloyd E.K., Wilson R.J., Oxford University Press, 1976 (1986).

[CHA 97] CHAPELON L., Offre de transport et aménagement du territoire : évaluation spatio-temporelle des projets de modification de l'offre par modélisation multiéchelles des systèmes de transport, Thesis, Tours, 1997.

[CHO 64] CHOW Y., CASSIGNOL E., *Théorie et application des graphes de transfert*, Dunod, Paris, 1964.

[COR 94] CORMEN T., LEISERSON C., RIVEST R., *Introduction à l'algorithmique*, Dunod, Paris, 1994.

[DOL 01] DOLLARS A., "Le patron du Maître", *Pour la Science*, no. 282, p. 92-94, 2001.

[EUL 1736] EULER L., *Solutio problemats ad geometriam situs pertitnentis, commemtarii Academia Scientiarum Imperialis Petropolitanae* 8,128-140 = Opera Omnia, vol. 7, 1-10, 1736.

[EUL 1758] EULER L., *Demonstreatio nonnullarum insignium proprietatum quibus solidra hedris planis inclusa sunt praedita. Novi Comm. Acad. Sci. Imp. Petropl.* 4 (1752-1753, publié en 1758), 140-160 = Opera Omnia (1), vol. 26, 94-108, 1758.

[EUL 1758] EULER L., *Elementa doctrina solidorum. Novi Comm Acad. SCI. Imp. Petropol. 4* (1752-3, published in 1758) 109-140 = Opéra Omnia (1), vol. 26, 72-93, 1758.

[FER 97] FERBER J., *Les systèmes multiagents. Vers une intelligence collective*, Paris, InterEdition, Paris, 1997.

[FRA 94] FRANKHAUSER P., *La fractalité des structures urbaines*, Anthropos, 1994.

[GEN 00] GENRE GRANDPIERRE C., Forme et fonctionnement des réseaux de transport : approche fractale et réflexions sur l'aménagement des villes, Besançon, 2000.

[HEA 76] HEADWOOD P.J., "Map Colour Theorem", *Quarterly Journal of Pure and Applied Mathematics 24* (1890), p. 332-338, in *Graph Theory* 1736-1936 Biggs N.L., Lloyd E.K., Wilson R.J., Oxford University Press, 1976 (1986).

[KAU 64] KAUFMANN A., *Méthode et modèles de la recherche opérationnelle*, volumes 1 and 2, Dunod, Paris, 1964.

[LAB 81] LABELLE J., *Théorie des graphes*, Modulo éditeur, Mont-Royal Québec, 1981.

[LAN 74] LANTNER R., *Théorie de la dominance économique*, Dunod, Paris, 1974.

[LAN 97] LANGLOIS A., PHILIPPS M., *Automates cellulaires, application à la simulation urbaine*, Hermès, Paris, 1997.

[LAU 91] LAUWERIER H., *Fractals, endlessly repeated geometrical figures*, Princeton University Press, Princeton, 1991.

[LHO 97] L'HOSTIS A., Images de synthèse pour l'Aménagement du territoire : la déformation de l'espace par les réseaux de transport rapide, PhD Thesis, Tours, 1997.

[LHU 76] LHUILLIER S.A.J., *Mémoire sur la Polyédrométrie*, Annales de Mathématiques 3 (1812-3), p. 169-189, in *Graph theory* 1736-1936 Biggs N.L., Lloyd E.K., Wilson R.J., Oxford University Press, 1976 (1986).

[MAN 77] MANDELBROT B., *The fractal geometry of nature*, New York, 1977.

[MIN 86] MINOUX M., BARTNIK G., *Graphes, Algorithmes, Logiciels,* Dunod, Paris, 1986.

[POI 76] POINSOT L., "Sur les polygones et les polyèdres", *J. Ecole Polytechnique 4* (Cah.10) (1810), p. 16-48, in *Graph Theory* 1736-1936 Biggs N.L., Lloyd E.K., Wilson R.J. Oxford University Press, 1976 (1986).

[PON 69] PONSARD C., *Un modèle topologique d'équilibre interrégional*, Dunod, Paris, 1969.

[ROB 00] ROBERT D., Le réseau routier français dans la dynamique des échanges de Marchandises de la France avec ses partenaires d'Europe Occidentale, PhD Thesis, Paris 1, 2000.

[ROS 91] ROSEN K.H., *Mathématiques discrètes*, Chénelière/McGraw-Hill, 1991 (1995, 1998).

[WIT 32] WITHNEY H., "Non separable and planagraphs", *Transaction of the American Mathematical Society,* 34 (1932), 339-362, 1932.

[XUO 92] XUONG N.H., *Mathématiques discrètes et informatique*, Masson, Paris, 1992.

Chapter 13

Practical Examples

The elements presented above are not the result of chance or simple theoretical research: they were necessary to solve a certain number of difficulties which the modeling group had encountered and for which we had sometimes developed empirical solutions or inferred the existence of a solution.

Below we initially present some of these intuitions or anticipations and then in the second part we will show not only the applicability but the great interest that lies in the results above within a large scale simulation of urban motor traffic, as well as in the study of its characteristics at the level of agents and cells, whose results can be used directly in urban planning.

13.1. Premises of multiscale analysis

13.1.1. *Cellular percolation*

The first example of partial use of cellulars and graphs carried out by the CESA research group goes back to 1992. The very simple model simultaneously employs a cellular frame to simulate urban expansion by percolation with or without the existence of free space and a mass transit system, such as the suburban train combining a graph to the cellular and therefore a bimodal approach with delay.

Chapter written by Philippe MATHIS.

In this example[1], the fastest means of transport was public transport with fixed stops of the train type (tunnel effect), from which there is normal diffusion, i.e. by 8-connected This type of modeling shows diffusions in successive blots, the development of the center and less quickly of the starting points of public transport (which might well have been highway exchangers), all the more slowly as they were further removed from the center, and a tendency for congestion as well as urban diffusion in the Parisian suburbs (Figure 13.1).

This type of modeling is perfectly appropriate for the analysis of stocks, but the process described remains very succinct. It is necessary to go beyond a simple, purely mechanical description of diffusion.

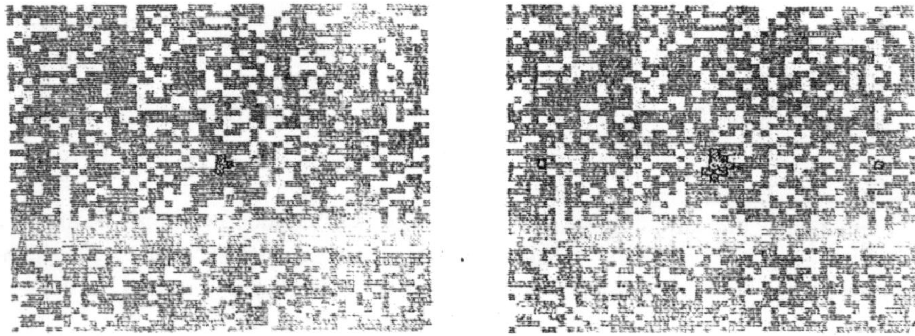

Figure 13.1. *Diffusion of housing by connectivity and according to public transport networks*

This type of cellular modeling must be supplemented by MAS modeling to simulate flows and we can then follow the definition of states at the local micro-level, which is that of one or several pieces of land (stocks in dynamic system modeling of the Forrester type [FOR 69]) *as well as the traditional level of overall optimization, and thus to detect variations, tensions, contradictions and the possible mechanisms of resolution thereof.*

The kind of spatial description enabled by this type of technique can be very fine.

1 Multimodal percolation Address at the Sixth World Conference on Transport Research, Lyon 1992.

13.1.2. *Diffusion of agents reacting to the environment*

The second preliminary application of these methods goes back to 1993, carried out in a different spirit in order to model the behavior of tourists in a sensitive natural environment: the crossing of the dune range on the Grande Côte at the mouth of the Gironde. This is a dune range whose structure is often disrupted by trampling and which required expensive dune restoration work, despite areas where entry was prohibited to tourists, which are marked in red in the diagrams of the color plates section.

After having observed and noted the behaviors of tourists during their journey to the range from the parking lots and back, and perhaps their stay on the accessible dunes, the objective of this simulation was to be able to simulate them in order to avoid invasive trampling of the dunes and the creation of many new paths.

The size of a represented cell must be more than 1 pixel, however, screen definition was too basic in 1993 to be able to display sufficient detail [MAT 92, 93], as shown by the modeling diagram (Figure 13.2; see the color plates section) describing the paths of tourists crossing the dune of the Grande Côte to reach the range situated at the bottom from the parking spots on the sides of the road or from the car park at the top of the diagram.

The diagrams in Figure 13.3 (see the color plates section) show two different hardness hypotheses, the paths used and the results on the walkways according to the behavior rules and the sensitivity of tourists: unconstrained movement of tourists through the alleys between the dunes, with the principle of least effort (the most direct and easiest path) between the parking on the edge of the main road and the edge of the sea, and the location on the range, naturally, without superposition of agents. The cells (used for the simulation) are of the size of a pixel due to the basic screen resolution in 1993.

In these preliminary examples to true MAS the interaction between the tourist and the physical state of the environment did not exist. The interaction between the tourists was taken into account only on the range where the selected location was the closest, but on a free cell.

In the next simulation we introduced stronger hardness on a path represented in yellow above in order to simulate a visible obstacle on the path, but without prohibition. We observe the detour taken by the tourists to reach the range. A current simulation would take into account a probabilistic behavior according to the type of user: age, accompanied by children of a certain age, outward or return journey, time of day, etc. In 1993, the data of study could not have been exploited to this level of details.

In the last simulation the bold line at the top represents the creation of a car park and a passage under the relatively busy access road, causing the concentration at this level. Another difficulty was added to the trip by involving a concentration of flows in a specific zone where dune erosion by trampling was considerable, and one of the objectives was to reproduce this aspect of tourist behavior.

13.1.3. *Taking relief into account in the difficulty of the trip*

In the following simulations for the same area Serrhini [SER 00] used MNT, i.e. a GRG to explore the influence of the slopes on the behavior of the same tourists by simulating different sensitivities to the effort required for climbing a sand dune.

The diagram in Figure 13.4 (see the color plates section) is drawn for an acceptable level of difficulty corresponding to a zero percent slope: tourists considered in this example as carrying items or infants, or as old people avoid any climbing, which perfectly describes the image in this figure.

In the second simulation the acceptable slope is changed to 14%, i.e. the double of the maximum slope of a highway, which is a considerable slope on sandy ground: 14 meters for every hundred meters is the height of a five storey building (Figure 13.5; see the color plates section).

With this level of difficulty accepted the routes do not change with respect to the previous assumption: only the same three paths cross the considered area.

Numerical exploration can then continue until a threshold is determined, from which percolation through the represented dune massif starts, with only the slope for an obstacle here.

With a threshold of 28% of slope acceptance, we note that practically all the paths, indeed, cross this space (Figure 13.6; see the color plates section). This slope is already very steep: it is a natural embankment. It is too much for old people and parents carrying young children, especially since the only constraint taken into account here is the slope and absolutely not the difficulty of the surface, such as sand, for example.

These two old examples show how the traditional and cellular-MAS approaches complement each other: the first is global and is an optimization procedure with strong assumptions including a perfect rationality and a total knowledge of the area considered. The second relates to an agent with limited rationality, but able to take into account all the "roughness" of the terrain and the "specificity" of behaviors.

13.2. Practical application of the cellular graph: fine modeling of urban transport and spatial spread of pollutant emissions

Taking into account the cellular and the attributes of cells makes it possible to pass from the micro level of agents to the macro-level of the spatial system, by making it possible, among other things, to evaluate "instantaneous stocks" that modify the functional valuations of networks and spaces in terms of speed, capacity, accessibility, but also constructability, etc. The software developed by Decoupigny [DEC 06] makes it possible to better understand the often significant difference between theoretical and practical performance of a system. We will see that its contributions, even at the theoretical level, are numerous and significant.

From the results of the previous chapter Decoupigny worked out an algorithmic method to transform a transport network represented by a graph into a cellular graph.

We developed the following considerations: it is obvious that such a network must be simulated on two levels, on the one hand, globally, as a problem of transport network optimization, which is done at the level of a PUM (Plane of Urban Movements), but also at the level of each arc and crossroads and that of the user. From this stems the need for cellular simulation to establish the various states of the network, its storage capacity (number of traveling cars as well as parked ones), its possible saturation and its capacity to enable the level of usability expected by a resident in his daily practice, and we add the characteristics of pollutant emissions of vehicles on a very fine spatial scale.

13.2.1. *The algorithmic transformation of a graph into a cellular graph at the level of arcs*

The first stage of the algorithm consists, as in the previous chapter, of transforming the arc by subdivision and then determining the dual without taking into account the external face, i.e. the ante primal, and to transform the faces into cells whose length is the average length of a car, that is, five meters.

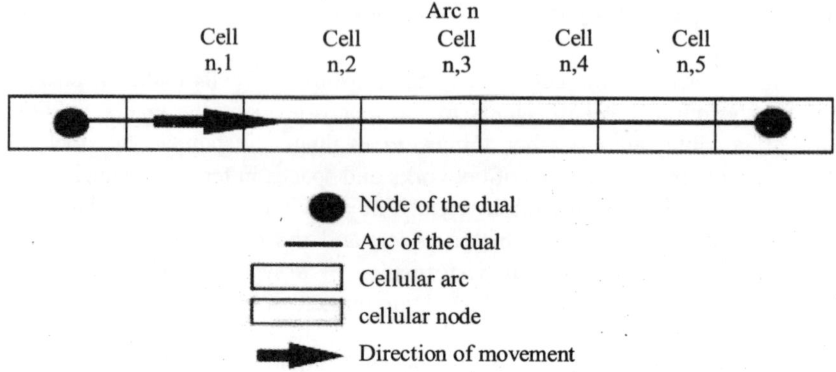

Figure 13.7. *Cellular arc*

The second stage as of the algorithms [DEC 06] makes it possible to supplement the cellular graph obtained in this manner by adding one or more lanes in the same direction and/or in the opposite direction, then possibly a right turn, as in the example below, or a left turn, etc.

It is even possible to create exclusive rights of way and reserved lanes with 30km/h, with parking on alternating sides [DEC 06].

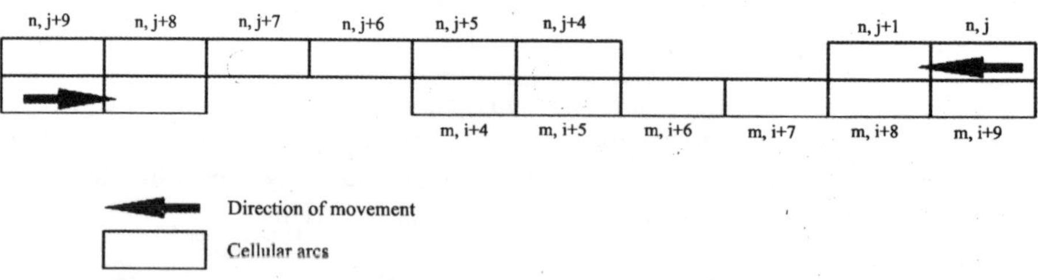

Figure 13.8. *Cellular arcs, two lanes and a right turn, and a cellular bayonet arc*

This example is a combination between contraction and widening, as all the examples above are additive combinations on the basis of the first transformation considering that two lanes in the same direction can be a directed 2-graph with possible lateral communication of the cells following the formal characteristics of the considered lane.

13.2.2. *The algorithmic transformation of a graph into a cellular graph at the level of the nodes*

The algorithmic transformation of nodes into cellular subgraphs or cellular nodes is more complex and generally includes two broad stages: the first determines the number of nodal cells, whereas the second determines the characteristics of spatial connection of the arcs at the considered nodes.

13.2.2.1. *Determination of the number of nodal cells*

This determination is based on the results of the previous chapter on the decomposition of a node into a subgraph and on the search for its dual, which we recall below.

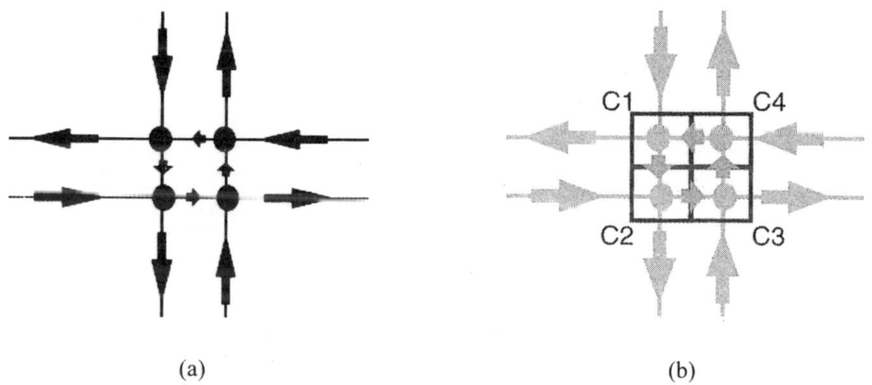

Figure 13.9. *(a) Nodal digraph; (b) nodal digraph and its ante primal*

As we may guess intuitively from the figures above, the number of cells is determined by the number of arcs adjacent to the considered node and, more precisely, by the number of lanes of each arc or, in other words, by the number of adjacent arcs to the considered node of the local p-graph.

Algorithmically this number will be at least equal to the number of arcs (or lanes) arriving at and leaving from the considered node, including the right and left turns mentioned in the definition of arcs. For obvious traffic reasons cells cannot be aligned except in the usual case of a crossroads with two one-way streets. In the general case the solution is a cellular subgraph with m rows and n columns of cells, or as close to it as possible in the particular case of a roundabout.

13.2.2.2. *Spatial connection of cellular arcs to the cellular node*

Decoupigny [DEC 06] determines the connection of the arcs to any of the four faces of a cellular sub-summit graph by analyzing the angle of the arc by taking the trigonometric circle as a basis.

The number of arcs (lanes) connected to each face in this way makes it possible to determine the m rows and n columns by considering each time the largest number of arcs at the each opposed lateral face, as we see in the figure below.

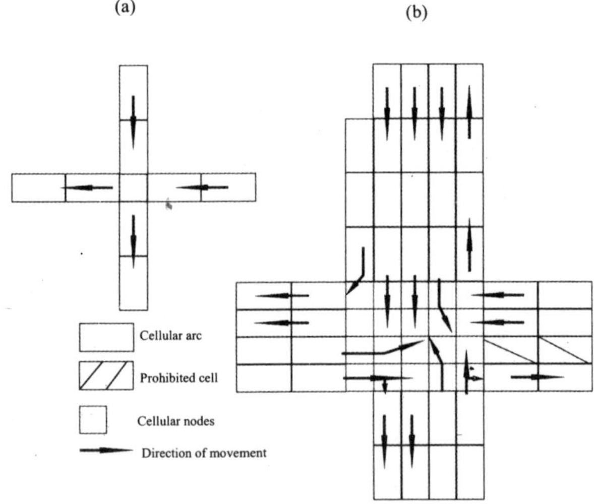

Figure 13.10. *(a) Simple node; (b) complex node*

For particular cases of interchanges and the roundabouts, as well as for a presentation of the algorithms see the work mentioned above.

The set makes it possible to automatically define a cellular graph on the basis of a transport network already transformed into a graph as shown in the figure and the details below.

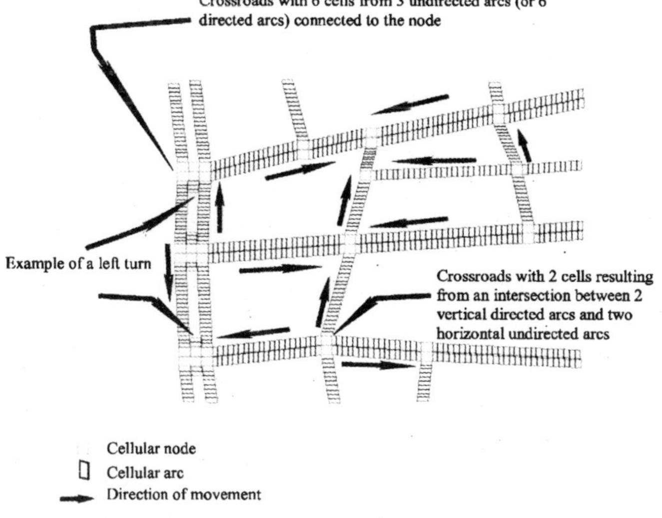

Figure 13.11. *Cellular graph: area of Tours*

Having shown the possibility of automatically representing the circulation network by using a cellular graph, the following stage consists of defining the behavior rules of the agents circulating in this network.

13.3. Behavior rules of the agents circulating in the network

"By assumption, traffic is composed of a discrete set of vehicles that share the same space and the same conditional objective, which is that of arriving at the destination point without accident. The journey time or the length of the path followed are secondary." The fundamental rule is thus to arrive at the destination point without accident.

Moreover, "a motorist exhibits self-centered behavior, since he adapts his behavior to his objective according to his immediate environment. He takes into account the motorists driving before him, the constraints imposed by space (width of the street, curve radius, etc.) and his capacity to evaluate these constraints. Consequently, he does not have an overall rational view of the traffic, since his rationality is limited by his field of vision and by his experience of the circulation conditions, in a generally known space at a given time".

On the basis of these fundamental hypotheses, Decoupigny defines, for a system of agents, at least three sets of strict, elementary and behavioral rules.

13.3.1. *Strict rules*

They combine spatial constraints with certain rules of the Highway Code, on the basis of which we define a behavior which is referred to as borderline. They stem from the fundamental rule.

The absolute constraint is of a physical nature. This is the speed limit of the vehicle, which in a curved trajectory of a crossroads is defined by the centripetal acceleration which is a function of the curve radius and the adherence of the vehicle. This limit implies a maximum speed of entry in a crossroad, and thus, the taking into account of the maximum negative acceleration defined by the vehicle's manufacturer, i.e. the stopping distance, which is maximum instantaneous speed at a distance from the crossroads.

The speed on the arc will thus be limited, on the one hand, by the maximum speed at the end of the arc and, on the other hand, by the speed of entry in the arc and by the capacity of maximum positive acceleration of the vehicle.

Moreover, the maximum speed on the arc will be constrained by a traffic light or a give way and/or stop sign at the crossroads and, thus, by zero speed of entry, lest the fundamental rule is not complied with.

Adhering to this fundamental assumption imposes complying with certain rules of the Highway Code, such as traffic lights, stop and give way signs in addition to the speed constraint of entry at a crossroads, as well as speed limits, respecting one-way streets and streets with assigned directions. However, these assumptions can be removed or modulated.

13.3.2. *Elementary rules*

They correspond to standard actions on behalf of the agents such as: the security distance depending on the braking capacity and relative speeds and therefore the free space between agents, the rules of starting and accelerating, following the vehicle in front and, thus, of the interaction between vehicles, of changing lanes, overtaking, parking, etc.

These rules are the same for everyone and are mainly acquired automatisms. They are defined to a limit. However, individual behaviors are variable according to the capacities of each driver.

13.3.3. *Behavioral rules*

They depend on the evaluation of space, experience, different visual and hearing perception capacities, as well as the nature of each agent. To simplify the model, three classes have currently been defined, but that can be modified.

Competitive or aggressive drivers are typically at the extremes of the acceleration and braking capacities and have shorter than average response times.

On the other hand, calm or careful drivers have slower reactions, higher braking distance, weaker accelerations and a reaction speed definitely lower than the previous category.

The intermediate category groups the largest number of agents and its characteristics are in effect average.

Together this means that even with only three groups, a behavioral differentiation among the agents is introduced.

On the basis of these basic elements and by integrating the data on pollutant emissions[2] Decoupigny has developed the TUREP software "URban Transport and Emission of Pollutants" [DEC 06].

13.4. Contributions of an MAS and cellular simulation on the basis of a graph representing the circulation network

The TUREP software, which not only uses the MAS, but also the attributes of cellular agents, leads to a certain number of results expected *a priori*. However, it also makes it possible to obtain much more general and theoretical results, if the applicability of the laws considered as general is specified.

13.4.1. *Expected simulation results*

First of all, the TUREP software shows the possibility of automatically passing from a graph representing a road network to a cellular graph. This makes it possible to simulate motor traffic of the agents thanks to a small number of behavioral rules and the description of phenomena which are permanently observed, but which are badly or not at all described by the traditional approaches of the four stages model type, like the stop and go, turnaround, the univocal influence of the vehicles upon

2 See Chapter 8.

each other, which the MAS makes it possible to simulate perfectly well, as the diagram below shows.

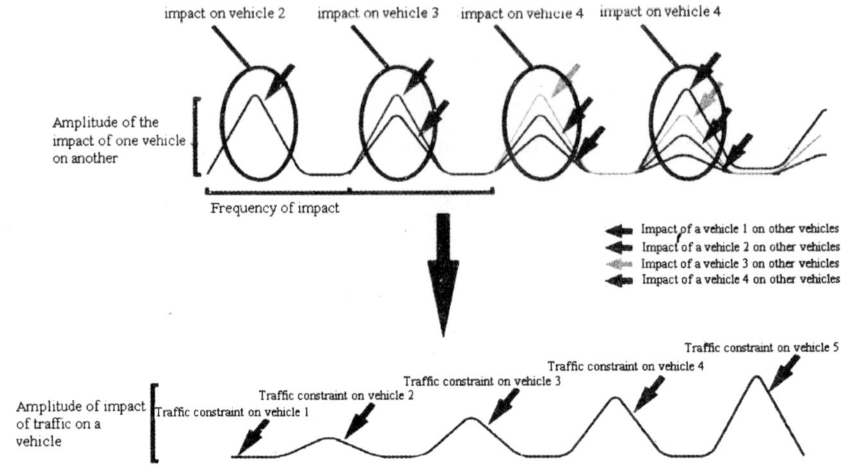

Figure 13.12. *Universal influence of the vehicles upon each other*

However, this also makes it possible to define the attributes of cells used as circulation support, which is a phenomenon that until now could have only been very coarsely approximated at the arc level after the phase of spatial allocation [MAT 05].

The results provided by the use of a cellular graph, a cellular MAS and agents has a totally different precision, it is micro simulation: space is defined by five meter-long cells, each attribute of which is calculated and mapped. Among these results appear the number of passes per cell, the speed and acceleration of vehicles, the total time of occupation, gas or particulate emissions, all this for a variable number of vehicles, from one to a hundred or even much more depending on the computer used.

13.4.2. *Limits of application of laws considered as general*

The most important results, confirming the intuition of the authors, are the descriptions of the limits at the level of microsimulations of the application of laws which are considered as general and, more precisely, the determination of the maximum flow in the graph, which no longer corresponds to the strict application of the Ford-Fulkerson theorem, the uselessness of the queuing theory, the non-applicability of the flow speed curve in its reciprocal form.

13.4.2.1. *Asymmetric application of the flow speed curve*

The simulations of Decoupigny show very clearly that a low speed is not necessarily the result of dense vehicle flow: there is a very clear network effect which does not enable high speed, especially, in an urban area. The consequence could be heavy pollutants emissions in zones with high density and low speed circulation. On the other hand, dense flow implies a speed that decreases the more flow is disturbed by turnaround, which are at the origin of the stop and go phenomena. However, these phenomena may just as well be the result of network effects, which appear dominant in an urban zone.

13.4.2.2. *The non-applicability of the Ford-Fulkerson maximum flow in a graph theorem*

In a fine representation of the circulation network, such as that made possible by the cellular graph, capacity is not limited by that of an arc, except for vehicle parking or an accident, but by the capacity of crossroads, network nodes, graph nodes. However, we saw that this aspect is not at all dealt with by traditional graph theory, for which the node is neutral. The simulations made with the TUREP software obviously reproduce what any driver knows and experiences systematically: congestion is initially and primarily caused by crossroads. Decoupigny shows that the capacity of crossroads does not merely depend on their physical shape, but that the rules of priority play a fundamental part in it by modifying the various degrees of freedom imposed on flows stemming from each incidental arc. It can also be seen in the example of an interchange that congestion can extend by percolation from node to node.

That is a very operational aspect for the layout of urban transport systems as well as for the forecast of pollutant emissions, without these two aspects necessarily being correlated.

It appears very clearly that the circulation rules are a primary influence on network flows and constitute an essential means of traffic control.

A perfectly clear illustration is given by the simulation of a speed bump in a street and the consecutive modification of pollutant emissions, whereas average speed remains almost identical, but the acceleration and braking phases change.

13.4.2.3. *The uselessness of the queuing theory for the simulation of urban circulation*

The precision with which crossroads and the use of simple agent behavior rules are reproduced renders the use of the queuing theory useless. We see them form naturally and develop in the same way, according to the priorities of each crossroads and the characteristics of the agents.

13.5. Effectiveness of cellular graphs for a truly door-to-door modeling

This type of approach is also perfectly adapted to the modeling of phenomena which had been ignored until now, such as the simulation of pedestrian movements in a city, the study of parking phenomena and initial and final paths.

Models which are currently qualified as door to door are only very partially so because the final paths are only estimated on average. Currently, 20% of urban car journeys are less than 1 km. However, the distances covered for parking and leaving the parking spot may be between 200 and 400 meters: the walking part is sometimes longer than the driving stretch! Another example is that it takes at least two hours to go from the University of Tours to the Ministry of Infrastructure, including 55 minutes by TGV for 230 km and more than an hour of initial and final trips on foot and in the subway over 10 km.

All multimeans journeys have a pedestrian component, which until now was neither modeled nor really simulated. Multilevel simulation using graphs and cellular graphs would make it possible to really do this.

Similarly, the simulation of semi-pedestrian paths and mixtures of means such as public transport, pedestrians, cyclists and rollerskaters is not presently carried out, so far for lack of adapted tools. The cellular graph must make it possible to mitigate these difficulties.

The problem of parking vehicles on the public road is not dealt with, neither with respect to the process of searching for a spot, nor with respect to the effects on the speed of circulation and the capacity of the roads, which is what the cellular MAS model makes it possible to simulate.

Finally, more traditionally, movements for multiple reasons are currently increasing and the traditional methods of optimization are not very adapted to solve them, whereas the use of the MAS and a search for a satisfactory individual sub-optimum make it possible to approach it more efficiently

13.6. Conclusion

After developing certain aspects of graph theory to obtain a cellular representation of arcs and nodes in order to be able to use cellular MAS and agents with a view to carry out very precise spatially spread micro-simulations, we have shown using some examples that these theoretical developments had to some extent been anticipated by experience.

Multiscale existed already, thus anticipating the theoretical developments and requiring them in order to really have a foundation. By using two examples of modeling we have shown that each of these characteristics and theoretical properties had their counterpart and their need in the theory which based on them the already used concrete possibility.

The work conducted since the publication of the French version of this book by Decoupigny within the laboratory, as we have rapidly mentioned above, made it possible to show the practical possibility of such a method and its efficiency in micro-modeling.

Indeed, the expected results have been obtained: algorithmic passage from the graph to the cellular graph, development of an MAS in this space and simultaneous use of cellular agents in order to arrive at flows, on the one hand, and, on the other hand, at stocks at the moment t characterized by the various attributes of cells. We have thus returned to the spirit of dynamic system models of Forrester [FOR 69], but finely spatially spread here.

However, most interesting and also, perhaps, least expected are the most general results limiting the fields of validity in micro-simulation of the law, which is a theory and theorem hitherto used systematically.

The use of TUREP enables a true current and perspective simulation of transport networks.

It makes it possible to test structural modifications of the network itself in terms of the priority rules, one-way streets and speeds, which the traditional technique of four stage models did not enable, since it is based on econometric relations and therefore on the fundamental condition "all other things being equal". This set is thus also a model of spatial perspective, which enables a more justified, more "objective", reproducible and more verifiable representation in the sense given by Waldo Tobler.

13.7. Bibliography

[DEC 06] DECOUPIGNY C. "Modélisation fine des émissions de polluants issues du trafic en milieu urbain" PhD Thesis, March 24, 2006, Tours.

[FOR 69] FORRESTER J.W., *Urban dynamics,* 1969 MIT, French translation *Dynamiques urbaines*, 1979, Economica, Paris.

[MAT 92] MATHIS P., LIARD V., "Percolation et diffusion dans une espace sensible", Address at the International Conference: Tourisme en Façade Atlantique, Royan 14, 15, and 16 October 1992.

[MAT 93] MATHIS P., LEROI B. and LIARD V., "Diffusion des touristes dans un espace dunaire sensible", ASRDLF-CESA Conference, Tours 1993.

[MAT 05] MATHIS P. (ed.), ESPON Project 1.2.1, Final Report, 2005, www.espon.lu.

[SER 00] SERRHINI K., "Intégration quantitative et multicritère du paysage lors de la détermination d'un aménagement linéaire", *Mappemonde*, 2000.

List of Authors

Manuel APPERT
UMR Espace
CNRS
University of Montpellier
France

Hervé BAPTISTE
CESA
Centre de recherche VST-ART
University of Tours
France

Laurent CHAPELON
UMR Espace
CNRS
University of Montpellier
France

Christophe DECOUPIGNY
CESA
Centre de recherche VST-ART
University of Tours
France

Fabrice DECOUPIGNY
UMR Espace
University of Nice
France

Ossama KHADDOUR
CESA
Centre de recherche VST-ART
University of Tours
France

Sébastien LARRIBE
Centre de recherche VST-ART
University of Tours
France

Alain L'HOSTIS
INRETS-TRACE
Lille
France

Philippe MATHIS
CESA
Centre de recherche VST-ART
University of Tours
France

Kamal SERRHINI
Département Génie des Systèmes Urbains
Compiègne University of Technology
France

Index

P

Q

R

S

T

U

V

W